This item must be returned or renewed by the last date shown below. The loan period may be shortened if it is reserved by another reader.   A fine will be due if it is not returned on time.

DATE OF RETURN

WITHDRAWN

UCL LIBRARY SERVICES
Gower Street London WC1E 6BT

Bioprocess Technology:
Modelling and
Transport Phenomena

BOOKS IN THE BIOTOL SERIES

The Molecular Fabric of Cells
Infrastructure and Activities of Cells

Techniques used in Bioproduct Analysis
Analysis of Amino Acids, Proteins and Nucleic Acids
Analysis of Carbohydrates and Lipids

Principles of Cell Energetics
Energy Sources for Cells
Biosynthesis and the Integration of Cell Metabolism

Genome Management in Prokaryotes
Genome Management in Eukaryotes

Crop Physiology
Crop Productivity

Functional Physiology
Cellular Interactions and Immunobiology
Defence Mechanisms

Bioprocess Technology: Modelling and Transport Phenomena
Operational Modes of Bioreactors

In vitro Cultivation of Micro-organisms
In vitro Cultivation of Plant Cells
In vitro Cultivation of Animal Cells

Bioreactor Design and Product Yield
Product Recovery in Bioprocess Technology

Techniques for Engineering Genes
Strategies for Engineering Organisms

Principles of Enzymology for Technological Applications
Technological Applications of Biocatalysts
Technological Applications of Immunochemicals

Biotechnological Innovations in Health Care

Biotechnological Innovations in Crop Improvement
Biotechnological Innovations in Animal Productivity

Biotechnological Innovations in Energy and Environmental Management

Biotechnological Innovations in Chemical Synthesis

Biotechnological Innovations in Food Processing

Biotechnology Source Book: Safety, Good Practice and Regulatory Affairs

BIOTECHNOLOGY BY OPEN LEARNING

Bioprocess Technology: Modelling and Transport Phenomena

PUBLISHED ON BEHALF OF :

Open universiteit　　and　　**Thames Polytechnic**

Valkenburgerweg 167　　　　　　Avery Hill Road
6401 DL Heerlen　　　　　　　　Eltham, London SE9 2HB
Nederland　　　　　　　　　　　United Kingdom

Butterworth-Heinemann Ltd
Linacre House, Jordan Hill, Oxford OX2 8DP

 PART OF REED INTERNATIONAL BOOKS

OXFORD LONDON BOSTON
MUNICH NEW DELHI SINGAPORE SYDNEY
TOKYO TORONTO WELLINGTON

First published 1992

British Library Cataloguing in Publication Data
A catalogue record for this book is
available from the British Library

Library of Congress Cataloguing in Publication Data
A catalogue record for this book is
available from the Library of Congress

ISBN 0 7506 1507 9

Composition by Thames Polytechnic
Printed and Bound in Great Britain by
Thomson Litho, East Kilbride, Scotland

Contributors

AUTHORS

Professor Ir. H.E.A. van den Akker, Technical University Delft, The Netherlands

Professor Dr Ir. J.J. Heijnen, Technical University Delft, The Netherlands

Dr C.K. Leach, Leicester Polytechnic, Leicester, UK

Dr R.F. Mudde, Technical University Delft, The Netherlands

TECHNOLOGY AND EDUCATIONAL ADVISORS

Dr G.M. Hall, Loughborough University of Technology, Loughborough, UK

Dr G. Mijnbeek, Bird Engineering bv, Schiedam, The Netherlands

Dr G. Subramanian, Loughborough University of Technology, Loughborough, UK

SENIOR TECHNOLOGY ADVISOR

Professor Ir. K. Ch. A.M. Luyben, Technical University Delft, The Netherlands

SCIENTIFIC AND COURSE ADVISORS

Professor M.C.E. van Dam-Mieras, Open universiteit, Heerlen, The Netherlands

Dr C.K. Leach, Leicester Polytechnic, Leicester, UK

ACKNOWLEDGEMENTS

Grateful thanks are extended, not only to the authors, editors and course advisors, but to all those who have contributed to the development and production of this book. They include Mrs A. Allwright, Miss K. Brown, Dr M. de Kok, Miss J. Skelton and Professor R. Spier.

The development of the BIOTOL project is funded by **COMETT, The European Community Action Programme for Education and Training for Technology**. Additional support is provided by the Open universiteit of The Netherlands and by Thames Polytechnic.

Project Manager   Dr J.W. James

The Biotol Project

The BIOTOL team

OPEN UNIVERSITEIT, THE NETHERLANDS
Prof. M. C. E. van Dam-Mieras
Prof. W. H. de Jeu
Prof. J. de Vries

THAMES POLYTECHNIC, UK
Prof. B. R. Currell
Dr J. W. James
Dr C. K. Leach
Mr R. A. Patmore

This series of books has been developed through a collaboration between the Open universiteit of the Netherlands and Thames Polytechnic to provide a whole library of advanced level flexible learning materials including books, computer and video programmes. The series will be of particular value to those working in the chemical, pharmaceutical, health care, food and drinks, agriculture, and environmental, manufacturing and service industries. These industries will be increasingly faced with training problems as the use of biologically based techniques replaces or enhances chemical ones or indeed allows the development of products previously impossible.

The BIOTOL books may be studied privately, but specifically they provide a cost-effective major resource for in-house company training and are the basis for a wider range of courses (open, distance or traditional) from universities which, with practical and tutorial support, lead to recognised qualifications. There is a developing network of institutions throughout Europe to offer tutorial and practical support and courses based on BIOTOL both for those newly entering the field of biotechnology and for graduates looking for more advanced training. BIOTOL is for any one wishing to know about and use the principles and techniques of modern biotechnology whether they are technicians needing further education, new graduates wishing to extend their knowledge, mature staff faced with changing work or a new career, managers unfamiliar with the new technology or those returning to work after a career break.

Our learning texts, written in an informal and friendly style, embody the best characteristics of both open and distance learning to provide a flexible resource for individuals, training organisations, polytechnics and universities, and professional bodies. The content of each book has been carefully worked out between teachers and industry to lead students through a programme of work so that they may achieve clearly stated learning objectives. There are activities and exercises throughout the books, and self assessment questions that allow students to check their own progress and receive any necessary remedial help.

The books, within the series, are modular allowing students to select their own entry point depending on their knowledge and previous experience. These texts therefore remove the necessity for students to attend institution based lectures at specific times and places, bringing a new freedom to study their chosen subject at the time they need and a pace and place to suit them. This same freedom is highly beneficial to industry since staff can receive training without spending significant periods away from the workplace attending lectures and courses, and without altering work patterns.

Contents

How to use an open learning text

An open learning text presents to you a very carefully thought out programme of study to achieve stated learning objectives, just as a lecturer does. Rather than just listening to a lecture once, and trying to make notes at the same time, you can with a BIOTOL text study it at your own pace, go back over bits you are unsure about and study wherever you choose. Of great importance are the self assessment questions (SAQs) which challenge your understanding and progress and the responses which provide some help if you have had difficulty. These SAQs are carefully thought out to check that you are indeed achieving the set objectives and therefore are a very important part of your study. Every so often in the text you will find the symbol Π, our open door to learning, which indicates an activity for you to do. You will probably find that this participation is a great help to learning so it is important not to skip it.

Whilst you can, as an open learner, study where and when you want, do try to find a place where you can work without disturbance. Most students aim to study a certain number of hours each day or each weekend. If you decide to study for several hours at once, take short breaks of five to ten minutes regularly as it helps to maintain a higher level of overall concentration.

Before you begin a detailed reading of the text, familiarise yourself with the general layout of the material. Have a look at the contents of the various chapters and flip through the pages to get a general impression of the way the subject is dealt with. Forget the old taboo of not writing in books. There is room for your comments, notes and answers; use it and make the book your own personal study record for future revision and reference.

At intervals you will find a summary and list of objectives. The summary will emphasise the important points covered by the material that you have read and the objectives will give you a check list of the things you should then be able to achieve. There are notes in the left hand margin, to help orientate you and emphasise new and important messages.

BIOTOL will be used by universities, polytechnics and colleges as well as industrial training organisations and professional bodies. The texts will form a basis for flexible courses of all types leading to certificates, diplomas and degrees often through credit accumulation and transfer arrangements. In future there will be additional resources available including videos and computer based training programmes.

Preface

Biotechnology essentially involves the exploitation of the chemical processes and products of living systems. Much of the recent growth in this area of enterprise has undoubtedly arisen from increased knowledge of the molecular mechanisms that underline the basic functions of living systems and from the discovery of restriction enzymes that have enabled the development of genetic engineering. Such advances are, however, only part of the story. Equally important has been the progress made in process technology which has facilitated the application of the new knowledge on a scale, and with suitable economics, to meet market requirements. It is the function of the technologist to turn possibilities which arise through new scientific knowledge into practical realities. In the context of biotechnology, the process technologist's role is vitally important in converting the possibilities arising from new biological discoveries into the products and processes in such diverse areas as healthcare, agriculture, food processing, environmental management and chemical manufacture.

Bioprocess technology has its roots in chemical technology but there are many important and fundamental differences. Almost invariably in bioprocess technology one works with lower concentrations of reactants, in an aqueous milieu and with strict limitations of chemical and physical parameters. We can, however, recognise, for both technologies, a number of underpinning principles and concepts that need to be understood if a logical and disciplined approach is to be made to solving technological problems. This text is designed to provide such understanding.

Central to all processes are issues relating to transport phenomena. How do we deliver reactants (nutrients) to the system effectively? What is the most efficient method to remove end products? Does the delivery of barely soluble gases such as oxygen pose any particular problems? How can heat be supplied or removed from the system to ensure an even and controlled temperature for the process. The first part of the text provides an in-depth discussion of the principles of mass and energy transfer.

Equally important is the ability to model systems. Before embarking upon the considerable expense of building a process, it is essential to understand relationships between, for example, cellular growth and product yield and be able to model this in order to make predictions about the likely performance of the process. The second part of this text provides insight into the strategy for modelling bioprocesses with particular emphasis on cellular and enzyme-based systems. Together, these two facets provide the basis for understanding bioreactor design and operation. They also form the underpinning for the downstream processes needed for product recovery. These latter topics are dealt with in other BIOTOL texts.

Although fundamental, the subjects covered in this text are by no means simple. It is in the nature of the discipline that there are certain pre-requisites required of the reader. If we are to apply the principles covered in a meaningful manner, then the relationships between the various parameters involved and the desired outcome needs to be understood. There is, therefore, a requirement for some mathematical acumen to handle these relationships. Similarly, in modelling cellular systems there is also a requirement for some fundamental knowledge of cellular metabolism. Essential mathematics include differential calculus and the ability to handle differential equations. An appendix (Appendix 5) describing the key manipulations and relationships used in differential calculus has been included and should provide useful reminders for those for whom calculus is not a way of life.

The contributors are at the forefront of contemporary bioprocess technology. They are to be congratulated in using their expertise to provide readers with the opportunity to gain the knowledge needed to contribute to this exciting and expanding area. Not least amongst their achievements has been the integration of the mathematical treatments with textual description and the provision of guidance and help to readers by the many and varied in-text activities. Readers are also enabled to check their understanding by the inclusion of self-assessed questions. The understanding will serve the reader well as a foundation in bioprocess technology.

<table>
<tr><td>Scientific and Course Advisors:</td><td>Professor M.C.E. van Dam-Mieras
Dr C. K. Leach</td></tr>
<tr><td>Senior Technology Adsvisor:</td><td>Professor Ir. K. Ch. A.M. Luyben</td></tr>
</table>

The importance of transfer processes and mathematical modelling in process biotechnology

The importance of transfer processes and mathematical modelling in process biotechnology

1.1 Introduction

The rapid development of contemporary biotechnology since the 1970s holds great commercial, environmental and social promise. Table 1.1 gives some examples of biotechnology processes and products. Already, many applications of biotechnology have been made in the production of therapeutics; in food production and processing; in environmental management and in the production of fine chemicals.

Industrial sector	Examples of products and processes
Pharmaceuticals	vaccines, antibiotics, diagnostics, steroids, alkaloids.
Food	dairy, meat and fish products, beverages (alcoholic, coffee, tea etc), bread, baker's yeast, food additives (antioxidants, colours, flavours), food supplements (amino acids, vitamins, starch products, glucose and high fructose syrups, enzymes, novel foods and fungi.
Agriculture	plant cell and tissue culture for quality assured stock, genetically improved varieties, bio-pesticides, ensilaging and composting, animal feedstuff, manufactured vaccines
Chemicals (bulk)	ethanol, butanol and organic acids; metal extraction.
Chemicals (fine)	enzymes, polymers (gums, agars) and perfumes.
Energy	ethanol, methane and biomass.

Table 1.1 Industrial sectors and biotechnology products and processes

Π Examine Table 1.1 carefully and mark on this Table those products and processes which involve the application of process technology be it for primary production, downstream processing or product transport and packaging.

You may not have been familiar with some of the products listed in Table 1.1. Nevertheless, a moments thought will have enabled you to realise that virtually all biological products will require the application of process technology. In most cases, the requirements of the market, the nature of the product and the needs to retain commercial viability, results in the application of process technology to processes carried out on a large scale. Examination of Table 1.1 could not fail to have impressed you not only by the great diversity and value of bioproducts, but also by the importance of process technology in enabling us to practically apply these bioproducts.

These advances have been made possible through increasing knowledge of how living systems function derived from the discoveries of the biological sciences. Especially important in this context have been the increased knowledge of how genetic material functions within organisms and the characterisation of restriction enzymes and genetic vectors. These latter two enable us to select desirable genetic elements and to introduce them into alternative organisms, a process referred to as genetic engineering. Genetic engineering is however only part of the story. Increased knowledge of cell physiology, molecular biology, biochemistry and developmental biology have all made contributions.

The development of biotechnology is not however merely a reflection of increased knowledge of biological systems. It has depended upon similar advances being made in process technology. This is especially true for processes involving the large scale cultivation of cells. Examples are the production, isolation and purification of metabolic products such as enzymes, antibiotics and antibodies on a large scale and the large scale use of enzymes as catalysts for chemical transformations. It is process technology that has enabled the biotechnological processes to be scaled up to make products on the desired scale and with the desired purity. Without the advances in process technology, many potential applications of biological discoveries will not be realised.

The purpose of this chapter is to outline briefly the importance of process technology in biotechnology and to contrast this with the process technology associated with chemical technology. It is also a function of this chapter to explain how the various aspects of bio-process technology have been divided up within the Biotol series of texts and to provide a context for the remaining chapters of this text.

1.2 A brief history

Process biotechnology is not new. Methods for producing alcoholic beverages, cheese and bread have been known for thousands of years. The early technology associated with these processes was however based on the empiricism of trial and error and tradition. During the 19th Century, the scientific basis of these processes emerged and began to be applied to fermentation processes. These developments were paralleled by increasing knowledge of the factors that governed the behaviour and performance of bio-processes.

The development of the science and technology associated with bio-processes has continued. Thus improvements were made to existing processes and new processes emerged. The production of the antibiotic penicillin was a particularly spectacular advance. The crucial issue arising from Fleming's 1921 discovery of penicillin was how to produce it on sufficient scale to meet the needs arising from infection. The production of penicillin remains one of the most technically demanding processes. There has also been continued development in the technology associated with food production.

The other demands on the continued development of bio-process technology arise through:

- the diversification of the biological systems used for biotechnological purposes;

- the diversification of the products from biotechnology;

- the commercial needs for larger scale processes and greater automation;

- the regulatory requirement to maintain good manufacturing practise and to provide quality assurance.

We can identify these demands as arising from the requirements of the consumer, the needs of the producer to remain competitive and from the regulatory issues arising from national and international regulatory bodies.

Thus this century, especially lately, has seen considerable advances. These advances have been made in the design of bioreactors and in downstream processing. Increasing emphasis has been laid on the development of mathematical models of the processes to enable calculation of optimal conditions. There has also been emphasis on continuous monitoring and control and on process stability.

Bio-process technology has many similarities with chemical process technology. In the next section we will examine why this is so but we will also mention in which ways the two technologies differ.

1.3 Biotechnology as a special form of chemical technology

A key feature of all living systems is that they carry out metabolism. By metabolism, we mean they take in one set of chemicals (nutrients) and produce a new set (biomass or other metabolic products). The range of chemicals that living systems can make is enormous. They include all of the many thousands of complex chemicals (eg proteins, carbohydrates, nucleic acids and lipids) which make up biomass and the wide variety of metabolic products (eg CO_2, ethanol, organic acids and secondary metabolics etc). Amongst these products are those that are of value as foods, medicines and fine chemicals. We can, therefore, regard biological systems as superb chemical factories. Biotechnology is concerned with making use of either the chemical products or the chemical transformations brought about by biological systems. In this sense, biotechnology has its parallels with chemical technology. But, however, it differs from chemical technology in some important respects. We can list these as:

- the range and complexity of the chemical transformations that can be achieved;

- the fact that bioproducts are made in an aqueous environment and often at relatively low concentrations;

- the high degree of stereospecificity of the chemical products made by biological systems;

- the sensitivity of the biological systems to physical and chemical parameters;

- the ability of the biological systems to reproduce and multiply.

Thus, although we might envisage that many of the principles which underpin chemical technology will be pertinent to biotechnology, we must anticipate some substantial differences between the two.

The special nature of biological systems as 'chemical factories' arises because of their ability to make a wide range of special catalysts (enzymes), their ability to integrate these individual catalysts into functional units to bring about sequential changes to chemicals (metabolic pathways) and to control these pathways and thus the amount of product made. A key function of biologists in biotechnology is to maximise the capability of the biological system to make the product. It is the function of the process technologist to develop processes which will optimise the manufacture and isolation of the product. As in chemical technology, we can therefore identify a number of important stages of concern to the process technologist. These are:

- the preparation of the reactants (feedstocks, ie the upstream processes);

- the optimisation of the conditions within the reactor to maximise product yield;

- the recovery of product(s) from the reactor (ie the downstream processes).

1.4 The division of bio-process technology

It is perhaps tempting to think that the study of bioprocess technology should be divided into upstream, bioreactor and downstream processes. Although some merit can be attached to this, it is not in the first instance the most effective approach. There are a large number of principles which underpin upstream, bioreactor and downstream processes and it is logical to establish these principles before embarking upon the specifics of each process stage. This text explains some of the fundamental principles underpining bio-process technology.

We can divide these principles into aspects of transport and modelling. In Chapters 2-7 we will examine transport phenomena. In the remaining chapters we will explore the methods available for modelling of bioprocesses. The companion Biotol texts 'Operational Modes of Bioreactors', 'Bioreactor Design and Product Yield' and 'Product Recovery in Bioprocess Technology' build upon these principles and examine upstream processes, bioreactors, downstream processes as well as aspects of process control.

1.4.1 The importance of transport phenomena

Think of the process by which oxygen travels from a stream of air being pumped into a bioreactor until it reaches the enzyme reaction sites inside a bacterium.

∏ Write down the sequence of oxygen transfers during this process.

You probably wrote down a sequence similar to this. The oxygen is transported into the bioreactor as part of the gas stream in a pipe. Within the bioreactor it would be dispersed into small bubbles which rise to the surface of the medium. During this process, oxygen diffuses across the air-medium interphase of the bubbles and becomes dissolved into the medium. Dispersion within the medium may be by diffusion or as a result of mechanical mixing of the medium by an impeller or some other device. Equally oxygen dissolved in the medium, diffuses across the medium and the cell surface into the cell.

We can see from this simple example that many different transport processes are taking place. Some, like the flow of the air in a pipe and the mixing of the medium we readily think of on a macro scale. Others, like the diffusion of individual molecules across the surface of a bubble or cell, we immediately tend to think of at the micro- (molecular) level.

The case of oxygen transport described above is an example of mass transport. No doubt you can think of many other examples of mass transport phenomena in bio-processes. Amongst these will be the transport of substrates and other nutrients into cells and the removal of metabolic end products (eg CO_2). You should not however confine your thinking to processes that take place within the bioreactor. Mass transfers also take place in many downstream processes (eg evaporation of solvent, removal of desired end products by dialysis, adsorption of products onto immobilised supports during chromatography etc).

Transport phenomena are not, however, confined to the transfer of mass. Heat can be transported too. This may take place, by conduction or convection. Heat is an important component in process technology. In most processes, temperature has to be maintained within a very small range in order to achieve optimal performance. Many fermentations generate significant quantities of heat and the process-engineer is faced with the problems of removing this heat. Similarly, it is often desirable to carry out downstream processing at temperatures well below ambient. The process engineer is again faced with heat removal. It should not, therefore, be surprising that a substantial portion of this text is devoted to discussion of both mass and heat transport phenomena.

We hinted earlier, in discussing oxygen transport, that transport phenomena can be treated either at a molecular (ie micro) level, or at a macro level. You should anticipate that in order to understand process technology, you will need to have knowledge of how to analyze transport at both micro and macro levels. Again, we will use the transfer of oxygen into cells in a bioreactor as an illustration. The availability of oxygen is often the rate-limiting step that governs growth rates and product yields.

A key question is, which of the various stages of oxygen transport (as a flow in a pipe, diffusion in bubbles, mixing of the medium, diffusion into cells) is rate-limiting. Is it the flow in the pipe or the mixing of the medium or diffusion out of bubbles or into cells that restricts the rate of oxygen availability to the cells? Clearly such a problem needs examining from both the macro and the micro levels.

The analysis of mass and heat transfer is based upon the principles of conservation of mass and energy (ie mass and energy cannot be destroyed). Application of the principles of conservation enable us to construct macrobalances for mass and energy. This text begins by explaining how macrobalances may be constructed, using a variety of situations to describe how macrobalances may be applied. We then move on to consider transport mechanisms and processes including molecular and convective (macro-) transport phenomena. An important tool in carrying out a transfer analysis, is a procedure known as dimensional analysis. We devote a chapter to explaining this technique and to providing experience of its application. We conclude this section of the text by examining heat and mass transport in depth. The knowledge you gain from these chapters will enable you to analyze a wide variety of processes relevant to biotechnology.

1.4.2 Importance of mathematical modelling

Consider the following process. It is proposed that we grow an aerobic micro-organism in a fermentor to produce an antibiotic. After growth, the micro-organism is to be separated from the culture medium by centrifugation. The spent culture medium containing the antibiotic is then to be acidified and the antibiotic is to be separated by ion exchange chromatography. It is subsequently to be eluted from the ion exchange resin and then the concentrated solution of antibiotic is to be filtered and then dried in a spray drier.

The process is to be conducted on a large scale to produce commercial quantities of the antibiotic.

Π You might find it useful to make a sketch of this process to refer to as you read the following discussion.

Before we embark on building such a process there are a large number of pieces of information that we need to know.

Π Make a list of at least 10 factors we will need to know before designing the process in detail.

We do not propose to try to guess all 10 items you listed, but high on your list should have been how much antibiotic the cells make and how big the market is for the antibiotics. Clearly, these two items will govern the overall size of the bioreactor needed. In turn, this will also govern the capacity of the centrifuge(s) needed to remove the cells; the size of the acid storage tank and the pump used to add it to the 'spent' medium; the size of the ion exchange chromatography column and of the spray drier. The capacity of the cells to make the antibiotic will be governed by the physical (pH, temperature) and chemical (nutrients etc) parameters. In principle we need to know which of these parameters are important and, therefore which need to be monitored and controlled. These factors too will influence the design and size of the bioreactor.

The size of the bioreactor is not the only key issue. The rate of growth of the cells and their growth yields on the substrate will govern the rate which we will need to feed in substrate and remove metabolic end products such as CO_2 and heat. The size and shape of the vessel will also influence a number of parameters such as the efficiency of O_2 transport into the vessel.

Even from this relatively simplistic description, it should be evident that the design of a biotechnological process is a complex matter especially if the desired end point is to produce a highly integrated process in which there is neither excessive over- or under capacity at each stage. Such a process might be developed on a trial and error basis but that is an extremely costly approach both in terms of time and equipment. A more profitable approach is to mathematically model the process. Then one can examine the consequences of changing a parameter without the expense of running costly experiments.

Central to this approach to bio-process design and development is to use chemical, biochemical and physiological knowledge to develop a model of a bioprocess and to use this model to describe growth and product formation. Chapters 8-11 will explain how mathematical modelling is carried out. We describe how a system may be defined and use the principles of the conservation of elements and energy, macroscopic balances and metabolic knowledge to construct simple models of bioprocess fermentations. In the final chapter, we examine the kinetics of single enzymes, metabolic pathways and metabolism in general to demonstrate how these kinetics may be applied to the mathematical description of fermentation processes.

1.5 Assumed knowledge

As we have explained, this text primarily concerns developing the reader's knowledge in two important areas which are necessary for the design and operation of efficient bio-processes. These are, transport phenomena and mathematical modelling. These two issues, although fundamental to the understanding of bioprocess technology, are not simple and demand some previous knowledge especially of important and basic physico-chemical principles (eg laws of thermodynamics; gas laws; conservation laws etc) . Competence in mathematics is also required.

1.6 Symbols

There are many different symbols used by different authors to represent quantities (eg mass, time, temperature) and processes (eg flow rates, reaction rates). In this text we describe the meaning of the various symbols used as we introduce them. Nevertheless it may prove difficult for you to remember them. Therefore, we have collected the symbols together into a glossary for easy reference. This glossary has been split into two sections. The first covers the symbols used in the discussion of transfer (transport) processes (Chapters 2-7); the second covers the symbols used in the Chapters (8-11) dealing with mathematical modelling. These are presented in Appendices 1 and 2.

Symbols are of course a form of shorthand. Even with a wide range of type faces and available symbols, these are still insufficient to go round all of the quantities and processes we wish to represent. Thus we have adopted D for diameter and D for diffusion coefficient. The context in which these are used should enable you to distinguish between them. Nevertheless you will need to read equations carefully.

Conservation laws of mass and energy

Conservation laws of mass and energy

2.1 Introduction

In this chapter we are going to apply the conservation laws of mass, and energy to enable us to write balances for the transport of mass, and energy. The ability to write such balances is of great importance.

Balances and conservation laws are tools widely used throughout physics, also play a role of paramount importance in describing transport phenomena both in nature (eg meteorology) and in the world of (bio-)chemical and environmental engineering. The basic principle of these balances and conservation laws is analogous to keeping a ledger (in book keeping) but for a particular physical quantity. The concept is especially useful in dealing with so-called 'conserved quantities': quantities like mass and energy which remain conserved during the transport process under consideration. Setting up balances accounts for the transport of the particular quantity over the boundaries of the control volume of interest. This technique of book keeping can be pursued on a very small scale, and then results in 'micro-balances'; it can also be applied to larger scales and leads to 'macro-balances'. In this chapter we will restrict ourselves to the latter case, ie to the macro-balances.

macro-balances

2.1.1 General scheme for setting up a balance

It is always a good idea to begin by giving a verbal description of the transport process that you wish analyse. To do this we need to:

- choose the 'conserved quantity', G, that is transported in the process of interest;

- choose the 'control volume', V;

- calculate the changes in the total amount of G in volume V during a small time interval Δt. There may be 'flow' into V, 'flow' out of V, and there may be production or loss of G inside V. Note that G simply represents the quantity one is interested in: for example, when calculating a temperature T, a thermal energy balance must be set up and consequently G stands for the thermal energy E.

We can write a verbal description of the balance:

$$\boxed{\frac{\text{increase of G inside V}}{\Delta t}} = \boxed{\begin{array}{c}\text{flow of G} \\ \text{into V (in } \Delta t)\end{array}} - \boxed{\begin{array}{c}\text{flow of G} \\ \text{out of V (in } \Delta t)\end{array}} + \boxed{\begin{array}{c}\text{net production} \\ \text{of G inside V} \\ \text{(during } \Delta t)\end{array}}$$

$$(E - 2.1)$$

We can convert this into a mathematical form.

The symbol ϕ will be used to indicate any kind of flow. Taking the limit $\Delta t \to 0$ results in the expression:

$$\frac{dG_{inV}}{dt} = \phi_{G,in} - \phi_{G,out} + P_G \qquad\qquad\qquad \text{(E - 2.2)}$$

where $\phi_{G,in}$ is the flow of G into the control volume of V, $\phi_{G,out}$ denotes the flow of G out of V, and P_G represents the net production of G inside V per unit of time (this will be negative for loss of G in V). If for a particular quantity G (eg, total mass M) the net production P_G always equals zero, then Equation 2.2 simplifies to:

$$\frac{dG_{inV}}{dt} = \phi_{G,in} - \phi_{G,out} \qquad\qquad\qquad \text{(E - 2.3)}$$

Effectively Equation 2.3 describes a conservation law (ie G is conserved).

Compare Equations 2.1 and 2.2. They are really saying the same thing except one says it in words, the other in mathematical symbols. You should note that there are no universally accepted conventions for these mathematical expressions. Some authors for example use superscripts, others subscripts. The system we have adopted here is however widely used.

You might find it helpful to pin up a piece of paper and write on it the various symbols we have used in this text together with a brief description of what they represent. We have included an extensive list of the symbols used in the appendix.

Now let us have a practice at setting up a balance. We have choosen a non-biochemical one to demonstrate that we can write balances for all sorts of things.

∏　Consider the amount of money (m) present in a bank (which is our 'control volume). Using Figure 2.1, see if you can write a balance for money.

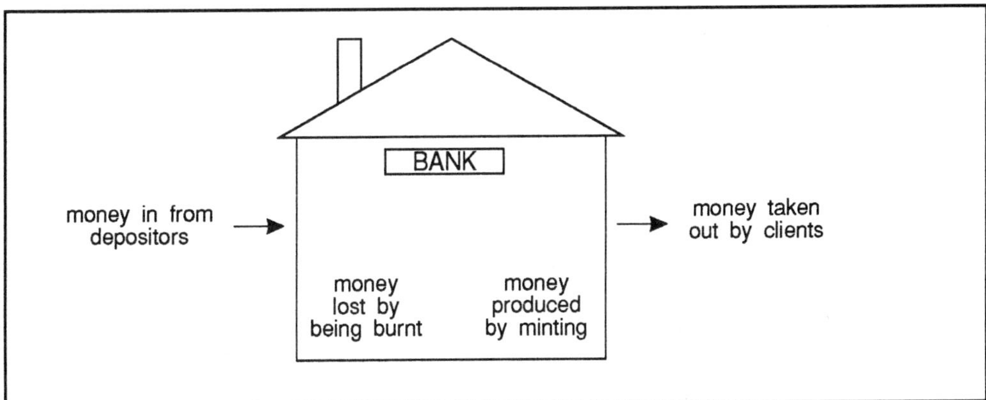

Figure 2.1 Change in money in a bank (see text).

We have formulated a balance thus:

$$\frac{dm}{dt} = \text{flow}_{in}^{via\ depositors} - \text{flow}_{out}^{via\ clients} + \text{production by minting - loss by burning}$$

$$\text{(E - 2.4)}$$

You might have written this in a slightly different form. For example:

$$\frac{dm}{dt} = \phi_{in} - \phi_{out} + P_{mint} - L_{burning}$$

where ϕ = flow, P_{mint} = production by minting, $L_{burning}$ = loss by burning.

The benefit of balances is that they enable analysis of processes and, if necessary, to adjust them. Thus they allow for an analysis of the weak and strong points of the process under consideration and thus are helpful in optimisation eg, a mass balance on a product can indicate losses at each stage of a purification process involving several steps.

∏ What would happen to Equation 2.4 if there was no production or loss of money in this bank?

The equation would simplify to a form as in Equation 2.3. Thus:

$$\frac{dm}{dt} = \phi_{in}^{via\ depositors} - \phi_{out}^{via\ clients}$$

(E - 2.5)

This is the case we find in an ordinary high street bank.

2.1.2 The mass balance

mass balance   The first type of balance that we will deal with extensively is a mass balance. Book keeping is done for the mass of substance A (named M_A) present in a 'control volume' V.

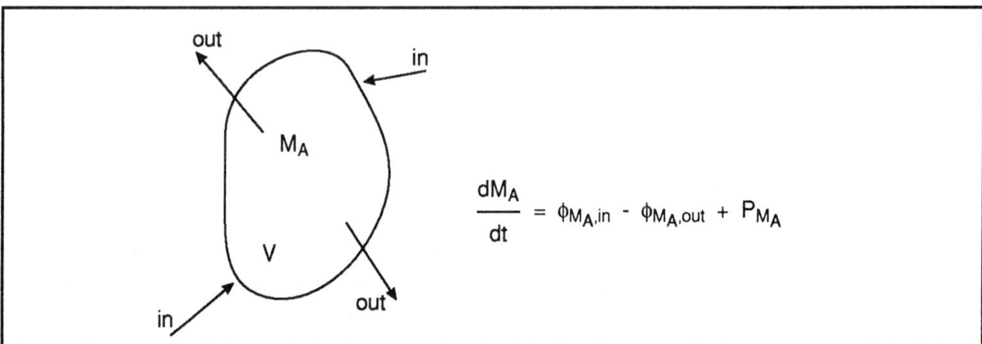

Figure 2.2 Representation of a macro-balance relating to a mass of substance A (M_A).

Let us examine the equation shown in Figure 2.2 a little more closely to learn something about the conventions that are used so that you can read these types of equations more easily.

Already we have learnt that we have used the symbol ϕ to represent a flow: the term $\phi_{M_A,in}$ is a short-hand way of saying the mass (M) of substance A (A) which flows (ϕ) into (in) the control volume. Likewise $\phi_{M_A,out}$ is a short-hand way of saying the mass (M) of a substance A(A) which flows (ϕ) out (out) of a control volume. P_{M_A} is a short hand way of saying the net production (P) of mass (M) of substance A (A).

If the system under consideration in Figure 2.2 consists of a single component there is no mass production term (P_{M_A}), since total mass itself is a conserved quantity. Consequently, Equation 2.5 becomes:

$$\frac{dM_A}{dt} = \phi_{M_A,in} - \phi_{M_A,out}$$

(E - 2.6)

In the case of a multi-component system, one has to put up several mass balances for the various substances. In addition, creation and/or loss of the various components is possible via chemical reactions inside the 'control volume' V (probably a reaction vessel) and has to be taken into account. In such a case it is convenient to introduce

concentrations   concentrations C (or mass densities).

Thus: for A: C_A = mass of A per unit volume = M_A/V; for B: C_B = mass of B per unit volume = M_B/V

(E - 2.7)

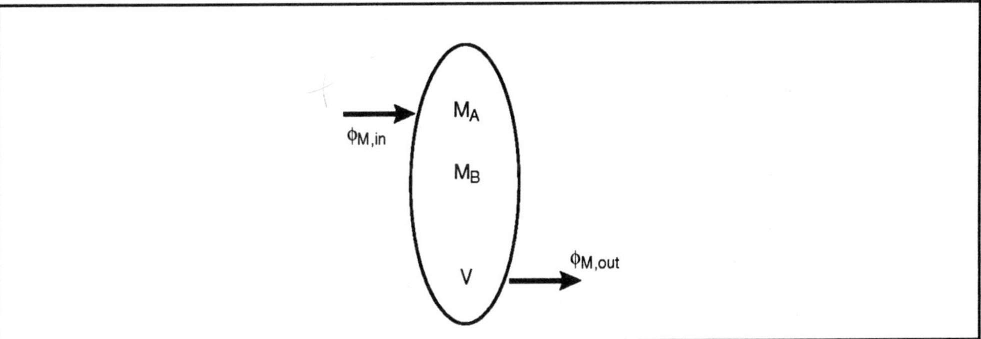

Figure 2.3 Representation of a mass-balance relating to a multi-component system.

mass density   For the total mass density $\rho = M/V$ the following relation holds:

$$\rho = \frac{M}{V} = \frac{M_A + M_B + \ldots\ldots}{V} = C_A + C_B + \ldots\ldots$$

(E - 2.8)

There may be several flows into and out of the reactor, each with its particular composition. Thus in general the concentrations of the various components of the in- and outflows must be specified as well. For example, the mass flow rate of component A in Figure 2.3 into the reactor is given by:

$\phi_{M_A,in}$

= mass of A which flows in per unit of time

= (concentration of A in the incoming flow) * (volume which flows in per unit time)

= $C_{A,in} \cdot \phi_{V,in}$

(E - 2.9)

with $\phi_{V,in}$, denoting the volumetric flow rate into the reactor.

The net-production rate per unit volume (r_A) is defined as:

$$r_A = \frac{P_A}{V}$$

$$(E - 2.10)$$

Now let us use these relationships to determine a mass balance for a substance A.

We can write overall that (for constant volume V):

$$\frac{dM_A}{dt} = \frac{d}{dt}(C_A V) = V\frac{dC_A}{dt}$$

$$(E - 2.11)$$

The flow into the system is:

$$\phi_{M_A,in} = \phi_{V,in} \cdot C_A$$

$$(E - 2.12)$$

The flow out of the system is:

$$\phi_{M_A,out} = \phi_{V,out} \cdot C_{A,out}$$

$$(E - 2.13)$$

But remember that the increase in a component in a defined volume V = the flow into the defined volume - flow out of the volume + the net production in the volume.

That is:

$$\boxed{\begin{array}{c}\text{change in amount}\\\text{in a defined volume}\end{array}} = \boxed{\begin{array}{c}\text{flow into}\\\text{the volume}\end{array}} - \boxed{\begin{array}{c}\text{flow out of}\\\text{the volume}\end{array}} + \boxed{\begin{array}{c}\text{net production}\\\text{within the volume}\end{array}}$$

Therefore, substituting into this equation yields:

$$V\frac{dC_A}{dt} = \phi_{V,in} \cdot C_{A,in} - \phi_{V,out} \cdot C_{A,out} + r_A V$$

$$(E - 2.14)$$

This can be rewritten as:

$$\frac{dC_A}{dt} = \frac{\phi_{V,in}}{V} \cdot C_{A,in} - \frac{\phi_{V,out}}{V} \cdot C_{A,out} + r_A$$

$$(E - 2.15)$$

Note that all the terms in Equation 2.15 have the dimension of amount/unit volume/unit time (in SI units kg m^{-3}s^{-1}).

A particular application of the mass balance applies when the reactor volume (V) is constant and $\phi_{V,in} = \phi_{V,out}$. This principle is important in analysing many bioprocesses. For example it is applicable to traditional Continuous Stirred Tank Reactors (CSTRs) and to the continuous operation of airlift and pumped airlift reactors.

CSTR

Now that we have established the principles behind mass balances, let us consider some examples.

Example 1:

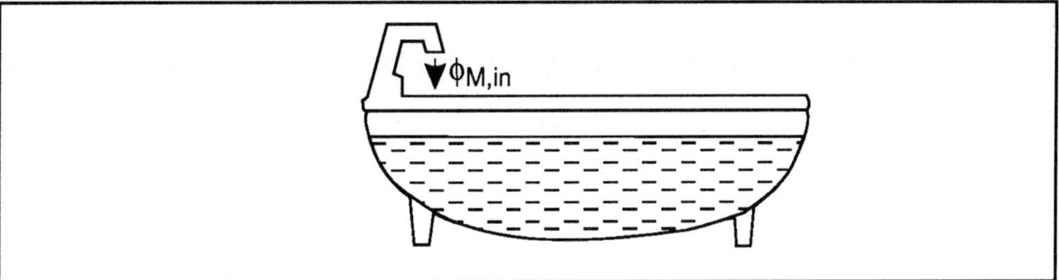

Figure 2.4 Example 1 - the filling of a bath (see text).

Consider the filling of a hot bath. There is no flow out of the bath and there is no production of water inside the bath. Thus the mass balance simplifies to:

$$\frac{dM}{dt} = \phi_{M,in} \; (= \text{constant})$$

(E - 2.16)

with the solution:

$$M(t) = \phi_{M,in} * t + \text{const}$$

(E - 2.17)

where const represents the mass of liquid at $t = 0$

Ⅱ Consider the kettle illustrated in Figure 2.5. See if you can write down a mass balance for the water in the kettle before reading on. Assume that the water in the kettle is boiling. You will need to consider both the liquid and vapour phases.

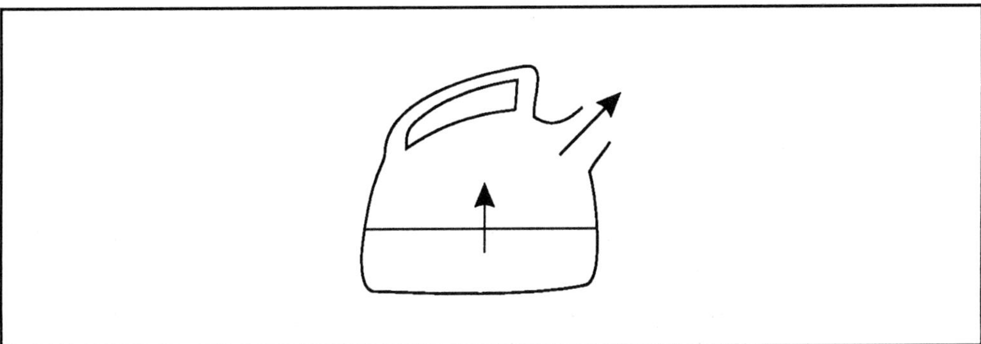

Figure 2.5 Water in a kettle. Arrows indicate movement of water (see text for details).

When the water inside the kettle of Figure (2.5) is boiling, there are two mass flows we have to take into account. The first one is the vapour flow ($\phi_{M,out}$) out of the kettle, while the second one is the evaporative mass flow ($\phi_{M,ev}$) from the liquid water into the vapour phase (inside the kettle). Thus for the liquid phase in the kettle the mass balance reads as:

$$\frac{dM_l}{dt} = - \phi_{M,ev}$$

(E - 2.18)

where M_l is the mass of the water in the liquid phase. Note that $\phi_{M,ev}$ is preceded by a minus sign since the water is lost from the liquid phase by evaporation.

For the vapour phase inside the kettle, the relationship in:

$$\frac{dM_v}{dt} = +\phi_{M,ev} - \phi_{M,out}$$

(E - 2.19)

where M_v is the mass of the water in the vapour phase. Note that $\phi_{M,ev}$ here is positive as it increases the mass of water in the vapour phase.

For the total mass of water inside the kettle the balance results in:

$$\frac{dM_{tot}}{dt} = \frac{d}{dt}(M_l + M_v) = -\phi_{M,out}$$

(E - 2.20)

where M_{tot} is the total mass of the water in the kettle.

Overall, there will be a loss of water from the kettle at a rate equal to the vapour flow.

⊓ Let us try another example. An ideally mixed tank (as in a continuous stirred tank reactor - CSTR) is filled with coconut-oil. There is a volumetric flow ϕ_V into, through and out of the tank. For an ideally mixed tank the composition of its contents is the same throughout the tank. This implies that at any time the composition of the outflowing liquid is similar to that of the tank contents. Now consider the following experiment. At $t = 0$ the feed to the inlet is suddenly switched to palm-oil. The question is how will the concentration of the coconut-oil in the outlet respond to this stepwise change at the inlet?

Use the following to help you. $C_{c,in}$ = concentration of coconut-oil in the inlet, $C_{c,out}$ = concentration of coconut-oil in the outlet. ϕ_V is the flow rate into and out of the vessel. Remember that $C_{c,out}$ = the concentration of coconut-oil in the tank = C_c.

We have represented this system in Figure 2.6.

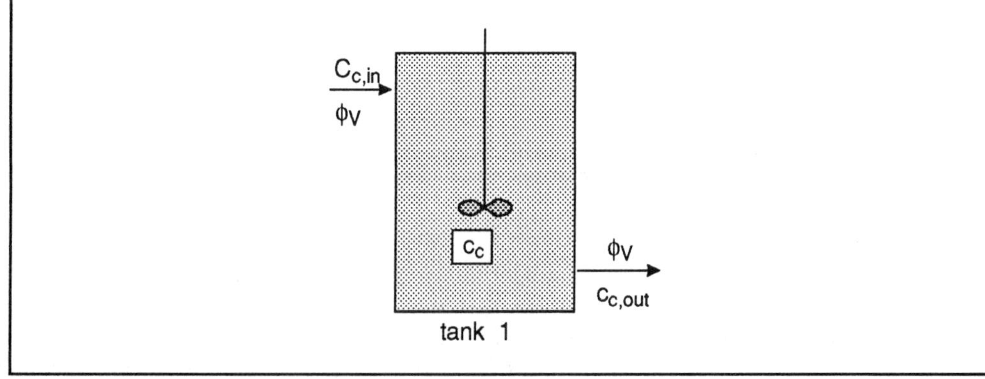

Figure 2.6 Ideally mixed tank (CSTR). See text for details.

Our response to this question is to first write a balance. Remember the change in coconut-oil in the tank = input of coconut-oil - output of coconut-oil + net production of coconut-oil.

$$\frac{dM_c}{dt} = \phi_V\, C_{c,in} - \phi_V\, C_c \text{ or } V\, \frac{dC_c}{dt} = \phi_V\, C_{c,in} - \phi_V\, C_c.$$

Since the concentration of coconut-oil in the inflow is 0 then we can deduce that $\phi_V\, C_{c,in} = 0$.

The mass balance for the coconut-oil therefore reads as:

$$V\frac{dC_c}{dt} = -\,\phi_V\, C_c \tag{E - 2.21}$$

Clearly the value of C_c changes with time.

Mathematically we can describe the value of C_c at time $t = 0$ as:

$$C_c\,(t{=}0) = C_{c,0} \tag{E - 2.22}$$

This enables us to mathematically solve Equation 2.21 as:

$$C_c(t) = A\ \exp\left(-\frac{\phi_V}{V}\,t\right) \tag{E - 2.23}$$

The integration constant A is found from the initial condition and equals $C_{c,0}$. Thus the response at the outlet is given by:

$$C_c(t) = C_{c,0}\ \exp\left(-\frac{\phi_V}{V}\,t\right) \tag{E - 2.24}$$

Thus the concentration of coconut-oil can be calculated if its initial concentration ($C_{c,0}$), the volume (V) of the vessel and the flow rate (ϕ_V) are known.

Note our convention: we have written $\exp\left(-\dfrac{\phi_V}{V}\,t\right)$ to represent the exponent to the power of $\left(-\dfrac{\phi_V}{V}\,t\right)$. Other ways of writing this is $e\left(-\dfrac{\phi_V}{V}\,t\right)$ or $e^{-\frac{\phi_V}{V}t}$.

Now that we have worked through some examples, see if you can work out the numerical case in SAQ 2.1.

SAQ 2.1

Coconut-oil is being pumped through an ideally mixed tank of the type illustrated in Figure 2.6. The volume of the tank is 4 l and the oil is being pumped into the tank at 2 l h^{-1}. At 2pm the coconut-oil feed into the vessel is replaced by a palm-oil feed with a similar flow rate as that of the coconut-oil. What will be the relative proportions of coconut-oil and palm-oil in the outflow at 3pm? (We suggested that you describe pure coconut-oil as having a concentration of 100% coconut-oil and pure palm-oil as having a concentration of 100% palm-oil).

2.1.3 Mole balances

It is not always appropriate to work with a mass balance. Sometimes it is more convenient to use the number of particles (ie the number of moles) of a substance as the property we want to book keep. Then we set up a mole balance. Remember that the mass of one mole of a substance is the molecular weight of the substance in grams. Thus 100g of a substance with molecular weight of 50 contains 2 moles. Basically we use the relationship:

$$\text{moles} = \frac{\text{mass (in grams)}}{\text{molecular weight}}$$

To demonstrate the use of mole balances the production of butanol and isopropanol in a pilot plant fermentor by a *Clostridium* (a bacterium) will be discussed in some detail. Under anaerobic conditions the *Clostridium* produces both products from glucose. The governing chemical reactions are:

$$\text{butanol production:} \quad C_6H_{12}O_6 \;\rightarrow\; C_4H_9OH \;+\; 2CO_2 \;+\; H_2O$$
$$\text{glucose} \qquad\qquad \text{butanol}$$

$$\text{isopropanol production:} \quad C_6H_{12}O_6 + H_2O \;\rightarrow\; C_3H_7OH \;+\; 3CO_2 \;+\; 3H_2$$
$$\text{glucose} \qquad\qquad \text{isopropanol}$$

Now let us put some numerical values on our system.

The volume of the fermentor is 200 l, while the concentration of glucose in the feed equals 30 g l^{-1}. It is found that when the volumetric flow rate of the feed is 27 l h^{-1}, no bio-mass is produced in the tank, ie there is no increase in the number of bacteria. Under these conditions the molar concentration of butanol in the outlet is found to be twice as high as the molar concentration of isopropanol in this flow, while there is no glucose in the outlet flow.

The sort of questions we would like to answer are: How large is the gas production under these conditions? What mass of isopropanol is produced per hour?

Since there is no growth of bacteria, the flow scheme is as depicted in Figure 2.7.

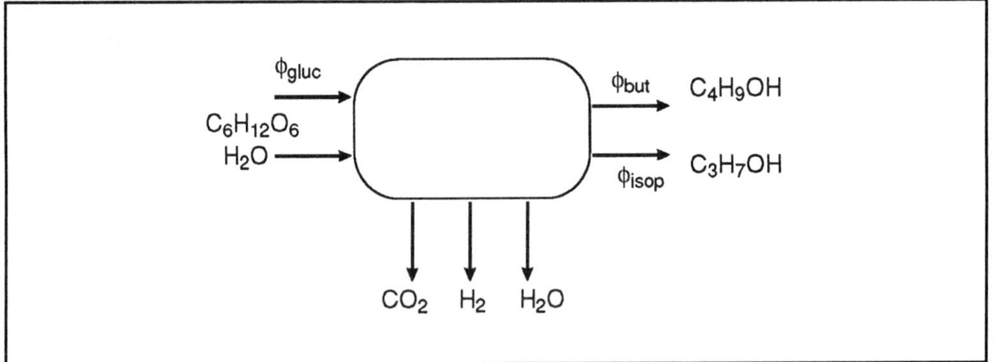

Figure 2.7 Mass transfer in a fermenting system. Where ϕ_{gluc} is the molar flow rate of glucose into the vessel, ϕ_{but} is the molar flow rate of butanol out of the vessel, ϕ_{isop} is the molar flow rate of isopropanol out of the vessel. See text for details.

mole balance

The molar flow of glucose into the tank (ϕ_{gluc}) is completely converted into a butanol flow (ϕ_{but}) and an isopropanol flow (ϕ_{isop}), since no glucose leaves the tank. Under the specified conditions the ratio of these flows equals 2:

$$\frac{\phi_{but}}{\phi_{isop}} = 2 \qquad\qquad\qquad\qquad (E - 2.25)$$

Furthermore, one molecule of glucose is converted into one molecule of butanol or one molecule of isopropanol (from the governing reactions), thus:

$$\phi_{gluc} = \phi_{but} + \phi_{isop} \qquad\qquad\qquad\qquad (E - 2.26)$$

The molar flow of glucose into the tank (ϕ_{gluc}) is calculated from the total volume flow ($\phi_{v,in}$), the glucose concentration ($C_{gluc,in}$) in this flow, and the molar mass of glucose (M_{gluc}), according to the equation:

$$\phi_{gluc} = \phi_{v,in}\frac{C_{gluc,in}}{M_{gluc}}$$

where ϕ_{gluc} is the molar flow of glucose; $C_{gluc,in}$ is the concentration of glucose in the flow in $g\ l^{-1}$; M_{gluc} is the molar mass of glucose (ie the molecular weight in grams).

Π Use the values provided in the description above for $\phi_{v,in}$, $C_{gluc,in}$ and the knowledge that the molecular weight of glucose is 180 to calculate the molar flow of glucose into the tank.

You should have come to a value of 4.5 moles h^{-1} since:

$$\phi_{v,in} = 27\ l\ h^{-1},\ C_{gluc,in} = 30g\ l^{-1} = \frac{30}{180}\ mol\ l^{-1}.$$

Therefore: $\phi_{gluc} = \dfrac{27\ x\ 30}{180} = 4.5\ mol\ h^{-1}.$

Π Now calculate the flow rate of butanol (ϕ_{but}) and isopropanol (ϕ_{isop}) out of the vessel.

You should have determined $\phi_{but} = 3.0\ mol\ h^{-1}$ and $\phi_{isop} = 1.5\ mol\ h^{-1}$.

We know that $\phi_{gluc} = \phi_{but} + \phi_{isop}$ (see Equation 2.26).

We also know that $\phi_{but} = 2\phi_{isop}$ (we told you that the molar concentration of butanol in the outflow was twice as high as the molar concentration of isopropanol).

So $\phi_{gluc} = 2\phi_{isop} + \phi_{isop}$

Therefore $4.5\ mol\ h^{-1} = 3\phi_{isop}$ and thus $\phi_{isop} = 1.5\ mol\ h^{-1}$ and $\phi_{but} = 1.5\ x\ 2 = 3.0\ mol\ h^{-1}.$

Π Now calculate the gas flow (in mol h^{-1}) out of the vessel. To do this you will need to go back to the reaction equations showing the conversion of glucose to butanol and isopropanol.

Your calculation should show you that the gas flow out of the vessel is 15 mol h^{-1}. If you did not get this answer, then the following explanation should help you understand where you have gone wrong.

The gas flow consists of a CO_2 flow and a H_2 flow. From the reaction equation, each mole of butanol is coupled to the production of 2 moles of CO_2. Similarly, for each mole of isopropanol produced, 3 moles of CO_2 are produced and 3 moles of H_2.

We can calculate that the total CO_2 flow must be $2\phi_{but} + 3\phi_{isop}$.

We previously calculated that $\phi_{but} = 3.0$ mol h^{-1} and $\phi_{isop} = 1.5$ mol h^{-1}.

Thus the total flow of CO_2 (ϕ_{CO_2}) = (6 + 4.5) mol CO_2 h^{-1}. Similarly for H_2, 3 x ϕ_{isop} = (4.5) mol H_2 h^{-1}.

The total gas flow (ϕ_{gas}) is therefore (6 + 4.5) + 4.5 = 15 mol h^{-1}.

Thus the total gas flow (ϕ_{gas}) is 15 mol h^{-1}.

Π Now calculate the mass production (P_{isop}) of isopropanol in g h^{-1} given that the molecular weight of isopropanol = 60.

You should come to the conclusion that 90 g of isopropanol are produced h^{-1}.

We know that the molar flow rate of isopropanol (ϕ_{isop}) is 1.5 mol h^{-1}.

If we multiply this by the molecular weight of isopropanol we will convert moles into grams.

Thus the mass flow rate of isopropanol ($\phi_{M,isop}$) = 1.5 x 60 = 90 g h^{-1}.

Since there is no inflow ($\phi_{M,isop,in}$) of isopropanol, the outflow of isopropanol ($\phi_{M,isop,out}$) must be equal to the production of isopropanol (P_{isop}) in the vessel.

ie: $P_{isop} = \phi_{M,isop,out}$. Therefore $P_{isop} = 90$ g h^{-1}.

It might be useful to write the values we have calculated for ϕ_{but}, ϕ_{isop}, ϕ_{gas} and P_{isop} onto Figure 2.7.

SAQ 2.2

Consider the aerobic and anaerobic metabolism of glucose by yeasts. These may be written as two overall reactions:

$$C_6H_{12}O_6 + 6O_2 \longrightarrow 6CO_2 + 6H_2O \qquad \text{aerobic}$$

glucose

$$C_6H_{12}O_6 \longrightarrow 2C_2H_5OH + 2CO_2 \qquad \text{anaerobic}$$

glucose   ethanol

The relative rates of these two processes depend on the availability of oxygen.

A tank which behaves ideally, is set up containing yeasts. Glucose, at a concentration of $30g \, l^{-1}$ is pumped into the tank at $60 \, l \, h^{-1}$. Given that ethanol is produced at 10 moles h^{-1} and that all of the glucose fed into the tank is converted to CO_2 or C_2H_5OH (ie there is no glucose in the outflow), calculate the flow rate of CO_2 out of the vessel. Assume that there is no net growth of yeast and that the molecular weight of glucose is 180. It might be helpful to make yourself a drawing of this set up.

2.2 Residence time distribution

residence time

The time that a certain package of fluid flowing into the tank stays in the tank (the so-called residence time) is not necessarily the same for all packages. On the contrary, for most pieces of equipment there will be variation in the residence times. It can be important to know the distribution of the residence time. For example, in a bioreactor the feed input has to stay long enough in the tank to allow the micro-organisms to consume the substrate, but not too long otherwise we may get substrate limitation or product-inhibition. Also consider sedimentation of solid particles in waste water: the residence times of all packages must have a lower limit in order to let the solids settle. Residence times can be determined using a mass balance approach. A first estimate of the residence time for a tank of constant volume V (operated so that $\phi_{V,in} = \phi_{V,out} = \phi_V$) can be found as follows. Consider the mass balance for component A:

$$V\frac{dC_A}{dt} = \phi_V \, C_{A,in} - \phi_V \, C_{A,out} + r_A V \tag{E - 2.27}$$

or alternatively:

$$\frac{dC_A}{dt} = \frac{\phi_V}{V} (C_{A,in} - C_{A,out}) + r_A \tag{E - 2.28}$$

What this means is that the rate of change in the concentration of A in the system is related to the volumetric flow through the system (ϕ_V), the volume of the system (V); the concentration of A in the inflow ($C_{A,in}$) and outflow ($C_{A,out}$) and the rate at which A is produced in the system.

∏ Using Equation 2.28, what would happen to the rate of change in the concentration of A if we doubled the volume of the vessel. Assume that all other factors remain constant and that A is not produced or destroyed in the system.

You should have concluded that the rate of change in the concentration of A $\left(\dfrac{dC_A}{dt}\right)$ would be halved.

This is because Equation 2.28 reduces to $\dfrac{dC_A}{dt} = \dfrac{\phi_V}{V} (C_{A,in} - C_{A,out})$ because $r_A = 0$ (A is not produced or destroyed).

Thus $\dfrac{dC_A}{dt}$ becomes inversely proportional to the volume of the vessel ie $\dfrac{dC_A}{dt} \sim \dfrac{1}{V}$. Thus by doubling V, we will halve the value of $\dfrac{dC_A}{dt}$.

dilution rate   The term $\dfrac{\phi_V}{V}$ in Equation 2.28 is quite important and is called the dilution rate. It is usually given the symbol D. The inverse of D is $= \dfrac{V}{\phi_V}$ and is called the mean residence time and given the symbol τ.

mean residence time

Thus $\tau = \dfrac{V}{\phi_V}$

∏ What would the mean residence time (τ) be if the volumetric flow into a system was 20 l h^{-1} and the volume of the system was 40 l?

The answer is 2 h because $\tau = \dfrac{40}{20} \dfrac{l}{lh^{-1}} = 2$ h.

∏ What would happen to the mean residence time (τ) if the flow rate was halved and what would be the dilution rate at this new flow rate?

The mean residence time would be doubled ie (4h) because ϕ_V would now be 10 l h^{-1}. Thus $\tau = \dfrac{40}{10} = 4$ h^{-1}.

The dilution rate D would be 0.25 h^{-1} as D $= \dfrac{1}{\tau}$ or $\dfrac{\phi_V}{V}$.

The rate of change in the concentration of A described in Equation 2.28 can be simplified in systems in which A is not being produced to:

$$\frac{dC_A}{dt} = \frac{\phi_V}{V} (C_{A,in} - C_{A,out})$$

$$(E - 2.29)$$

Now that you have understood the relationships between the terms dilution rate, residence time, flow rate and volume, let us see if we can find an explanation for the term residence time.

plug-flow

Consider a small fluid volume that flows into a pipe of volume V. The element is part of a volume flow ϕ_v and flows undistorted through the pipe ('plug-flow'). This is represented in Figure 2.8.

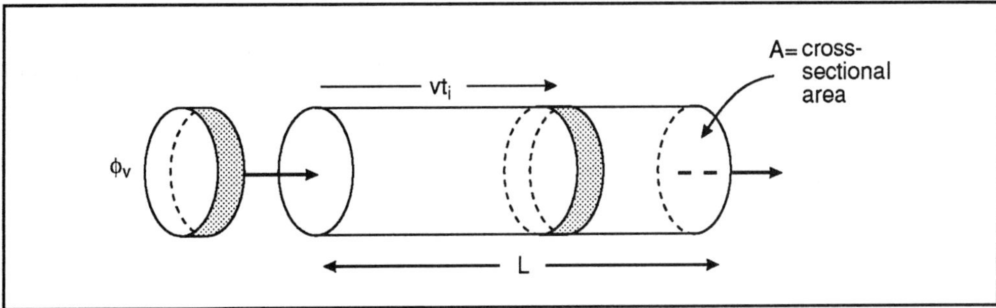

Figure 2.8 Plug-flow (ϕ_v = flow rate, v = velocity of fluid, L = length of pipe). See text for details.

All elements have the same velocity (v) which equals $\dfrac{\phi_v}{A}$ where A is the cross-sectional area of the pipe and ϕ_v is the volumetric flow rate.

At time t = 0, the element is just inside the pipe. After a time interval (t_i), it has travelled over a length l (l will of course be velocity * time = v * t_i). To travel over the whole length of pipe it will take a time $t = \dfrac{L}{v}$ where L = length of pipe and v = velocity of the fluid.

We could also write this as $t = \dfrac{A*L}{A*v}$

But A*L (cross-sectional area * length) = the volume V of the vessel. Also we defined the velocity $v = \dfrac{\phi_v}{A}$. Therefore, $\phi_v = A*v$.

So the time it takes for the fluid element to flow the whole length of the pipe
$$\tau = \frac{L}{v} = \frac{A*L}{A*v} = \frac{V}{\phi_v}.$$

As here there is only one major flow line and the fluid moves with a constant, undistorted velocity (v), the 'mean' residence time is in fact the 'minimum' residence time. However, when a vessel as shown in Figure 2.10 is used the 'minimum' residence time will be completely different from the 'mean' residence time. Such a time period of course measures the time the fluid element was in the pipe - hence the phrase 'mean residence time'.

dimensionless time measurement

Thus far we have defined residence time (τ) in terms of the time an element remains in a system. It has the units of time (eg hours, minutes etc). Another useful way of expressing residence time is to use a dimensionless time measurement given the symbol θ where $\theta = t/\tau$.

Essentially what θ does is to count how many time intervals equal to the residence time (τ) have passed in time t. The relationship between θ and residence time is important so try and remember it!

In the section above, we described a system through which everything was passing at the same velocity. In real life, this is very rarely the case. Some mixing or drag by the walls of the vessel almost inevitably occurs. Thus for most systems there is not a single residence time for all elements within the system. More usually, there is a distribution of residence times.

residence time distribution

To find the residence time distribution for a particular apparatus one performs an experiment in which one changes the input in a known way and measures the response at the exit of the device. The two most commonly used input signals are the 'step' and the 'pulse'. There are several ways to characterise the residence times. These are by terms known as:

- the E function;

- the C function;

- the F function.

2.2.1 The E-function

E-function

The definition of the E-function (exit age distribution) is:

$E(\theta)d\theta$ = volume-fraction of the outflow with dimensionless residence time θ between θ and $\theta + d\theta$. Remember that $\theta = \dfrac{t}{\tau}$ where t = time and τ = mean residence time. It is a ratio of two times namely actual time: mean residence time and is, therefore, dimensionless.

Thus a method for determining $E(\theta)$ would be: take a sample of the outflow and determine the fraction in it with a residence time θ between θ and $\theta + d\theta$ and do this for all θ's. An example of an E-function is given in Figure 2.9. Note that if one integrates $E(\theta)$ over the interval $(0,\infty)$, the integral equals 1, since this integral calculates the volume fraction in the sample with $\theta < \infty$ which is obviously 1. (A residence time ∞ means that the element never leaves the device). Mathematically this can be expressed by:

$$\int_{\theta}^{\theta_1} E(\theta)\ d\theta = \text{volume fraction of the sample of the outgoing flow with residence time}$$

between θ and θ_1. Thus:

$$\int_{0}^{\infty} E(\theta)\ d\theta = 1$$

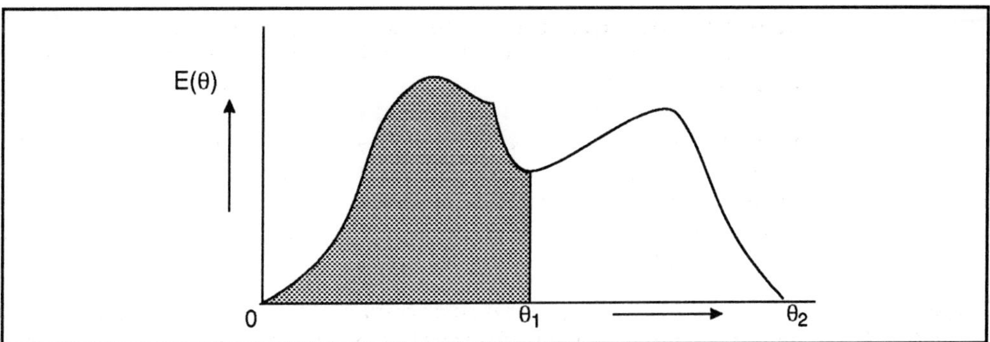

Figure 2.9 Plot of E-function versus θ ($\theta = t/\tau$ where t = time τ = residence time).

Flow processes in a sedimentation tank

Consider the flow of a process fluid into a sedimentation tank.

In Figure 2.10 the flow lines are drawn in a sedimentation tank of length L. The flow comes in with a velocity v and thus one expects for the minimum residence time τ_{min} = L/v.

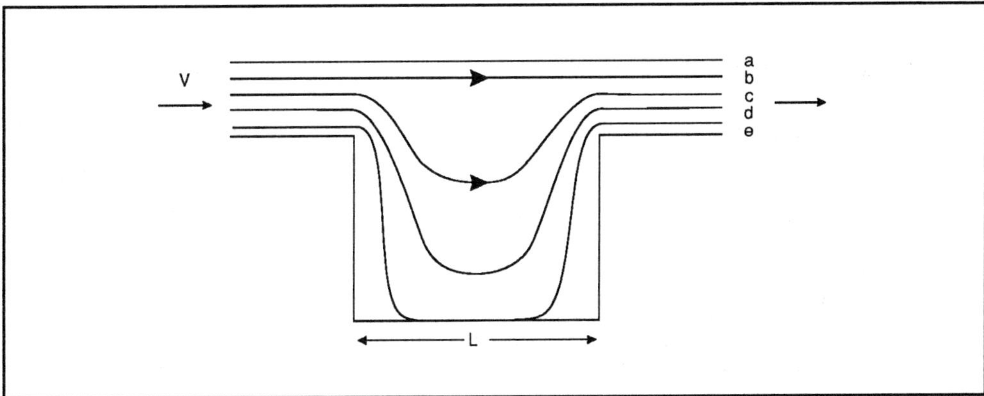

Figure 2.10 Stylised flow times in a sedimentation tank. a-e = flow lines.

Ⅱ Which of the flow lines (a-e) is most likely to give the minimum residence times (τ_{min})?

a and b because these go straight through the vessel. Packages of liquid following flow lines c, d and e take a longer route through the vessel and consequently they have increasingly longer residence times. Flow line e is likely to give the longest residence time. There will be a range of residence times. The shortest time any element will be in the vessel will be:

$t = \tau_{min}$ where τ_{min} = minimum residence time

All other residence times will be longer. Thus for the tank described above, a plot of E (θ) against θ will be similar to that shown Figure 2.11. Remember that $\theta = \dfrac{t}{\tau}$.

where τ = mean residence time

$\prod$ Examine Figure 2.11 carefully. From this figure, what is the value of τ_{min} in terms of τ.

In Figure 2.11, no fluid leaves the tank before $\theta = 0.5$.

That is $\theta_{min}\left(=\dfrac{t}{\tau_{min}}\right) = 0.5$. As $\theta = \dfrac{t}{\tau}$, the minimum residence time $\tau_{min} = \dfrac{1}{2}\tau$.

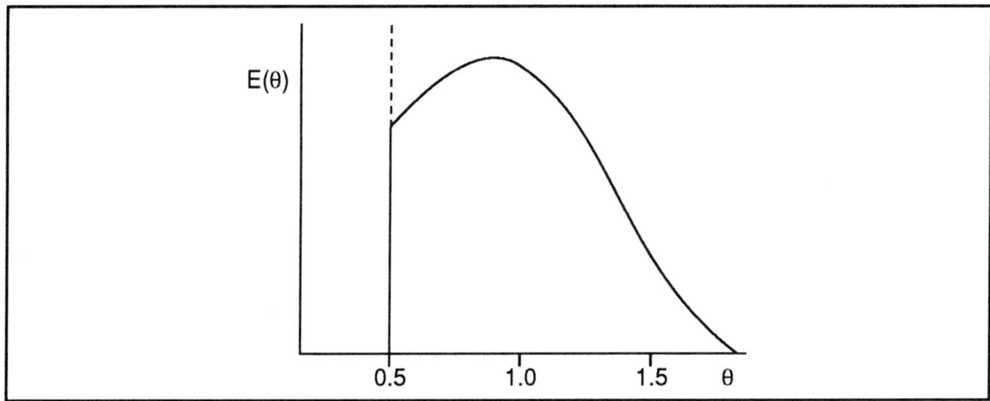

Figure 2.11 E-function versus residence time in terms of θ. (NB $\theta = 1$ when $t = \tau$ and τ = mean residence time).

$\prod$ Draw onto Figure 2.11 an E-function against θ graph for: 1) a shallower tank than that described in Figure 2.10; 2) a pipe in which all the fluid flows at the same rate (ie plug-flow).

A shallower tank means there will be a smaller range of flow paths. This will result in a narrow range of residence times (θ). Thus graphically:

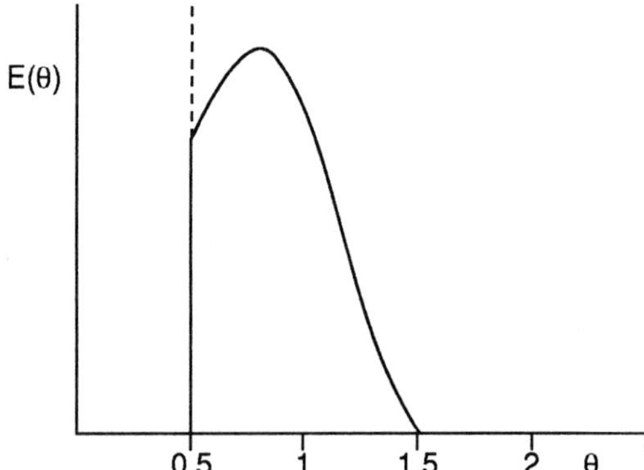

For a pipe in which there is no mixing and in which all fluid flows at the same rate, all elements in the system will have the same residence time and thus will be equal to τ (also τ_{min}). Graphically:

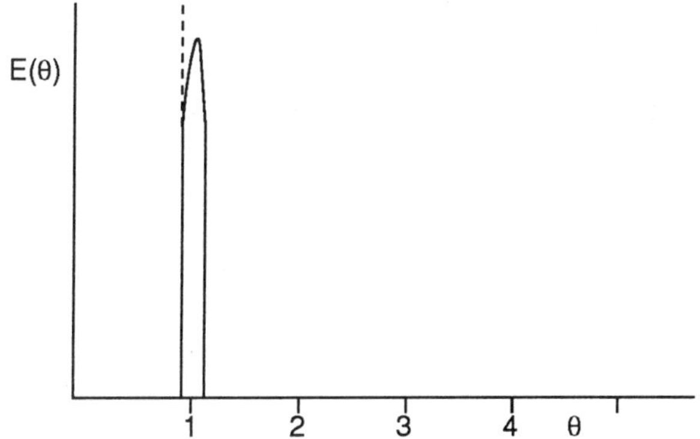

2.2.2 The C-function

One can look at the phenomenon of the residence time distribution in another way. Rather than taking one sample at a fixed moment in time, the response at the exit of a pulse or tracer material at the entrance can be followed. Then find the so-called C-function (C from concentration of tracer). Figure 2.12 shows two graphs in which the concentration of the material A (C_A) is plotted against time. It will be clear that the amount of tracer which is injected as a pulse in the flow of the system (graph A) is present for a larger period of time in the outflow (graph b). In the first graph a is present in the inflow into the system for a short period (ie it passes into the system as a short pulse). The second graph plots the concentration of A (C_A) in the outflow over a period of time.

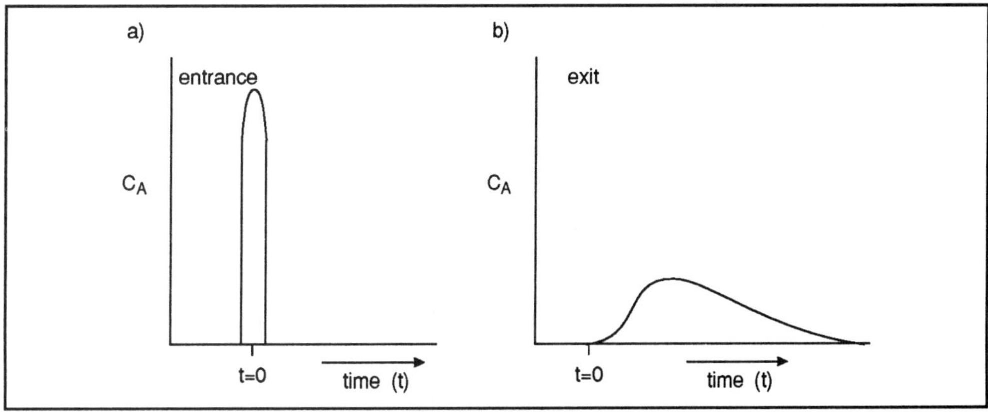

Figure 2.12 Concentration of tracer A versus time at entry a) and exit b) of a vessel (see text for details).

Mathematically we can treat this in the following way.

C-function   The total amount of tracer A injected in the vessel (with volume V) can be conceived as the product of the volume V and a concentration $C_{A,0}$ as if the injected tracer is distributed over the volume V instantaneously. The total amount of tracer is, therefore, V $C_{A,0}$. The C-function is defined as:

$$C(\theta) = \frac{C_{A,out}\,(\theta)}{C_{A,0}}$$

$$(E - 2.30)$$

with $C_{A,out}\,(\theta)$ denoting the concentration of tracer in the exit flow at dimensionless time θ. The mass balance (for $t > 0$) for the tracer material reads as:

$$\frac{dM_A}{dt} = \phi_V\,C_{A,in} - \phi_V\,C_{A,out} = -\phi_V\,C_{A,out}$$

$$(E - 2.31)$$

where $C_{A,in} = 0$ for $t > 0$.

Integration of the left-hand side of Equation 2.32 from $t = 0$ to ∞ renders:

$$\int_0^\infty \frac{dM_A}{dt}\,dt = -M_A^{tot} = -VC_{A,0}$$

$$(E - 2.32)$$

This can easily be understood: at $t \to \infty$ all of the tracer will have left the vessel.

Thus combining Equation 2.31 and Equation 2.32 gives:

$$VC_{A,0} = \int_0^\infty \phi_V\,C_{A,out}\,dt$$

$$(E - 2.33)$$

Rearranging Equation 2.33 results in:

$$1 = \int_0^\infty \frac{\phi_V\,C_{A,out}}{V\,C_{A,0}}\,dt = \int_0^\infty C(\theta)\,d\frac{t}{\tau} = \int_0^\infty C(\theta)\,d(\theta)$$

$$(E - 2.34)$$

The means of finding the C-function is to follow the concentration of a tracer at the exit after a pulse of tracer has been generated at the inlet.

An example of such a C-function is shown below.

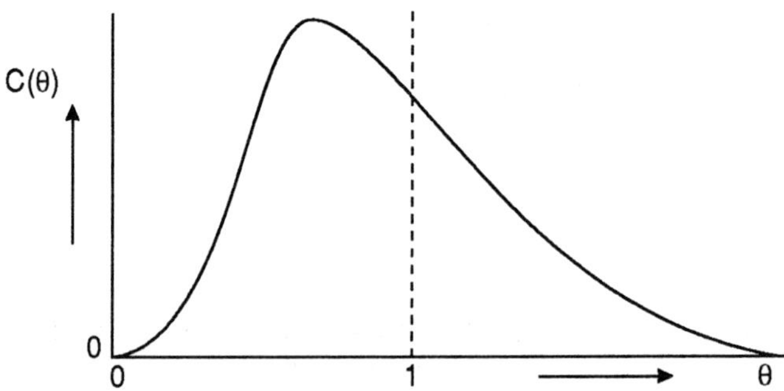

It can be shown that for a steady-state, the functions $C(\theta)$ and $E(\theta)$ are equal.

2.2.3 The F-function

Single tank

Another kind of distribution function found from observing the time behaviour of the concentration in the outlet flow is the F-function. The F-function is defined in the same way as the C-function, but as the response to a step-function at the inlet (Figure 2.13).

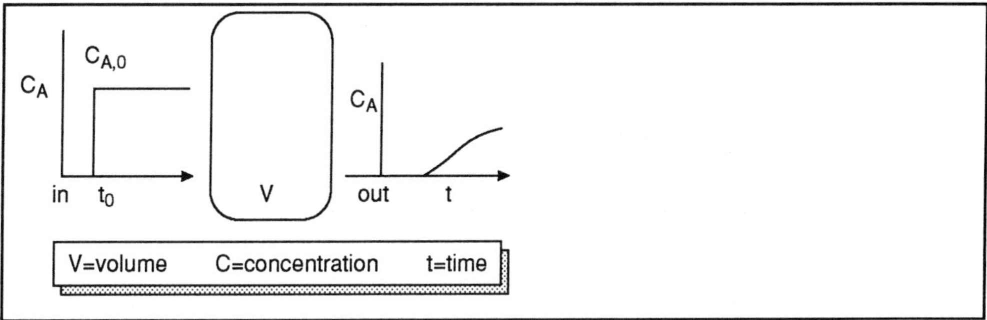

Figure 2.13 Relationship between a step-function at the inlet and the response at the outlet. Note that there is a step up in the concentration in the inlet, but it rises in a curve in the outlet.

∏ If for a fluid flowing through a pipe, all fluid elements were flowing with the same velocity (plug-flow) what would the concentration profile in the outlet look like?

You should have decided that it would show a step-wise increase. This step-wise increase would appear at τ (ie the residence time) after the step-wise increase in the inlet. Thus:

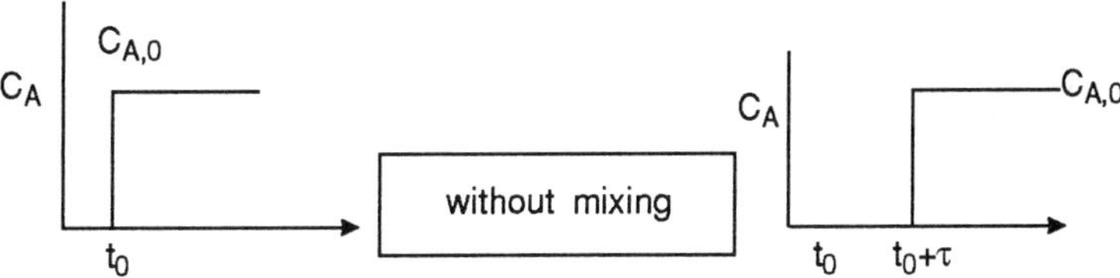

We will be examining plug-flow again later.

In most cases, however, the fluid passing into a vessel is being thoroughly mixed and so the more usual relationship is:

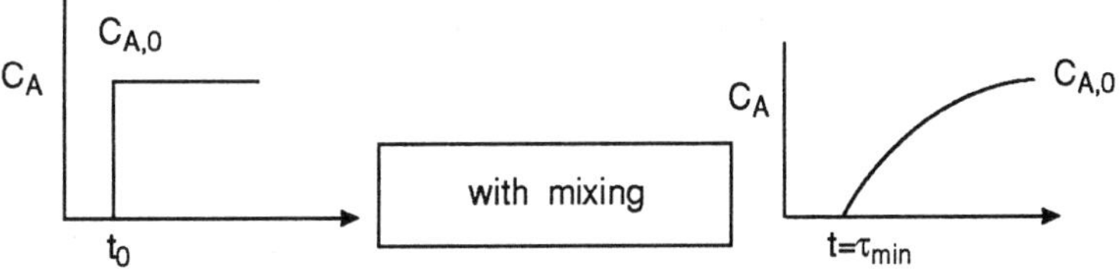

The shape of the outlet profile will of course reflect flow behaviour and the size of the vessel. Mathematically we can handle this in the following way.

We define the F-function by the equation:

$$F(\theta) = \frac{C_{A,out}(\theta)}{C_{A,0}}$$

$$(E - 2.35)$$

F-function   The F-function obviously approaches 1 for $t \rightarrow \infty$, since eventually the concentration in the outlet becomes $C_{A,0}$:

$$F(\theta) \rightarrow 1 \text{ for } \theta \rightarrow \infty$$

$$(E - 2.36)$$

A common form of $F(\theta)$ is drawn in Figure 2.14.

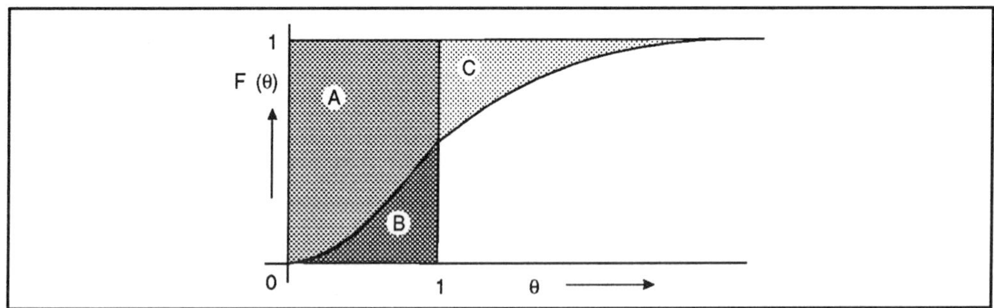

Figure 2.14 Relationship between F-function and residence time (see text for description of the significance of A, B and C).

A simple way to control the reliability of the experimentally derived form for the F-function is obtained by writing down the mass balance for the tracer and integrating the expression over $(0,\infty)$.

$$\frac{dM_A}{dt} = \phi_V C_{A,0} - \phi_V C_{A,out}$$

$$(E - 2.37)$$

thus:

$$\int_0^\infty \left(\frac{dM_A}{dt}\right) dt = \int_0^\infty \phi_V (C_{A,0} - C_{A,out}) \, dt$$

$$(E - 2.38)$$

The left-hand side equals $V\,C_{A,0}$. Since after ∞ time, concentration in the vessel of volume V will be $C_{A,0}$. Thus the total mass of A in the vessel will be $M_A = V\,C_{A,0}$. Thus Equation 2.38 can be written as:

$$1 = \frac{\phi_V}{V} \int_0^\infty \left(1 - \frac{C_{A,out}}{C_{A,0}}\right) dt = \int_0^\infty (1 - F(\theta)) \, d\theta$$

$$(E - 2.39)$$

Thus the sum of the areas A and C in Figure (2.14) is exactly 1: ie A + C = 1.

On the other hand, the sum of the areas A and B is also 1 (the square A + B has sides of length 1). Thus the area B equals the area C no matter what the $F(\theta)$ figure looks like.

The physical meaning of $F(\theta)$ is that it equals the volume fraction in the outgoing flow with a residence time smaller than θ, since if one looks at dimensionless time θ all tracer that flows out has entered the device between 0 and θ.

From the definition of $C(\theta)$ and $E(\theta)$ it is clear that since $C(\theta)d\theta$ is the volume fraction in the exit flow with a residence time between θ and $\theta + d\theta$, the following relation holds:

$$F(\theta) = \int_0^\theta C(\theta')\,d\theta' = \int_0^\theta E(\theta')\,d\theta' \tag{E - 2.40}$$

where the second equality only holds for a steady-state.

$\prod$ Let us again consider the ideally stirred tank we considered in Section 2.1.2, but rather than calculating the concentration of coconut-oil at the exit we now calculate the palm-oil concentration, thus deducing the F-function, since at $t = 0$ the inlet is suddenly switched from coconut-oil to palm-oil.

The mass balance for the palm-oil reads as:

$$V\left(\frac{dC_{palm,out}}{dt}\right) = \phi_V\,(C_{palm,in} - C_{palm,out}) \tag{E - 2.41}$$

with the initial condition:

$$C_{palm,out} = 0 \text{ at } t = 0 \tag{E - 2.42}$$

Note $C_{palm,out}$ = the concentration of palm-oil in the outflow. Remember this will also equal the palm-oil concentration in the reactor.

Let the inlet concentration be $C_{palm,0}$ for all $t \geq 0$. The subscript palm can now be dropped and Equation 2.41 can be written as:

$$\frac{dC_{out}}{dt} = \frac{\phi_V}{V}\,(C_0 - C_{out}) \text{ for } t \geq 0 \tag{E - 2.43}$$

$$C_{out}(0) = 0$$

The solution of Equation 2.43 is:

$$\frac{C_{out}}{C_0} = 1 - \exp\left(-\frac{\phi_V}{V}\,t\right) \tag{E - 2.44}$$

This relationship is an important one since it enables us to calculate the concentration of a component in an outlet after a step-function increase in the inlet if we know the volumetric flow rate, the volume of the vessel, the time and the concentration of the component in the inlet.

Let us assume we have a 40 l tank through which water is passed at a rate of $40 \, l \, h^{-1}$.

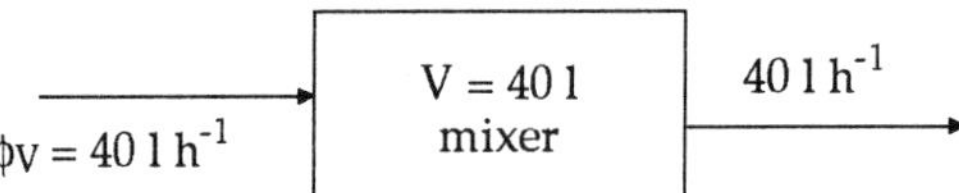

At a particular time, the water inlet is replaced by a 10% w/v sucrose solution at a flow rate of $40 \, l \, h^{-1}$.

What will be the sucrose concentration in the outflow:

1) 30 minutes after the switch at the inlet?

2) 1 hour after the switch at the inlet? Use Equation 2.44 to help you.

We have defined $F(\theta) = \dfrac{C_{out}(\theta)}{C_0}$ (Equation 2.35) and $F(\theta) = \displaystyle\int_0^\theta C\,(\theta)d\theta$ (Equation 2.40)

Using Equation 2.44 we can therefore establish that:

$$F(\theta) = 1 - \exp(-\theta) \quad (\text{Since } \theta = \frac{t}{\tau} = \frac{\phi_v}{V}\, t) \tag{E - 2.45}$$

$$\text{and } C(\theta) = \frac{dF}{d\theta} = \exp(-\theta) \tag{E - 2.46}$$

Equations 2.44, 2.45 and 2.46 are important. They describe changing concentrations of substances in the outflow from a vessel in relation to the volume of the vessel, flow rate and time.

Let us now turn our attention to a more complex situation in which two vessels are coupled together in series (eg a two stage process).

Two tanks in series

We will consider two perfect tanks coupled together in series as shown in Figure 2.15. By the term perfect tank, we mean that material entering a tank is instantaneously evenly mixed throughout the tank's volume (ie each tank's contents are homogenous).

For simplicity we have arranged it such that both tanks are of equal volume V.

The volumetric flow rate ϕ_v is constant, but at $t = 0$ suddenly the input is changed from $C = 0$ to C_0 ('step'). What do the F- and C-functions look like?

To answer this question we have to calculate the response at the exit of the second tank, ie $C_2(t)$. In order to find $C_2(t)$, we first have to calculate the response at the exit of the first tank, that is $C_1(t)$. Thus two mass balances are required.

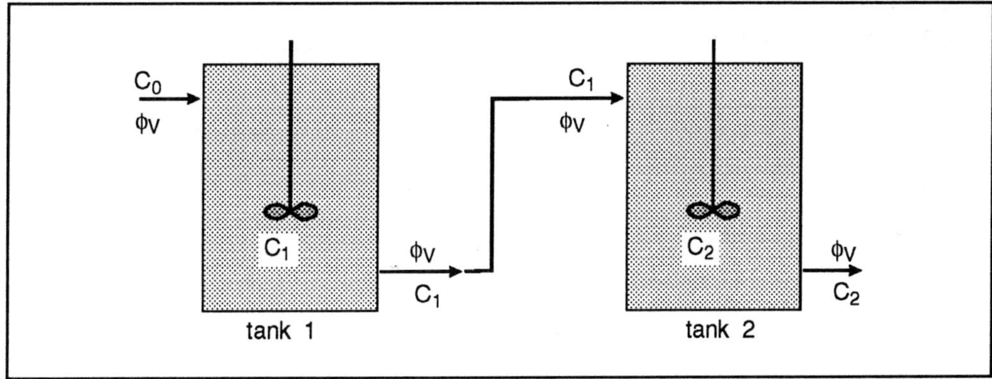

Figure 2.15 Two tanks coupled in series C_0 = concentration in the in flow to tank 1. C_1 = concentration in tank 1 and its out flow. C_2 = concentration in tank 2 and its out flow. ϕ_v = volumetric flow rate.

Tank 1:

$$V \frac{dC_1}{dt} = \phi_v \, (C_0 - C_1) \text{ with } C_1\,(0) = 0 \tag{E - 2.47}$$

Tank 2:

$$V \frac{dC_2}{dt} = \phi_v \, (C_1 - C_2) \text{ with } C_2\,(0) = 0 \tag{E - 2.48}$$

The solution of Equation 2.47 is:

$$C_1 = C_0 \left(1 - \exp\left(-\frac{\phi_v}{V} t\right) \right) \tag{E - 2.49}$$

Equation 2.49 is used to eliminate $C_1(t)$ from Equation 2.48 for tank 2. The resulting equation has as solution for $C_2(t)$ (as is seen from substitution of the solution for C_1 and C_2 into Equation 2.48):

$$C_2\,(t) = C_0 \left\{ 1 - \left(1 + \frac{\phi_v}{v} t \right) \exp\left(\frac{-\phi_v}{V} t \right) \right\} \tag{E - 2.50}$$

Thus the F-function is (taking $\tau = \dfrac{2V}{\phi_v}$)

$$F(\theta) = \frac{C_2\,(\theta)}{C_0} = 1 - (1 + 2\theta) \, \exp\,(-2\theta) \tag{E - 2.51}$$

and the C-function:

$$C(\theta) = \frac{dF}{d\theta} = 4\theta \, \exp\,(-2\theta) \tag{E - 2.52}$$

Plug-flow in a tube

Consider again the flow through a tube of length L as described by Figure 2.8. The fluid was thought to be in plug-flow, ie all fluid elements move with the same velocity v. Now at $t \geq 0$ a tracer with concentration C_{in} is continously injected into the volume flow ϕ_v. What is the F-function in this case?

The answer is quite simple. A fluid element which enters the tube will leave it after a time $\tau = L/v$; or, expressed in terms of the tube volume and volume flow, after $\tau = V/\phi_v$. Thus the response to the step input is the same step function, but shifted over a time V/ϕ_v. But this is just the definition of the residence time τ and the response at the exit of the tube equals C_{in}, but with a delay of just one τ. In other words:

$$F(\theta) = \begin{cases} 0 & \text{for } \theta < 1 \\ 1 & \text{for } \theta \geq 1 \end{cases}$$

(E - 2.53)

In the previous Section 2.1 we learnt the basics of mass balances. In this Section 2.2 we have used the principles of mass balances to explore residence time distribution within vessels. The determination of residence time is an important element in bioprocess engineering since it enables us to calculate, for example, the possible contact time between a substrate and a catalyst. We have learnt that, in practice, systems often behave far from ideal and we have learnt that at least 3 concepts exist (E, C and F) to describe the residence time distribution for a piece of apparatus. Before moving on to another aspect which applies the laws of conservation, attempt the following SAQs.

SAQ 2.4

A solution containing 5g glucose l^{-1} is fed into a tank at time t=0. One hour later, the concentration of glucose in the outflow of the tank contains 2g l^{-1}. If the flow rate is 20 l h^{-1}, what is the volume of the tank?

SAQ 2.5

Below is a drawing of a steam generator. Water is supplied via pipe A, and removed as steam via 3 separate pipes (B, C and D). If the steam from B, C and D are condensed, they yield 2 l h^{-1}, 1 l h^{-1} and 5 l h^{-1} respectively.

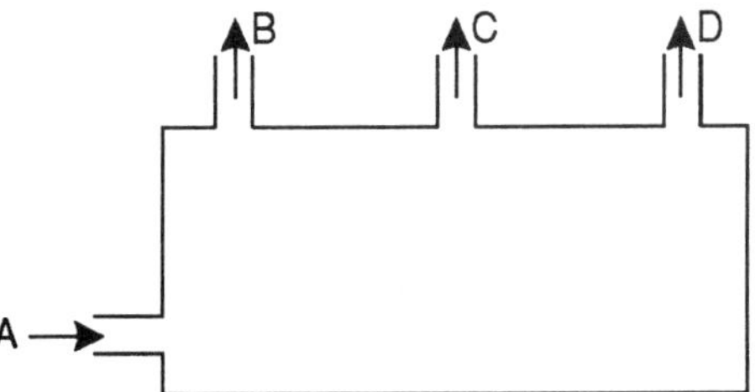

1) Write a mass balance for water in the steam generator and calculate the rate at which water must be pumped in via pipe A.

2) If the volume of the steam generator is 50 l, what is the mean residence time for water in this vessel.

SAQ 2.6

Below are two graphs, a) represents the concentration of glucose in the inflow entering a tube over the period 0-60 minutes, b) represents the concentration of glucose in the outflow leaving a tube 0-60 minutes.

The tube is 20m long and has a cross-sectional area of 10cm^2.

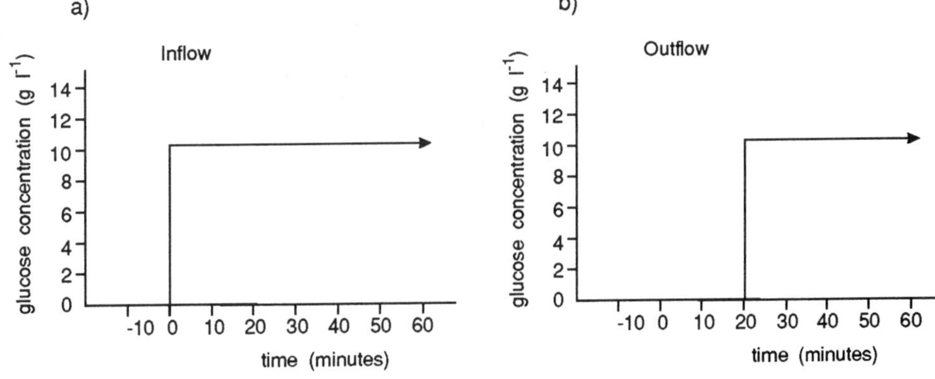

1) Is the fluid in the tube showing plug flow characteristics?

2) What is the residence time (τ) of the fluid in the tube?

3) What is the value of θ at 20 minutes after the step up in glucose in the inlet?

4) What is the value of θ at 40 minutes after the step up in glucose in the inlet?

5) What is the F(θ) value 10 minutes after the step up in glucose in the inlet?

6) What is the F(θ) value 40 minutes after the step up in glucose?

7) Calculate the flow rate of the fluid through the tube.

SAQ 2.7

1) The mean residence time of compound A in two tanks coupled in series is 2 h. If the two tanks are of equal volume, each being 10 l, what is the volumetric flow rate?

2) For the system described above, what will the value of the F-function be after 1 h^{-1}?

<table>
<tr><td>

SAQ 2.8

</td><td>

Air containing 20 vol % O_2 is pumped into a reactor at a flow rate of 20 m^3 h^{-1}. 95% of the oxygen is removed in the reactor by a culture of micro-organisms. The overall chemical reaction taking place can be represented by the reaction.

O_2 + carbon substrate $\rightarrow CH_3CHOHCOOH$

(Assume $CH_3CHOHCOOH$ is non volatile)

1) Draw up a mass balance for oxygen in the vessel.

2) Assuming a steady state is reached and there is no growth of the micro-organisms in the vessel, what is the volumetric gas flow out of the reactor?

3) Calculate the rate of oxygen consumption in the reactor in mol h^{-1} (assume 1 mol of oxygen = 24.0 l of O_2 at the temperature and pressure used in the reactor).

</td></tr>
</table>

2.3 Summary of mass and mole balances

So far in this chapter we have introduced the concepts of the macro-balance with special reference to the mass and mole balance. We have also applied a macro-balance approach to explore the behaviour of fluids in systems in terms of residence time. In the next section we will examine another type of macro-balance: the energy balance.

Now that you have completed Sections 2.1 and 2.2 you should be able to:

- apply mass balances to determine changes in amounts of substances in fluid systems using supplied data;

- apply mole balances to determine flow rates of components in fluid systems from supplied data;

- calculate residence time functions (F, C, E) from supplied data.

Now let us move on to consider the energy balance.

2.4 Energy balances

2.4.1 Energy forms

In the previous sections we dealt with mass balances. Another useful balance is the
energy balance. Now the quantity G that is subjected to the process of book keeping is
no longer mass, but instead we concentrate on energy E. Energy manifests itself in a
variety of forms. When a balance is set up for energy, not only must all transports of
energy be accounted for, but also all conversions of energy from one form into another
must be included. The variety of manifestations of energy is quite large and there are
many conversion processes. However, a useful list of different forms of energies and
expressions for their respective concentrations is given in Table 2.1.

energy balance

Read through this table and make sure you are familiar with each.

manifestation	concentration of energy (expressed per unit mass)
kinetic energy	$\frac{1}{2}v^2$
potential energy	gz (gravitation)
internal energy	u (thermodynamics)
pressure energy	p/ρ
enthalpy	$h\ (= u + p/\rho)$

Table 2.1 Different forms of energy. v = velocity, p = pressure, g = acceleration due to gravity, ρ = density, z = height, h = enthalpy, u = internal energy.

In most applications of transport phenomena the list given above suffices, and the energy of a given system is a sum of all these terms written as (again per unit mass):

Total energy = internal energy + pressure energy + kinetic energy + potential energy.

$$e = u + p/\rho + \frac{1}{2}v^2 + gz \qquad\qquad (E - 2.54)$$

Make a note of the following convention:

- lower case letters refers to energy per unit mass;

- upper case (capital) letter refer to the total energy of the system.

2.4.2 Energy balance

The general energy balance (macro-balance) can be written as:

energy change in a system	=	flow of energy into the system	−	flow of energy out of the system	+	flow of heat into the system	+	work done on the system	+	energy produced within the system

Mathematically, this can be written as:

$$\frac{dE}{dt} = d\frac{(\rho V e)}{dt} = \phi_{M,in}\,e_{in} - \phi_{M,out}\,e_{out} + \phi_q + \phi_w + P_e \qquad\qquad (E - 2.55)$$

Note that the energy per unit mass (e) acts like a concentration and the structure of the balance is the same as in the case of a mass balance involving concentration (see Equations 2.2 and 2.14). On the left-hand side the rate of change in the total energy E (= ρVe) is written. The terms on the right-hand side denote: the flow of energy into the system via a mass flow (which carries along a certain energy concentration e_{in}), the energy flow out via a mass flow out of the system (with energy concentration e_{out}), the flow of heat ϕ_q into the system (heat supplied to the system has a positive sign, withdrawal of heat has a negative sign), the amount of work performed on the control volume per unit of time ϕ_w (for example due to the action of a pump or compressor) and the energy production P_e inside the volume, respectively. The production can have several forms. It can stem from heat created in a metal by dissipation of an electrical current, or from an exothermic chemical reaction etc. Let us consider some examples of conversions that are possible between various forms of energy.

∏ Before reading on, see if you can write down the energy conversions which take place when a tennis ball is dropped onto the floor.

If a tennis ball is dropped onto the floor, the ball loses potential energy on its way down and gains kinetic energy. After it has bounced on the floor, the kinetic energy is converted into potential energy until no kinetic energy is left any more and the ball has reached its 'highest' position. Of course, after a few bounces the ball lies down on the floor and all kinetic and potential energy is gone. Both forms are now completely converted into heat (internal energy). The reason for this is that during every bounce some of both energy forms is converted to heat. Not only will the ball warm up but also, to a lesser extent, will the ground. After several bounces, all of the mechanical energy will be converted to heat and the ball will no longer bounce.

∏ Now consider a steady flow of water through a horizontal tube due to a pressure difference. Using Equations 2.54 and 2.55 and Figure 2.16 show that pressure energy can be converted to internal energy under the following conditions.

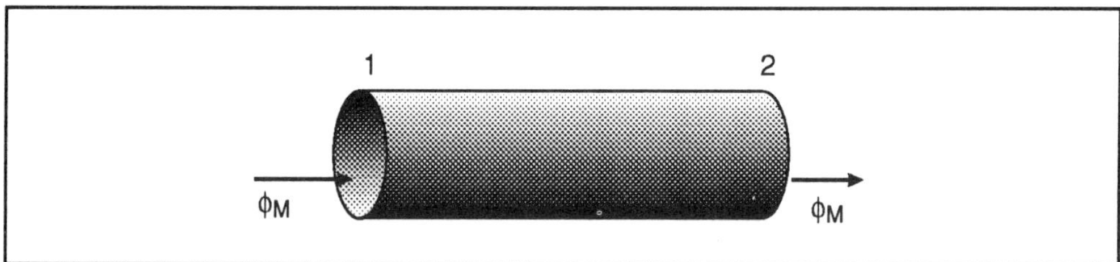

Figure 2.16 The flow of water through a horizontal tube.

Assume that there is no heat exchange with the surroundings, that the flow rate is constant and that there is no pump operating within the section of tube under consideration. To work out the energy balance over the tube you will need to work out the internal energy in position 1 and position 2.

We will explain how we would work this out. Since we are considering a steady-state, the mass flow in, must equal the mass flow out: $\phi_{M,in} = \phi_{M,out} = \phi_M$. Now we can write down an energy balance over the pipe (ignoring the possibility of heat exchange with the surroundings, thus $\phi_q = 0$). Since it is a steady flow, the dE/dt-term in Equation 2.55 is zero. Furthermore, there is no action of a pump (thus $\phi_w = 0$) and there is no production of energy within our 'control volume'.

Using Equation 2.55, the flow rate of energy at position 1 can be expressed as:

$$= \phi_M(u_1 + p_1/\rho_1 + \frac{1}{2}v_1^2 + g\,z_1).$$ Note that subscript 1 refers to position 1.

Likewise the flow rate of energy at position 2 is given by:

$$= \phi_M \left(u_2 + p_2/\rho_2 + \frac{1}{2}\, v_2^2 + g\, z_2\right)$$

(Note that subscript 2 refers to position 2)

Thus we can write the energy balance over the tube as:

$$\frac{dE}{dt} = 0 = \phi_M \left(u_1 + \frac{p_1}{\rho_1} + \frac{1}{2}\, v_1^2 + gz_1\right) - \phi_M \left(u_2 + \frac{p_2}{\rho_2} + \frac{1}{2}\, v_2^2 + g\, z_2\right) \qquad \text{(E - 2.56)}$$

ie 0 = (energy flow in at position 1) - (energy flow out at position 2).

We can, however, take this derivation even further.

For a horizontal tube we have $z_1 = z_2$. Furthermore, to a good approximation water is incompressible, ie $\rho_1 = \rho_2$. If we further assume that the tube has a constant cross-sectional area, it is found that $v_1 = v_2$.

Prove to yourself that $v_1 = v_2$ from a mass balance. Start with $\phi_{M1} = \phi_{M2}$. When you have attempted this, compare you approach with ours.

Since Mass (M) = Volume (V) x density (ρ)

then $\phi_{V1}\, \rho_1 = \phi_{M1}$ and $\phi_{V2}\, \rho_2 = \phi_{M2}$

Thus $\phi_{V1}\, \rho_1 = \phi_{V2}\, \rho_2$

But $\phi_{V1} = v_1\, A_1$ and $\phi_{V2} = v_2\, A_2$

Thus $v_1\, A_1\, \rho_1 = v_2\, A_2\, \rho_2;\ A_1 = A_2$ and $\rho_1 = \rho_2$

Thus $v_1 = v_2$

Using all these equalities simplifies Equation 2.56 to:

$$u_1 + \frac{p_1}{\rho} = u_2 + \frac{p_2}{\rho} \qquad \text{(E - 2.57)}$$

Expressed in another way, during the flow from entrance to exit, pressure energy is converted into internal energy. From standard thermodynamics it is known that for a system at constant density.

$$u_1 - u_2 = c_V\,(T_1 - T_2) \qquad \text{(E - 2.58)}$$

where c_v is the specific heat at constant volume (J kg^{-1} K^{-1}).

Thus from Equation 2.57 and Equation 2.58 we can derive the relationship:

$$T_2 - T_1 = \frac{p_1 - p_2}{\rho c_v} \qquad \text{(E - 2.59)}$$

It shows that the temperature of this water at the outlet will be higher than at the inlet.

SAQ 2.9

The pressure of a fluid entering a tube is 10^6 N m^{-2}. 200 meters further down it leaves the tube at a pressure of 10^5 N m^{-2}. Assume that the density of the fluid is 10^3 kg m^{-3}, its specific heat is 1×10^3 J kg^{-1} K^{-1}, its temperature at the inlet is 300 K and the fluid is incompressible. The mass flow rate is 1 kg s^{-1} and the radius of the tube is 10 cm. What will the temperature (in K) of the fluid when it leaves the tube?

SAQ 2.10

1) Consider the flow of water ($c_v = 4.2 \times 10^3$ J kg^{-1} K^{-1}, $\rho = 10^3$ kg m^{-3}) at a mass flow rate $\phi_M = 1$ kgs^{-1} through a tube with radius $r = 1$ cm due to a pressure difference $\Delta p = 2 \times 10^5$ Nm^{-2}. What is the temperature difference (in °C) across the tube if all pressure energy is converted into internal energy?

2) What would be the temperature change if the outflow was 200 m lower than the inlet. (Assume all other parameters are the same as in 1) and that the loss of potential energy is converted to heat. Take $g = 9.81$ m s^{-1}.

dissipation

This kind of conversion from mechanical energy (ie kinetic, potential and pressure energy) into 'thermal energy' (ie chaotic kinetic energy of the individual molecules) is called dissipation, the mechanism for this being friction.

2.4.3 Internal energy and thermal energy

heat balance
thermal energy balance

Up to now we have only looked at the total energy, but sometimes it is more convenient to look at a certain kind of energy like for instance the internal energy U. Then we formulate what can be called a heat balance:

change in internal energy	=	inflow of internal energy	-	outflow of internal energy	+	heat transfer	+	production of thermal energy

$$\frac{dU}{dt} = \frac{d(\rho V u)}{dt} = \phi_{M,in}\, u_{in} - \phi_{M,out}\, u_{out} + \phi_q + \text{production} \qquad (E - 2.60)$$

The term on the left-hand side obviously represents the change of total internal energy U in our control volume, while the first two terms on the right-hand side represent the difference in the flow of internal energy into and out of the system due to mass flow. The third term on the right (ϕ_q) accounts for the flow of heat into the system and the last term ('production') stands for the production of internal (or thermal) energy inside the system due to dissipation (friction) and work performed on the system. In a steady-state Equation 2.60 is usually written in a somewhat different form. First of all, in a steady-state the mass flow rates in and out are equal: $\phi_{M,in} = \phi_{M,out} = \phi_M$. Further, the production term in Equation 2.60 is split in two parts. The reversible conversion of

friction

pressure energy is represented by: $\displaystyle\int_{in}^{out} p\, d\frac{1}{\rho}$ (this is in fact the pdV (work) term in thermodynamics), while the irreversible conversion, ie the dissipation of mechanical energy is written as $\phi_M\, e_{fr}$, where e_{fr} is the amount of mechanical energy per unit mass that is dissipated into thermal energy (heat). Thus the production term is equal to:

$$-\phi_M \int_{in}^{out} pd\frac{1}{\rho} + \phi_M \, e_{fr}$$

We can replace the production term of Equation 2.60 by this relationship. For the steady-state version of 3.7, $\frac{dU}{dt} = 0$. Thus we can write:

$$0 = \phi_M\left[u_{in} - u_{out} - \int_{in}^{out} pd\frac{1}{\rho}\right] + \phi_M \, e_{fr} + \phi_q \qquad \text{(E - 2.61)}$$

This is called the thermal energy balance.

2.4.4 Mechanical energy

If we write down the steady-state balance for the total energy from Equation 2.55 and subtract from this the thermal energy balance (Equation 2.61) we would be left with what is known as the mechanical energy balance.

There are various final forms that this can be written in, but a common one is:

$$0 = \phi_M\left[- \int_{in}^{out} \frac{1}{\rho}\,dp + \frac{1}{2}\left(v_{in}^2 - v_{out}^2\right) + g\left(z_{in} - z_{out}\right)\right] - \phi_M \, e_{fr} + \phi_w \qquad \text{(E - 2.62)}$$

Equation 2.62 is called the Bernoulli equation.

The term ϕ_w is the amount of work done on the system and serves as a production term, to cover mechanical energy inputs other than those already considered and $e_{fr} =$ frictional energy.

This is quite a complex equation so do not try to memorise it. You should, however, be able to use this equation and remember that the mechanical energy balance is the difference between the total energy balance and the thermal energy balance.

Now that we have covered the principles underpinning energy balances, in the next section we will examine some applications of energy balances.

2.4.5 Application of energy balances

We will again use the example of the kettle we used earlier and consider two conditions: 1) in which the water is not boiling and 2) in which the water is boiling.

We will refer to Figure 2.17. Remember that here we are focusing on an energy balance.

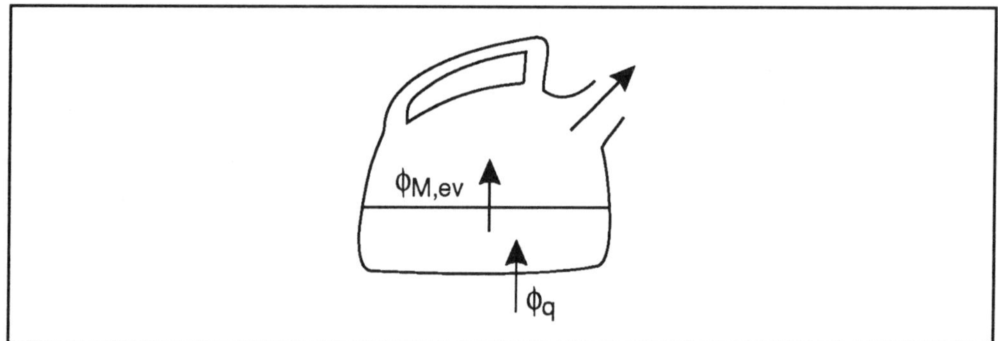

Figure 2.17 Energy balance in a kettle (see text for details).

Water not boiling

As long as the water does not boil, the mass flow from the liquid to the vapour can be neglected. Thus the thermal energy balance over the liquid volume reads as (thermal expansion being ignored):

$$\rho V \frac{du}{dt} = \phi_q \tag{E - 2.63}$$

we remind you that ϕ_q is the flow of heat into the system, ρ = density, V = volume and ρV = mass.

If we further approximate by assuming that over the entire temperature regime the specific heat c_V is constant Equation 2.63 becomes:

$$\rho c_V V \frac{dT}{dt} = \phi_q \text{ with } T = T_0 \text{ for } t = 0 \tag{E - 2.64}$$

because $\frac{du}{dt} = c_V\frac{dT}{dt}$ (see Equation 2.58).

Simplifying by taking the heat input ϕ_q as constant results in the solution:

$$T(t) = T_0 + \frac{\phi_q}{\rho c_V V} t \tag{E - 2.65}$$

ie the temperature (T) at time t is equal to the initial temperature (T_0) plus an increase due to the heat input represented by the second part of the Equation 2.65.

This is of course a very useful relationship. It means if we know the temperature of a liquid in a vessel and we know its specific heat, density and volume (or mass), we can calculate the heat input needed to raise the temperature in a given time. Try it by attempting SAQ 2.11.

SAQ 2.11	Consider a vessel which holds $50m^3$ of water ($c_V = 4.2 \times 10^3$ J kg^{-1} K^{-1}, $\rho = 10^3$ kg m^{-3}) and we seek to raise its temperature by 0.1 °C every second. What rate of heat input would be required?

Water boiling

When the water is boiling, $\phi_{M,ev}$ can no longer be neglected. If we write down the thermal energy balance for the liquid, there are two things we have to keep in mind. Firstly, the liquid volume V is no longer constant. Secondly, not only the flow $\phi_{M,ev}$ leaves the liquid but there is also a phase change with an associated density change from ρ_l to ρ_g. Thus the term $\phi_{M,ev} \cdot \frac{p}{\rho}$ has to be taken into account as well. So, we will concentrate on h, which

enthalpy

is called enthalpy rather than on u alone. From thermodynamics we know that $h = u + \frac{p}{\rho}$

The following derivation is quite complex but, as you will see at the end, the conclusion is quite straight forward.

The enthalpy balance equation (ignoring kinetic and gravity effects) for the liquid phase can be written as:

$$\frac{d\,(\rho_1 V_1 h_1)}{dt} = -\,\phi_{M,ev}\,h_g + \phi_q \qquad\qquad (E\text{ - }2.66)$$

(Note subscript l = liquid and g = gas phase).

where the first term on the right-hand side contains the enthalpy of the gas. From a mass balance over the liquid volume we find:

$$\frac{d\,(\rho_1 V_1)}{dt} = -\,\phi_{M,ev} \qquad\qquad (E\text{ - }2.67)$$

Combining Equation 2.66 and Equation 2.67 yields:

$$\rho_1 V_1 \frac{dh_1}{dt} = -\,\phi_{M,ev}\,(h_g - h_1) + \phi_q \qquad\qquad (E\text{ - }2.68)$$

heat of
evaporation

The difference h_g -h_1 is called the heat of evaporation and is written as h_g -h_1 = Δh_{ev}. Furthermore, since the water is boiling, the temperature and pressure are constant and thermodynamics tells us that h_1 is constant. Consequently, Equation 2.68 becomes:

$$\phi_{M,ev} = \frac{\phi_q}{\Delta h_{ev}} \qquad\qquad (E\text{ - }2.69)$$

$\phi_{M,ev}$ is the flow of mass from liquid to vapour phase. Thus equation 2.69 indicates that increasing the heat input via the heating device only serves to evaporate the liquid faster.

SAQ 2.12

1) A vessel holds 20 m^3 of a liquid (c_v = 2 x 10^3 J kg^{-1} K^{-1}, ρ = 10^3 kg m^{-3}). If heat is supplied at a rate of 2 x 10^6 J s^{-1} at what rate will the temperature rise? (Assume that the liquid does not expand).

2) If the liquid in the vessel is described in 1) is brought to its boiling point, what will be the rate of evaporation if the heat of evaporation is 10^7 J kg^{-1} and the heat is supplied at a rate of 2 x 10^6 J s^{-1}.

In SAQ 2.12, you should have demonstrated that you can calculate the rate at which a liquid will warm up given information concerning its volume, density and specific heat (c_v). This is of course a useful calculation in biotechnology as we often need to warm fluids. Also you will have demonstrated to yourself that you can calculate the rate at which a fluid will evaporate by boiling given a particular energy input.

We can use analogous calculations for cooling a liquid.

SAQ 2.13	A liquid has a $c_v = 2 \times 10^3$ J kg^{-1} K^{-1} and a density (ρ) = 10^3 kg m^{-3} and a volume of 10 m^3. If we wish to be able to cool this liquid at a rate of 10 °C per hour, at what rate does the refrigeration unit have to remove heat to achieve this rate of cooling?

Finally in this section, let us do a calculation which has obvious relevance to biotechnology.

SAQ 2.14	We have a bioreactor which has a volume of 10 m^3. Heat is generated at a rate of 6×10^4 J min^{-1} from the metabolism of the micro-organisms in the culture. Heat is lost through the walls of the vessel at a rate of 2×10^3 J s^{-1}. At the same time, some of the water of the culture is evaporating (at a rate of 0.01 kg min^{-1}).

The specific heat (c_v) for the culture = 4.0×10^3 J kg^{-1} K^{-1} and its density (ρ) = 10^3 kg m^{-3} and the heat of evaporation of the liquid from the culture = 10^6 J kg^{-1}. From this information:

1) Draw up a thermal energy balance for the bioreactor.

2) Use this balance to determine if the vessel will warm up or cool down. (Assume a batch system).

3) Calculate the capacity of the heating/cooling device needed to maintain the reactor at constant temperature.

4) The reactor described above is now to be operated as a continuous flow process. Culture media is fed into the vessel at a rate of 10 m^3 h^{-1} and a temperature of 20°C. The culture is removed from the vessel at the same rate, but the temperature of this outflow is 30 °C. What will be the additional heat requirement to maintain the reactor at a constant temperature. (Assume all other parameters described above remain constant).

Summary and objectives

In this chapter we have examined mass and energy balances. We have learnt how to write general mass and energy balances which will enable us to calculate a wide variety of parameters.

Now that you have completed this chapter you should be able to:

- write mass and mole balances;

- apply mass balances to determine changes in the amount of substances in fluid systems;

- apply mass and mole balances to determine rates of product formation from supplied data;

- calculate residence time functions (F, C, E) from supplied data;

- write general energy balances;

- calculate temperature rises in a fluid from data relating to volume, fluid density and specific heat;

- calculate heat flow from data relating to temperature changes, specific heat, fluid density and energy input;

- determine temperature gradients and heat flow through a system from supplied data.

Transport mechanisms of mass, energy and momentum

Transport mechanisms of mass, energy and momentum

3.1 Introduction

In Chapter 2 we described how to set up mass, mole and energy balances and to use these balances to calculate flow rates of energy and mass. We did not, however, discuss the mechanisms by which mass and energy were transported. It is this aspect of transport phenomena which is the topic of this chapter.

Transport can be considered at two different levels, namely the macro level and the micro (molecular) level. We begin this chapter by giving a description of transport at these two levels. This will enable us to discuss transport at these two levels and it will also enable us to discuss transport coefficients.

3.2 Transport mechanisms and processes

3.2.1 Convective transport

The type of transport we came across in the preceding chapter is, in fact collective transport: 'fluid elements', containing very large numbers of molecules (order 10^{15} or more), are carried along by a flow. The successive passage of fluid elements resembles a train of vans, each with its own cargo. Similarly, each fluid element may have its own concentration, eg of some component molecules and/or of thermal energy ('temperature'). This transport of properties by flow is called convective transport.

types of flow Flow can occur due to an external driving force, such as a pressure difference applied to a pipe or pipe element, and then has a forced character. It can also occur 'spontaneously', for example as the result of temperature differences inside the fluid which are related to some local heat supply or heat generation; in this way density differences arise (or even bubble formation when boiling occurs) and thus gravity induces flow (free convection). A common example of the latter type of flow is that of air shimmering above the surface of a highway on a hot summer day or above the radiator of a central heating system.

Thus in convective transport we are looking at the movement of large numbers of molecules. We have described this transport mathematically in terms of mass flow rate (ϕ_M in kg s^{-1}). For convective energy transport we have also described this transport in terms of mass flow rate, while regarding the energy content of mass as being a function of mass such that $\phi_q = \phi_M\, e$ (with e = J kg^{-1}) where e is the concentration of energy (expressed as energy per unit of mass).

3.2.2 Molecular transport

In addition to the above convective transport of molecules and properties, molecules also travel individually. This molecular traffic finds its origin in the criss-cross thermal motion of the individual molecules. Each molecule has its own translational, vibrational and rotational velocities. Note that on the average these velocities are related to temperature via the Boltzmann constant (k). Each molecule also carries with it its own momentum and kinetic/thermal energy, and is able to transfer them to other molecules by means of collisions. Notwithstanding the fact that this so-called molecular transport goes back to individual motions, it can quite satisfactorily be expressed in 'mean quantities', ie in quantities like density, concentration and temperature. This section is explicitly dealing with such a phenomenological description rather than with fundamental theories on collisions and motions of molecules. Let us, however, consider some examples to develop at least an insight into the molecular transport mechanisms.

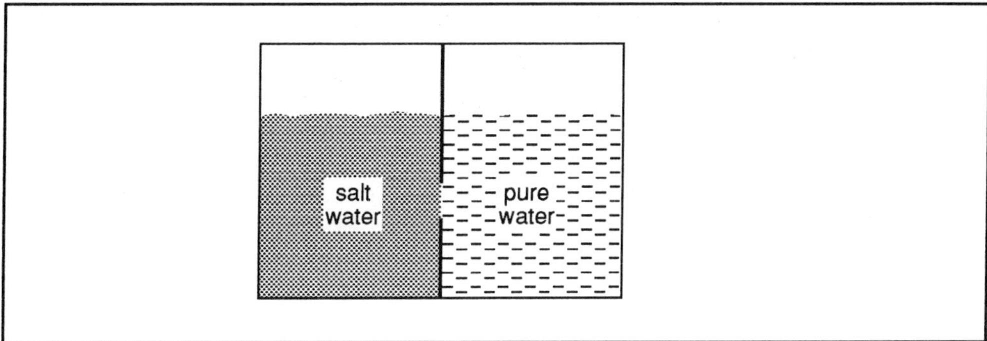

Figure 3.1 Example of molecular mass transport.

Π Consider the following experiment where a reservoir is divided into two sections by a partition wall with a small hole in it (Figure 3.1).

The right-hand side of the reservoir is filled with pure water, the left-hand side with salt water. The water levels on both sides of the partition wall are equal so that no static pressure difference and hence no flow and convective transport occurs. What change from this initial condition would you expect with time?

If we wait long enough, both sides will contain salt water and this situation does not spontaneously return into one side being pure water and the other side being salt water again. How can this be explained? Let us focus on the small hole in the partition wall (see Figure 3.2). From both sides a real bombardment with water molecules takes place. As many molecules find their way from the left to the right as vice versa, otherwise a level difference between the two sides would be established (as the two parts are really two 'communicating vessels'). If we look, however, at the salt molecules only (for the sake of simplicity ionisation has been ignored), the story is quite different. Initially there is only a bombardment with salt molecules on the hole from the left-hand side, with the result that individual salt molecules will move from the left to the right while there is no compensating salt flow from the right to the left! This net flow from the left to the right continues until the concentrations on both sides are equal (a state of equilibrium). Obviously the longer the distances the molecules have to travel, the longer it will take before concentration differences will have disappeared thence, concentration gradients (rather than concentration differences) determine the net transport rates resulting from molecular motion.

molecular
diffusion
The above described phenomenon is known as molecular diffusion (often referred to simply as diffusion) and finds its origin in concentration gradients. We may expect the

resulting mass flow to be a simple function of the concentration gradients: the larger the difference the stronger the flow, and the larger the distance between the places of high and low concentration the weaker the flow. We have represented molecular diffusion in Figure 3.2.

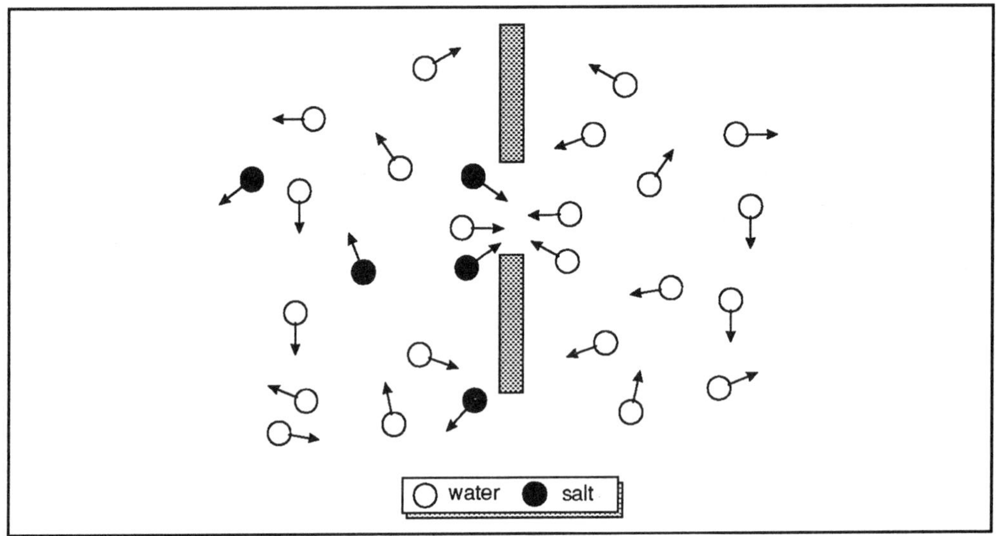

Figure 3.2. Model of molecular diffusion. Ionisation is ignored for the sake of simplicity.

molecular transport of heat

An analogous reasoning can be set up for molecular transport of heat. Now, the driving 'force' is a temperature difference. Temperature itself is related to an average of the kinetic energy of a very large number of molecules.

⊓ Imagine a copper rod the left-hand side of which has a higher temperature than the right-hand side. This means that on the average the molecules on the left-hand side have a greater kinetic energy than those on the right-hand side. What will happen to this temperature difference with time?

Now molecules at both sides of the 'partition' plane collide (this happens in a solid between neighbouring molecules which vibrate about a fixed position). The effect of such a collision is that the faster molecules lose kinetic energy and this is gained by the slower ones. Thus, on the left-hand side the temperature drops a little, whereas on the right-hand side it increases a bit. This continues until the temperatures at both sides are equal, and obviously the net transport of energy (heat) becomes zero (a state of equilibrium). We conclude therefore that a temperature difference in a solid gives rise to an energy flow, namely a flow of kinetic energy of the molecules. This mechanism for heat transport by individual molecules is called conduction.

conduction

Seen from a macro-scale point of view there is simply a flow of heat. This possible mechanism for heat transport is not restricted to solids, but is also conceivable in liquids and gases. In the latter media the transport of heat does not take place through collisions of vibrating adjacent molecules, but occurs mainly during the seemingly chaotic motions of 'hot' and 'cold' molecules: a 'hot' one diffuses so to say from the left to the right, where there are too few fast, hot molecules. Cold molecules diffuse in the opposite direction.

It is clear that convective transport usually is a much more efficient way of transporting heat (or mass) than molecular transport. Although the velocity of any fluid element is much lower than the velocity of an individual molecule (convective velocities of water, for instance, are of the order of 1 m/s whereas the mean velocity of a water molecule at room temperature is of the order of 600 m/s), in convective transport all molecules travel collectively and massively, whereas in the concept of molecular transport molecules behave like individuals going their own way, mutually independent of each other. The molecules move randomly in all directions, while their net displacements are seriously restricted by their mutual collisions. In addition, for convective transport it is the quantity itself that counts (for instance thermal energy), while for molecular transport only differences (or more precisely gradients) in this quantity are important. Usually these differences are at least one order of magnitude smaller than the absolute value of the quantity under consideration.

3.3 Transport coefficients

In Section 3.2.2 we have seen that the diffusive flows due to molecular mobility are generated by 'driving' gradients, for example a gradient in concentration (C) results in diffusion. The greater the difference the stronger the flow and the greater the distance between the regions of high and low concentration, the smaller the flow. This can be summarised in the simple mathematical expressions described by Fick's and Fourier's laws.

3.3.1 Fick's law

Fick's law   The mass diffusion flux in, say, the x-direction (ie the mass flow rate per unit area), for a concentration gradient of component A in that x-direction is given by Fick's law.

$$\phi''_{M,A} = - D \frac{dC_A}{dx}$$

(E - 3.1)

diffusion coefficient   The superscript '' of ϕ indicates that the flow is taken per unit area. Note the minus-sign, expressing the fact that mass diffuses from a region of high concentration to a region of low concentration. The transport coefficient D is called the diffusion coefficient, it has SI units of $m^2 s^{-1}$. There are some limitations to the validity of Fick's law. First of all, it applies to two-component systems only. Further, it is not generally valid in a medium in which there is a concentration gradient for substance A but not for the other components. Under these conditions, it is possible that there will be a pressure gradient as well, and hence a convective flow. We shall come back to this type of situation in more detail later (Chapter 7).

3.3.2 Fourier's law

A similar relation to that of Fick's law holds for the molecular transport of heat and is described according to Fourier's law:

$$\phi''_q = - \lambda \frac{dT}{dx}$$

(E - 3.2)

thermal conductivity   Again the minus-sign is necessary to show that the heat flows from a region of high temperature to a region of low temperature. The transport coefficient λ is called the thermal conductivity coefficient (SI units $J s^{-1} m^{-1} K^{-1}$). Fourier's law can be written in an alternative form. If variations in the density ρ and the specific heat at constant pressure (c_p) may be ignored we may consider $\rho c_p T$, which actually denotes a concentration of thermal energy, rather than temperature T. Then Equation 3.2 becomes:

$$\phi_q'' = - \frac{\lambda}{\rho c_p} \frac{d(\rho c_p T)}{dx} = - \alpha \frac{d(\rho c_p T)}{dx} \qquad\qquad (E - 3.3)$$

thermal
diffusion
coefficient

where α is called the thermal diffusion coefficient and now has the same dimension as the diffusion coefficient D (m^2/s). This is of course a consequence of writing the heat (= energy) flux in terms of a gradient in the energy concentration, on the analogy of writing the mass flux in terms of a gradient in the mass concentration C_A.

Π In principle we can write a general equation that represents Equations 3.1 and 3.2. Try this on a separate piece of paper and then compare it with our version.

A generalised form of these two equations is:

$$\text{diffusion flux of } X = - \text{ diffusion coefficient of } X * \frac{d\,[\text{ concentration of } X\,]}{dx}$$

Thus X could be mass or thermal energy.

3.3.3 Application of Fick's and Fourier's laws

Π Let us work through an example. Consider the copper rod in Figure 3.6. Both ends are at a fixed temperature. What is the steady-state the temperature profile along the rod and how large is the heat flow through the rod? The rod is thermally insulated from its surroundings.

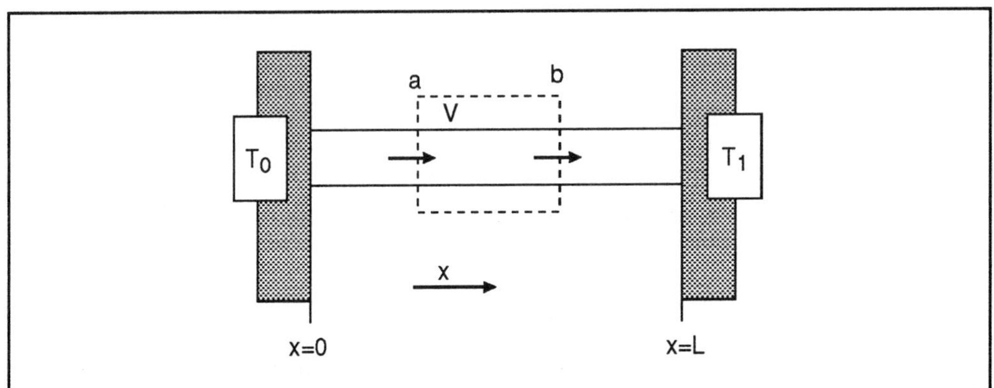

Figure 3.3 A copper rod with each end held at a specific temperature (see text for further details).

Solution:

Choose a control volume V as in Figure 3.3. Since there is no production of heat inside V, the thermal energy balance in the steady-state reads as: $\phi_{q,in} = \phi_{q,out}$. As the rod is insulated, the only possible flow of heat through the rod is due to conduction; thus the heat flows per unit area of the rod into and out of V can be written as:

$$\phi_q''\,(x = a) \;=\; -\,\lambda\left[\frac{dT}{dx}\right]_{x=a}$$

(E - 3.4)

and

$$\phi_q''\,(x = b) \;=\; -\,\lambda\left[\frac{dT}{dx}\right]_{x=b}$$

(E - 3.5)

These are derived from Equation 3.2 - Fourier's law.

Consequently, we find that the derivative of the temperature has the same value at x=a and at x=b. Since a and b are arbitrarily chosen it follows that:

$$\frac{dT}{dx} = C_1 = \text{constant and is independent of } x$$

(E - 3.6)

The solution of Equation 3.6 is:

$$T(x) = C_1\,x + C_2$$

(E - 3.7)

which describes the relationship between temperature and position along the rod.

The two constants C_1 and C_2 are determined via the boundary conditions:

x = 0: T(0) = T_0 (ie the left-hand end of the rod);

x = L: T(L) = T_1 (ie the right-hand end of the rod).

By substituting these boundary conditions into Equation 3.7, we find:

$$T(x) \;=\; \frac{T_1 - T_0}{L}\, x \,+\, T_0$$

(E - 3.8)

(where x = relative position along the rod).

Using Equations 3.8 and 3.2 the heat flow through the tube can be calculated from:

$$\phi_q \;=\; A\,\phi_q'' \;=\; A\left[-\lambda\frac{dT}{dx}\right] \;=\; -\,\frac{\lambda}{L}\,A\,(T_1 - T_0)$$

(E - 3.9)

Note that the heat flow is a consequence of a temperature gradient, $(T_1 - T_0)$, the 'driving' force for the flow.

Note also that the temperature gradient $= \dfrac{T_1 - T_0}{L}$.

> **SAQ 3.1**
>
> A bar similar to the rod displayed in Figure 3.3 is held at one end at a temperature of 390 K and at 300 K at the other. If the bar has a cross-sectional area of 0.1 m^2; a length of 1 m; a thermal conductivity coefficient of 1 J s^{-1} m^{-1} K^{-1} (or W m^{-1} K^{-1}) and is thermally insulated, what is:
>
> 1) the temperature 30 cm from the hottest end;
>
> 2) the heat flow through the bar?
>
> 3) if the specific heat of the rod is 1 . 10^3 J kg^{-1} K^{-1} and its density is 2 . 10^3 kg m^{-3} what is its thermal diffusion coefficient (α)?

Π Let us work through a more complex example.

A long narrow tube (length 25 cm, inner diameter D = 1 cm), partially filled with solid naphthalene, is connected to the open air, see Figure 3.4. The tube length L above the naphthalene is 20 cm. It is possible to measure the diffusion coefficient of naphthalene in an experiment involving measurement of changes in the level of naphthalene in the tube with an accuracy of 1 mm. Our task is to estimate the time required for such a measurement using the following data.

diffusion coefficient in air	$D = 7.10^{-6}$ m^2 s^{-1}
density of solid naphthalene	$\rho = 1150$ kg m^{-3}
molar weight	$M = 106$ kg kmol^{-1}
vapour pressure at 20°C	$p^* = 0.05$ mm Hg
gas constant	$R = 8.314$ J mol^{-1} K^{-1}

To estimate the time required, we set up a mass balance for the solid naphthalene. Obviously, there is no production and no flow in. On the other hand, the control volume is now not a constant since the level (z) of the naphthalene is decreasing.

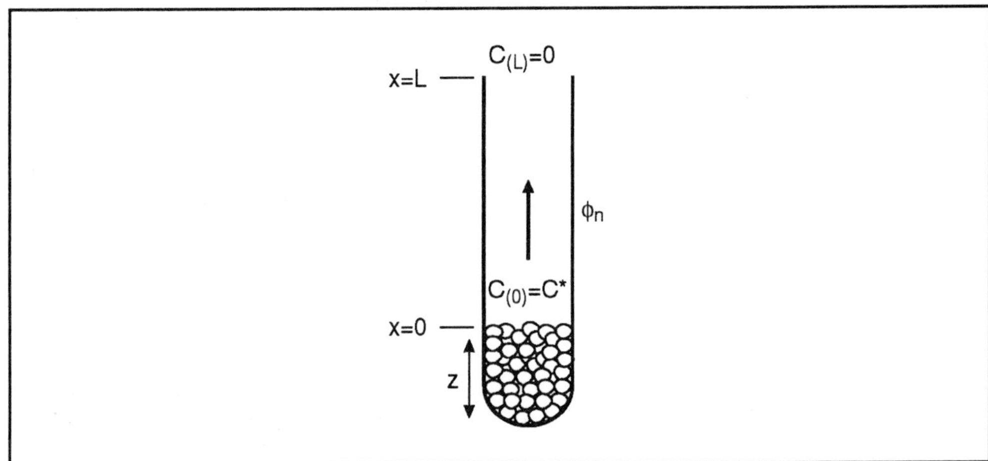

Figure 3.4 Illustration of the naphthalene diffusion experiment (see text for details).

The rate change in mass of solid naphthalene in the tube ($\frac{dM}{dt}$) is related to the change in volume of the naphthalene. This in turn is related to the surface area (A) and the height (z) since $M = \rho \times V = \rho A z$. But the density ρ of the naphthalene and the surface area of the tube are constant.

Thus $\dfrac{dM}{dt} = \rho A \dfrac{dz}{dt}$

This must also equal the mass outflow of naphthalene (remember none is being produced or added to the tube).

Thus $\rho A \dfrac{dz}{dt} = -\phi_{M,out} = -A\, \phi''_{M,out}$ where ϕ''_M = flow per unit area

But we know from Fick's law that the mass flow per unit area is related to the concentration gradient by the relationship

$$\phi''_{M,out} = D\,\frac{dC}{dx} \text{ (from Equation 3.1)}$$

and $A\phi''_{M,out} = AD\,\dfrac{dC}{dx}$

Thus $-\rho A\,\dfrac{dz}{dt} = AD\,\dfrac{dC}{dx}$ and also $-\phi_{M,out} = \dfrac{dM}{dt}$ $\qquad$ (E - 3.10)

The mass flow out of the control volume is because of diffusion. To proceed we need the concentration profile in the vapour/air phase down to the level of the solids. Let us approximate the transport of naphthalene through the air up to the edge of the tube by the steady-state solution, ie we suppose that the diffusion is so slow that the change in the level is small and that the concentration at the mouth of the tube ie $C_{(L)} = 0$.

The concentration at a point (x) along the tube can therefore be approximated by:

$$C(x) = \frac{C_{(L)} - C_{(0)}}{L}x + C_{(0)} = -\frac{C^{*}}{L}x + C^{*} \qquad \text{(E - 3.11)}$$

in which $C_{(L)} = 0$ and $C_{(0)}$ is the concentration of napthalene just above the solid and equals the saturation concentration of napthalene (C^{*}). Thus $C_{(0)} = C^{*}$. Substituting Equation 3.10 we obtain:

$$\frac{dz}{dt} = -\frac{DC^{*}}{\rho L} \qquad \text{(E - 3.12)}$$

Now all we need to do is calculate C^{*}. From the ideal gas law it is found that the equilibrium concentration C^{*} of the vapour just above the naphthalene is given by:

$$C^{*} = M\,\frac{p^{*}}{RT} \qquad \text{(E - 3.13)}$$

in which p* is the vapour pressure just above the solid naphthalene. Combining Equation 3.12 and Equation 3.13 yields for Δt:

$$\Delta t = \Delta z \frac{\rho R T L}{M D p^*} = 1.13 \times 10^8 \text{ s} = 3.6 \text{ years!}$$

(E - 3.14)

whereΔz = 1 mm. Our conclusion therefore is that measuring the decrease of the solids level with an accuracy of 1 mm is not a practical way of measuring the diffusion coefficient of naphthalene, although in theory it can be done.

This type of calculation is quite useful in determining how fast material, especially solvents, including water, might evaporate. Materials with high vapour pressure (p*) will, of course, evaporate quicker in part because the concentration gradient ($\frac{dC}{dx}$) will be greater. Similarly a shorter tube would also speed up the evaporation by increasing the concentration gradient.

Finally in this section let us take a look at a third example.

∏ Consider the following situation. We wish to measure at some position in a pipe the local pressure in a flow of some fluid in which an aggressive substance B is dissolved. We do not want to have dilute B diffusing into a side branch which is connected to the pressure sensor, since this might damage our sensor. So, what shall we do? Figure 3.5 illustrates the system under investigation.

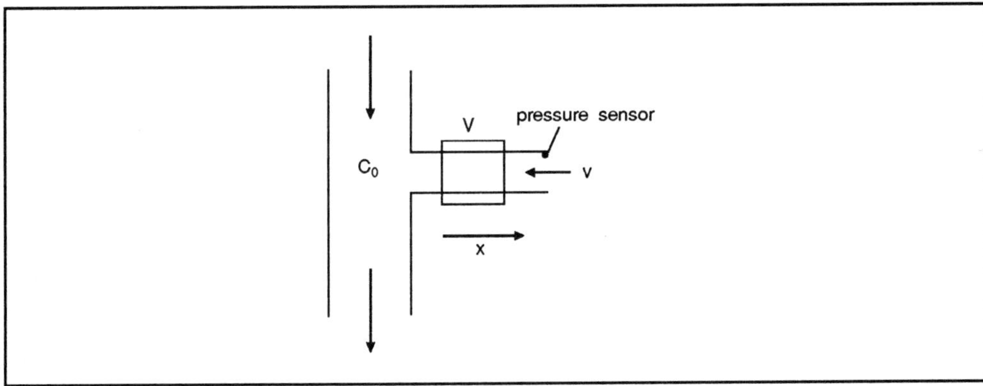

Figure 3.5 Illustration of the problem posed into protecting a pressure sensor (see text for details).

Suppose, we apply a small mass flow (velocity v) through the side branch towards the tube; the question then is whether this flow will be able to balance the diffusion of the solute?

Thus, what we want is for v to be large enough so that a steady state is reached. The desired mass balance over V in Figure 3.5 for the aggressive component B in a steady-state can be written as:

$$0 = \phi_{M,in}^{B} - \phi_{M,out}^{B}$$

(E - 3.15)

Thus the mass flow is constant, independent of the coordinate x along the side branch. Since we do not want any mass of B to enter the sensor, the net mass flow of B must equal zero. Now, ϕ_M^B at any position x along the side branch is composed of two terms, namely a diffusive part and a convective part. B enters the side branch by diffusion and is carried out by convection.

Thus ϕ_M^B = diffusion - convection.

From Fick's law, the diffusive part $= - AD \dfrac{dC_B}{dx}$

The convective part is related to the concentration of B, its velocity out of the pipe and cross-sectional area, ie $= (A.v)C_B$.

$$\text{Thus } \phi_M^B = - AD \dfrac{dC_B}{dx} - (A.v)\, C_B \qquad\qquad (E - 3.16)$$

Under the conditions we are operating there is no net flow hence

$$- AD \dfrac{dC_B}{dx} - (A.v)\, C_B = 0$$

$$\phi_M^B = \text{diffusion + convection} = - A\, D \dfrac{dC_B}{dx} - (A.v)\, C_B = 0 \qquad\qquad (E - 3.17)$$

Due to the boundary condition $C_B = C_0$ at $x = 0$, Equation 3.17 has the solution

$$C_B(x) = C_0 \exp\left(-\dfrac{vx}{D}\right) \qquad\qquad (E - 3.18)$$

Thus we can calculate the concentration of B at position x from the concentration of B in the main tube, the diffusion coefficient (D) and the velocity of the flow in the side tube (v).

SAQ 3.2	Using an arrangement of pipes shown in Figure 3.5, acid is being run through the main down pipe. A sensor we wish to protect is situated 10 cm away from the junction in the side arm. Given that the diffusion coefficient of the acid is $10^{-6}\,\text{m}^2\text{s}^{-1}$, calculate the velocity (v) of fluid we need to inject through the side arm to prevent the concentration of acid at the sensor from rising above 10^{-6} of the concentration in the down pipe.

3.4 Momentum and transport phenomena

3.4.1 Momentum

We need to consider another important aspect of transport. So far we have only considered transport in terms of the transport of mass and energy (heat). There is however a third quantity which can be transported. This is momentum.

momentum

Consider for example what happens when you stir your cup of coffee with a spoon. The movement (momentum) of the spoon transfers movement (momentum) to the liquid. The transfer of momentum is important in many biotechnology processes. For example the impeller used in a bioreactor transfers some of its momentum to the culture. Likewise some of the momentum of a bubble of CO_2 produced in a culture is transferred to the culture. It is, therefore, important that we understand momentum. We will learn in the following sections that transport of momentum can be treated in an analogous way to mass and energy transport. Momentum, like mass and energy is conserved.

Momentum is composed of two components, mass and velocity. In other words momentum = mass x velocity. The SI units of momentum is kg m s^{-1} or (preferably) Newton seconds (Ns). We can regard velocity as the 'concentration' of momentum. Thus if we double the velocity of a fixed mass we double the momentum; if we halve the velocity than we halve the momentum.

Since momentum = M * v then if we replace M by volume density $\rho \left(= \dfrac{\text{mass}}{\text{volume}} \right)$; then ρv = volumetric concentration of momentum. The volumetric concentration of momentum has the units of Ns m^{-3}.

∏ How can we change the momentum of a body?

Essentially we have to change the velocity. We can of course change velocity by applying a force. From Newton's second law of mechanics.

Force = Mass * acceleration (F = M *a) where acceleration is the rate of change of velocity.

A force (F) can therefore change velocity v of a mass (M) and can thus add or subtract momentum by an amount MΔv. We can therefore regard force as a source or sink of momentum.

We can represent this in the following way:

$$\text{force} \quad \underset{\text{subtract}}{\overset{\text{add}}{\rightleftarrows}} \quad \text{momentum}$$

A common unit for force is kg m s^{-1} s^{-1}. The SI unit of force is Newton (N) or Newton second per second (Ns s^{-1}).

∏ We have so far described the units of momentum as kg m s^{-1} or Ns and force as kg m s^{-1} s^{-1} or Ns s^{-1}. Can you determine the units of velocity in terms of Newtons?

Since velocity = ms^{-1} and Newton = kg m s^{-1} s^{-1}

Then $\dfrac{\text{N}}{\text{kg s}^{-1}} = $ ms^{-1} = velocity

Thus velocity = Ns kg^{-1}

Thus we have established the following:

Momentum = mass x velocity (kg m s^{-1} or Ns)
Volumetric concentration of momentum = density x velocity (Ns m^{-3})
Force = mass x acceleration (kg m s^{-1} s^{-1} or N or Ns s^{-1})
Velocity = distance ÷ time (m s^{-1} or Ns kg^{-1})

If you think carefully you will realise we can make an analogy between momentum and mass and energy balances. In energy balances we showed that the flow of energy (energy flux) was:

= flow of mass * energy concentration

= ϕ_M . e   where e = energy concentration

By analogy the flow of momentum (momentum flux)

= ϕ_M * (momentum concentration)

Since momentum concentration = velocity

Then momentum flux = ϕ_M . v

Momentum is produced and/or destroyed by forces. We will not however explore macrobalances for momentum in this chapter. We will however examine the transport of momentum.

3.4.2 Newton's law

Π Look back at the discussion of Fick's law and Fourier's law in Sections 3.3.1 and 3.3.2. Can you write down the generalised formula we derived for mass and energy transfer in terms of concentration gradients and transfer coefficients?

The general relationship we came to was:

$$\text{diffusion flux of X} = - \text{diffusion coefficient of X} * \frac{d\,[\text{ concentration of X }]}{dx}$$

where X could be mass or thermal energy

The exact relationships were:

$$\text{for mass } \phi''_{M,A} = -D \frac{dC_A}{dx} \qquad (E-3.1)$$

$$\text{for thermal energy } \phi''_q = -\alpha \frac{d\,(\rho c_p T)}{dx} \qquad (E-3.3)$$

From the above discussions on molecular transports of mass and energy we may suggest that in the case of a spatial gradient in momentum concentration (ρv), the individual molecules may provide a net transport of momentum from the region with higher momentum concentration to the region with lower concentration. This is indeed the case. For a certain class of liquids and gases a simple relation between this molecular transport of momentum and the prevailing differences in momentum concentrations again applies. It is known as Newton's law and reads for constant density as:

$$\phi_{p,xy}^{''} = -v\,\frac{d\,(\rho v_y)}{dx} = -\eta\,\frac{d\,(v_y)}{dx}$$

$$(E - 3.19)$$

These equations need some explanation. Let us work through them from left to right.

$\phi_p^{''}$ is the flow of momentum per unit area

Now remember that force and velocity have a direction (ie they are vectorial).

Now consider the three axis.

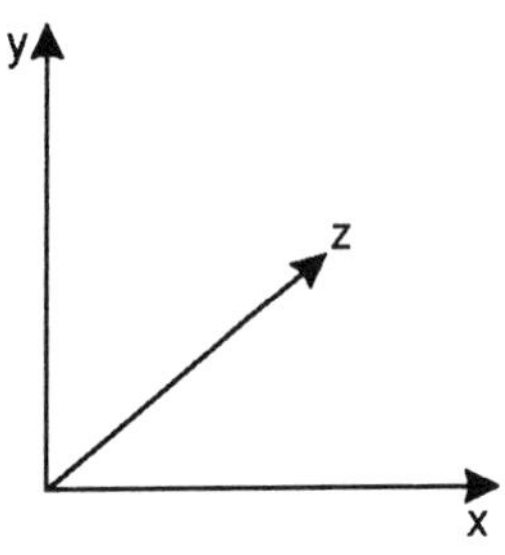

The subscript xy indicate that there is a variation in the y-component of the velocity and that this variation is between points with different x-value. The minus-sign is included to express that momentum flows from high to low momentum concentration regions. The coefficient η, called the coefficient of dynamic viscosity (for short the viscosity coefficient). It is expressed in SI-units of Ns m^{-2} and represents the fluid's potential of passing a momentum flux. A high value of η means that the exchange of momentum between fluid elements or layers as a result of molecular interactions is strong. Under these conditions the momentum concentrations of neighbouring fluid elements or layers may not differ too much, and velocity gradients may remain small. The coefficient v (= η/ρ) in Equation 3.19 denotes the coefficient of kinematic (kinetic) viscosity. The SI units of the latter are m^2 s^{-1}.

dynamic viscosity

kinematic viscosity

From this description you should have realised that the transport of mass, energy and momentum can be treated analogously.

⫿ We have so far met with five transport coeffficients (diffusion coefficient *D*; thermal conductivity coefficient λ; thermal diffusion coefficient α; coefficient of dynamic viscosity η and kinematic viscosity v). It would be quite a good test of your knowledge to see if you can write down the units of these coeffcients. You can compare your response with the units provided in the table of symbols in the appendix.

Now let us move on to another aspect of transport and introduce the concept of dimensionless numbers.

3.4.3 Prandtl number

In many applications both a velocity gradient and a temperature (or a concentration) gradient occurs. An example of such a more complicated situation is a flow of a fluid in a tube with the temperature of the wall of the tube being higher than the temperature of the fluid at the pipe entrance. On its way along the tube length the fluid gradually warms up owing to heat transfer from the tube wall. The direction of the driving force of the heat transfer, and hence of the heat flow is perpendicular to the direction of the velocity. It turns out that the transfer of heat from the wall to the fluid depends on the ratio of the kinetic viscosity (v) to the thermal diffusion coefficient (α). The larger the ratio the larger the heat flow. This ratio is named the Prandtl number (Pr) after the German scientist L. Prandtl:

$$Pr = \frac{v}{\alpha} \qquad\qquad\qquad\qquad\qquad (E - 3.20)$$

(Remember that $\alpha = \dfrac{\lambda}{\rho c_p}$ - see Equation 3.3).

Notice that Pr is dimensionless. The value of Pr is a measure of the relative ease with which velocity (or momentum) differences and temperature (or thermal energy) differences are smoothed out owing to the molecular interactions. The meaning of Pr and the dependence of the heat transport on Pr will be treated further later.

$\prod$ What is a dimensionless number?

All physical properties can be defined in terms of units, eg for thermal conductivity coefficient (λ) these are W m^{-1} K^{-1}. These units can be reduced to three fundamental dimensions, those of mass (M), length (L) and time (t); where appropriate temperature (T) is accepted as a dimension. In a dimensionless number, such as the Prandtl number, the properties making up the number have divisions which cancel out. Prove this for the Prandtl number using Table 3.1.

This should have proven easy since $Pr = \dfrac{v}{\alpha}$ and since v has the dimensions of m^{2}s^{-1} and α also has the dimensions of m^{2}s^{-1} the units cancel out. Therefore Pr is dimensionless and we can regard this as simply a ratio.

Before the end of this text we will come across many such dimensionless numbers. In each case they are ratios between two fundamental process. Let us examine another example of a dimensionless number.

3.4.4 Schmidt number

We can think of a combination of flow and mass transport in a similar way where the concentration gradient is perpendicular to the direction of the convective transport.

For example consider a flow through a semi permeable tube immersed in a solution containing a high concentration of solute. Thus:

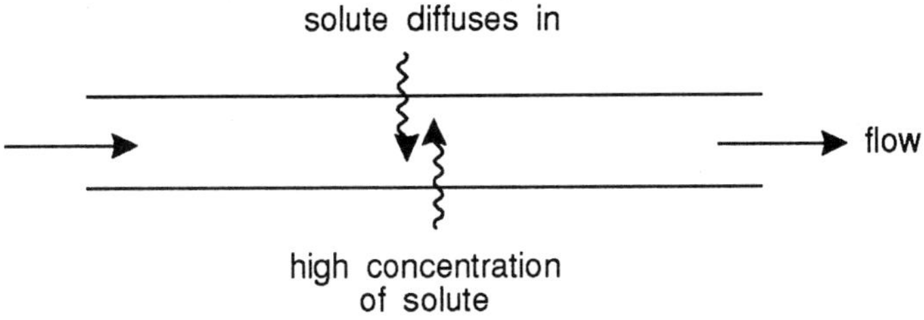

Schmidt number Now the transport of mass depends on the ratio of v and D. This ratio is named the Schmidt number (Sc) after the German scientist Schmidt and is given by the equation:

$$Sc = \frac{v}{D}$$

(E - 3.38)

Sc is without dimension, too. In its turn it measures the ability of the molecules of a fluid for smoothing the differences in velocities and concentrations, respectively. The dependence of mass transport on Sc will be treated more explicitly later.

All the coefficients introduced in this section along with the dimensionless numbers Pr and Sc, reflect processes on a molecular scale. These molecular interactions are highly affected by the composition of the molecules, the size of the molecules and the charge distribution within the molecules. The coefficients and the dimensionless numbers are therefore physical properties specific for each chemical compound. To get, however, an impression of the relevance of the various 'diffusive processes' it is useful to consider the orders of magnitudes of the coefficients D, α, λ and v. They are summarised, together with the Prandtl number and the Schmidt number, in Table 3.1.

	D $(m^2\,s^{-1})$	α $(m^2\,s^{-1})$	λ $(W\,m^{-1}\,K^{-1}$ or $J\,s^{-1}\,m^{-1}\,K^{-1})$	v $(m^2\,s^{-1})$	Pr (-)	Sc (-)
gas	10^{-5}	10^{-5}	10^{-2}	10^{-5}	1	1
liquid	10^{-8}-10^{-9}	10^{-7}	10^{-1}	10^{-5}-10^{-6}	10-10^2	10^2-10^3
solid	10^{-11}-10^{-12}	10^{-1}-10^{-4}	1-10^3	-	-	-

Table 3.1 Summary of the values of the coefficients D, α, λ, v and Prandtl and Schmidt numbers for gases, liquids and solids.

∏ As a way of revising what the symbols at the head of the columns stand for we suggest you re-examine Sections 3.4 and 3.5 and write out, at the bottom of each column the name of each coefficient.

You should come to the following conclusions.

D = diffusion coefficient derived from Fick's law.

α = thermal diffusion coefficient from Fourier's law.

(Note $\alpha = \dfrac{\lambda}{\rho c_p}$ if specific heat coefficient c_p and density ρ remain constant).

λ = thermal conductivity coefficient from Fourier's law.

v = kinematic (kinetic) viscosity coefficient from Newton's law.

Pr = Prandtl number = $\dfrac{v}{\alpha}$.

Sc = Schmidt number = $\dfrac{v}{D}$.

In attempting to predict the behaviour of large scale processes, especially in terms of:

•   the distribution (transport) of molecules and fluids within a system;

- the distribution (transport) of energy in the form of thermal or mechanical energy.

we can anticipate that the coefficients that have been derived in this chapter will be of fundamental importance.

The coefficients give us a measure of the ratio of transport. Dimensionless numbers give us a measure of the ratio of two fundamental processes. We will learn later that dimensionless numbers provide us with a useful way of characterising a system.

For example the Prandtl number gives us a ratio of the kinematic viscosity coefficient to the thermal diffusion coefficient. We will continue discussion of dimensionless numbers and dimensional analysis in the next chapter. Test your knowledge of what you have learnt in this chapter by attempting the following SAQs. It should give you good practice in handling units.

SAQ 3.3

1) If the temperature gradient in a system is 100 K m^{-1} and the thermal conductivity coefficient is 1 J s^{-1} m^{-1} K^{-1}, what is the flow of heat per unit area?

2) A linear concentration gradient of A of 1 kg m^{-3} m^{-1} has been estabished in a system. If the mass flow rate of A per unit area is 10^{-6} kg s^{-1}, what is the diffusion coefficient of A?

3) Given that for a liquid the kinematic viscosity coefficient is 10^{-5} m^{2}s^{-1} and the density of the liquid is 1000 kg m^{-3}, and the velocity gradient is 1 m s^{-1} m^{-1}, what is the momentum flux per unit area of the liquid?

4) What is the ratio of the kinematic viscosity coefficient to the diffusion coefficient known as?

5) At a given value of a kinematic viscosity, an increase in the thermal coefficient will increase or decrease the Prandtl number?

SAQ 3.4

1) If a Schmidt number is divided by a Prandtl number, we generate another dimensionless number called the Lewis number (Le). From your knowledge of Schmidt and Prandt numbers, what does the Lewis number measure?

2) The Schmidt number of a system has been determined as 10^2 and the Prandtl number has been also determined as 10^2. If the thermal diffusion coefficient has been determined as 10^7 m^2 s^{-1}, what is the Lewis number and diffusion coefficient of the system?

3) Is a culture of micro-organisms with a high Prandtl number more likely or less likely to result in localised overheating than a system with a low Prandtl number?

4) Is a system with a high Schmidt number more likely or less likely to have gradients of solute concentration than a system with a low Schmidt number.

5) From your answers to 3) and 4) would biotechnologists prefer systems with high or low Prandtl and Schmidt numbers.

Summary and objectives

This chapter has been quite brief. Nevertheless we have covered some very important concepts.

We explained that transport can be divided into two broad categories, convective and molecular transport and have derived a number of important transport coefficients. These include the diffusion coefficient (D), thermal conductivity coefficient (λ), thermal diffusion coefficient (α), the coefficient of dynamic viscosity (η) and the kinematic viscosity (ν). We demonstrated some analogies between these coefficients. Important to this development has been to show that momentum can also be transported.

At the end of the chapter we introduced the concept of dimensionless numbers by discussion of Prandtl and Schmidt numbers. We pointed out that there were ratios of two different coefficients each associated with a fundamental transport process.

Now that you have completed this chapter you should be able to:

- calculate gradients (temperature, concentration) from supplied data;

- use a variety of transport coefficients to calculate flux;

- use the concept of velocity as a concentration of momentum;

- describe what is meant by a dimensionless number.

Dimensional analysis

Dimensional analysis

Many of us run into difficulty in conducting calculations because we forget about the units we are operating in. How often have we produced an answer that is 100 times too big or too small because one of the items in the equation we were using was in cm and not m and we forgot to adjust for it? There is a requirement for dimensional consistency. Dimensional analysis is an algebraic treatment of the symbols of units considered independently of their magnitude. Dimensional analysis drastically simplifies the task of fitting experimental data to equations where a completely mathematical treatment is not possible. Dimensional analysis is based on the concept of dimensions and dimensional formula. In this chapter we will explain the technique involved in dimensional analysis and explore its advantages and disadvantages.

4.1 The meaning of dimensional analysis

The technique of dimensional analysis is driven by the need for dimensional consistency and the constraint it places on the functional relationship between variables. The technique enables us to group a number of variables in a problem to form dimensionless groups.

dimensionless groups

We have already encountered some examples of these in the previous chapter, namely Prandtl, Schmidt and, in SAQ 3.4, Lewis numbers. In each case they are the quotients (ratios) of two physical quantities .

Π From what you have learnt at the end of Chapter 3, you should be able to describe the two physical mechanisms involved in each of these numbers. Write these down on a piece of paper and check your response with ours given below.

The Prandtl (Pr) number $\left(= \dfrac{v}{\alpha} \right)$ is the ratio of kinematic viscosity to thermal diffusion; Schmidt (Sc) number $\left(= \dfrac{v}{D} \right)$ is the ratio of kinematic viscosity to mass diffusion; Lewis (Le) number $\left(= \dfrac{\alpha}{D} \right)$ is the ratio of thermal to mass diffusion.

Reynolds number

Let us now introduce another dimensionless number, the Reynolds number (Re). This is given by:

$$Re = \frac{\rho v^2}{\eta(v/d)} \tag{E - 4.1}$$

where ρ = density; v = velocity; and d = a typical length scale of the flow; v/d = a measure of the velocity gradient; η = coefficient of dynamic viscosity. We will not go into further details here except to explain that like the other dimensionless numbers the Reynolds number is the ratio between two forces, one coming from the tendency of a mass to continue on its way (inertia), the other from the viscous forces (the friction between neighbouring layers). Thus we could write Re as:

$$Re = \frac{\text{'inertia pressure'}}{\text{'viscous stress'}}$$

The point we are attempting to ensure you have grasped is that the ratio of two fundamental mechanisms or forces can often be described by a dimensionless number.

Let us see if we can generate dimensionless number for ourselves. To do this we will use a system we have already analysed. You may recall that we considered the following situation in Chapter 3 (see Section 3.3.3). We wished to measure, at some position in a pipe the local flow of some fluid in which an aggressive substance B is dissolved. We did not want to have this B diffusing into the side branch that is connected to the sensor, since this might damage our sensor (see Figure 4.1). We had to decide what to do.

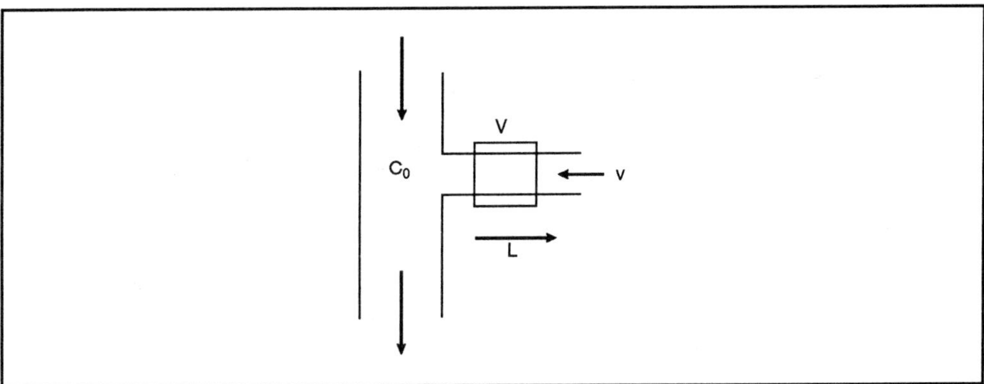

Figure 4.1 Flow and pipe arrangement described in the text.

Our approach to this was that we considered the competition between the mass flow via diffusion and a counteracting convective mass flow. The solution we came to was that the concentration of B at x distance from the mouth of the side pipe:

$$C_B(x) = C_0 \, \exp\left(- \frac{vx}{D}\right)$$

We can write this in a more general form of:

$$\frac{C_L}{C_0} = \exp\left(- \frac{vL}{D}\right) \tag{E - 4.2}$$

where C_L is the concentration at a distance L from the entrance; C_0 is the concentration at the entrance; v = velocity of the liquid in the tube; D = diffusion coefficient; L = distance between position C_L and C_0.

The group $\frac{vL}{D}$ is another example of a dimensionless group. It is called the Péclet number.

Péclet number

Like the other dimensionless groups we have met, the Péclet number is a quotient of two physical mechanisms (convective and diffusive transport).

Thus:

$$Pe = \frac{vL}{D} = \frac{vC}{D\,C/L} = \frac{\text{convective transport}}{\text{diffusive transport}} \qquad (E - 4.3)$$

The use of dimensionless numbers is a powerful tool in technology. It helps to reduce the number of parameters one has to take into account. Furthermore, most of the dimensionless numbers have a physical meaning: they compare the strength of two mechanisms as is clearly seen in the Péclet number we discussed above. One can try to reverse the way of reasoning and start from the important mechanisms of the problem under consideration. Then construct the dimensionless numbers by comparing the strength of the mechanisms and find out which are the important ones. For instance, consider our diffusion-convection problem. We see the two mechanisms for mass transport of the aggressive component:

$$\text{diffusion:} - A\,D\,\frac{dC}{dx} => - A\,D\,\frac{\Delta C}{L} \approx - A\,D\,\frac{C}{L}\ (\text{kg s}^{-1}) \qquad (E - 4.4)$$

$$\text{convection: A v C (kg s}^{-1}) \qquad\qquad\qquad\qquad\qquad\qquad\qquad\qquad (E - 4.5)$$

$$\text{Dimensionless group:}\ \frac{AvC}{AD\,C/L} = \frac{vL}{D} \equiv Pe \qquad (E - 4.6)$$

Thus now we know that Pe will determine our solution, although we do not know exactly how. All we know is that C_L and C_0 are a function of the Péclet number.

That is:

$$\frac{C_L}{C_0} = f(Pe) \qquad (E - 4.7)$$

To determine C_L from a value of C_0 we have to know what f is. Therefore to determine C_L we have to use:

$$\frac{C_L}{C_0} = \exp\,(- Pe)\ \text{since:}$$

$$\frac{C_L}{C_0} = \exp(- \frac{vL}{D})\ \text{and}\ \frac{vL}{D} = Pe$$

4.2 Advantages of using dimensional analysis

Apart from reducing the number of parameters and grouping the variables into quotients of mechanisms, there is another advantage of using dimensionless numbers. They help to compare similar situations.

Suppose we know all about one experiment like that described in Figure (4.1). Can we use dimensional analysis (eg Péclet numbers) to determine how we have to change v to obtain the same ratio $\dfrac{C_L}{C_0}$ if the experiment is done with another aggressive substance (denoted by C^1) with a different diffusion coefficient? Assume that the diffusion coefficient of C^1 is D^1 and that $D^1 = 8D$, where D is the diffusion coefficient of substance B, and that L is the same in each case. Try this before reading on.

The answer is yes, we can use the Pe number.

since $\dfrac{C_L}{C_0} = f\,(Pe)$ and $\dfrac{C_L^1}{C_0^1} = f\,(Pe^1)$

and we see if $\dfrac{C_L}{C_0} = \dfrac{C_L^1}{C_0^1}$ then:

$f\,(Pe) = f\,(Pe^1)$

Thus $Pe = Pe^1$

Thus $\dfrac{vL}{D} = \dfrac{v^1\,L}{D^1}$

We gave the condition that $D^1 = 8D$ and L is constant.

Thus $\dfrac{v}{v^1} = \dfrac{D}{D^1} = \dfrac{D}{8D}$

ie $v^1 = 8v$

Thus, in the case of the second substance we need to increase the gas velocity (v), in the side arm by a factor 8.

<table>
<tr><td>

SAQ 4.1

</td><td>

We set up an experiment similar to that described in the intext activity above. In this case we are dealing with two gases. Gas A has a diffusion coefficient D^A and Gas B has a diffusion coefficient $D^B = 2D^A$.

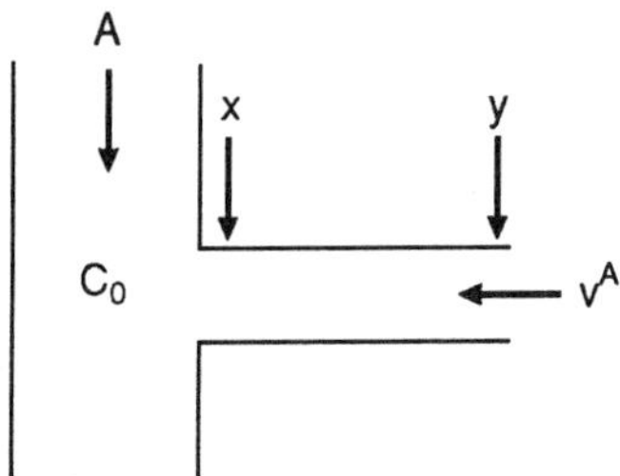

Gas A is passed down the tube and diffuses into the side arm A.

Another gas is pumped with velocity v^A (where $v^A = 1$ cm s^{-1}) along the side arm. A concentration gradient of A is thus set up such that the concentration of A at position x is C_0^A and at position Y is C_y^A. The distance x-y is 45 cm.

Gas B is passed down a similar piece of apparatus except in this case x and y are 180 cm apart. What you are asked to calculate is the velocity v^B of the gas that needs to be pumped into the side arm to maintain a concentration of gas B such that:

$$\frac{C_y^B}{C_0^B} = \frac{C_y^A}{C_0^A}$$

where C_0^B is the concentration of B at the neck of the side arm.

</td></tr>
</table>

The result obtained for SAQ 4.1 is a powerful one, for it shows us that by using this type of technique we are able to generate specific experimental results. It is particularly useful for scaling up laboratory experiments. Moreover, in a number of occasions the geometry is far too complicated to allow for an exact solution and there is not much left to do other than finding the dimensionless numbers and do laboratory experiments on models. We must however be cautious. If a change in regime takes place during scale up (for example turbulent flow is replaced by laminar flow see Chapter 5) then dimensional analysis breaks down. Let us explain this a little further. You have probably met with the term regime in everyday life to describe a government holding power, especially a totalitarian one. In process technology, we use the term analogously to describe the factor which governs (limits) the performance of a system. It could, for example, be the the nature of the flow, oxygen transfer rates, heat transfer rates etc. If a change in regime (for example at small scale oxygen transfer rates govern performance but on a large scale heat transfer becomes rate limiting on scale up, then dimensionless analysis breaks down.

4.3 Generalisation of the technique of dimensional analysis

Now that we have examined a specific example of dimensional analysis using Péclet numbers, we can ask the important question. How do we set about expressing a problem in terms of dimensionless groups. Below we describe a sequence of stages we must go through to enable us to express a problem in terms of dimensionless groups (Table 4.1). It is this sequence of stages which is called dimensional analysis.

Stage 1	Choose the quantity you want to consider: A
Stage 2	Determine the physical quantities B, ... P that determine the chosen one.
Stage 3	Write this in the form $A \sim B^{\beta} \times C^{\gamma} \times D^{\delta} \times ... P^{\pi}$.
Stage 4	Now make sure that the dimensions used in the expression are homogenous (eg do not mix cm and m). Then compare all the dimensions that occur. In examining for dimensions, only allow basic dimensions like length (m), mass (kg), time (s), temperature (K). This means for example that concentration is a derived dimension per cubed length ($kg\ m^{-3}$) or energy concentration $Jm^{-3} = kg\ m^{-1}\ s^{-2}$ (A table of units and some of their conversions to basic dimensions is provided in the appendix). Solve the resulting algebraic equations for β ... π.
Stage 5	Use the functions β ... π to group the quantities A, B to P into dimensional numbers.

Table 4.1 Stages in dimensional analysis.

This may seem a little abstract so let us work through some examples so that we have practice on how to carry out this process.

Example 1:

Again we consider the diffusion-convection problem discussed in SAQ 4.1. First we choose C_L (ie the concentration at distance L from the entrance) as the quantity we want to 'know'. (This is equivalent to Stage 1, Table 4.1).

Stage 2

Then use the quantities which determine C_L. Remember that $C_L = C_0 \exp(-\frac{vL}{D})$. Thus the quantities which determine C_L are: C_0 = concentration at the entrance (0); D = the diffusion coefficient; v = the velocity of the liquid in the tube; L = the distance between the position of C_L and C_0.

(ie C_L is a function of C_0, D, v and L)

Thus $C_L = f(C_0, D, v, L)$ (E - 4.9)

Stage 3

$$C_L \sim C_0^{\alpha} \times D^{\beta} \times v^{\gamma} \times L^{\delta}$$ (E - 4.10)

Stage 4

Substitute in the units for C_0, D etc. Thus:

$$\left[\frac{kg}{m^3}\right]^1 \sim \left[\frac{kg}{m^3}\right]^{\alpha} \cdot \left[\frac{m^2}{s}\right]^{\beta} \cdot \left[\frac{m}{s}\right]^{\gamma} \cdot [m]^{\delta}$$ (E - 4.11)

The basic dimensions that occur are: length (m), mass (kg) and time (s). We can obtain 3 equations for the 4 unknown powers α, β, γ, δ from Equation 4.11.

We can do this by examining each unit in turn. Let us use kg first. On one side of the equation we have kg^1, on the other kg^α

Thus $1 = \alpha$

$\prod$ See if you can do the same thing with m and s.

You should have come to the conclusion for s that the unit balance is:

$0 = -\beta - \gamma$ Thus $\beta = -\gamma$

For m, we have the unit balance of

$-3 = -3\alpha + 2\beta + 1\gamma + 1\delta$

But $\gamma = -\beta$ and $\alpha = 1$

Thus $-3 = -3 + 2\beta - 1\beta + 1\delta$

$0 = \beta + 1\delta$

thus $\beta = -\delta$

Thus $\alpha = 1, \beta = -\delta, \gamma = -\beta = \delta$

Substituting this into Equation 4.10 yields:

$$C_L \sim C_0^1 . D^{-\delta} . v^\delta . L^\delta \qquad\qquad\qquad (E - 4.12)$$

or grouped in dimensionless numbers (stage 5 of Table 4.1):

$$\frac{C_L}{C_0} \sim \left[\frac{vL}{D}\right]^\delta \sim [Pe]^\delta \qquad\qquad\qquad (E - 4.13)$$

Of course this does not mean that C_L/C_0 equals Pe to some unknown power δ. The analysis tells us which group is a function of certain other groups:

$$\frac{C_L}{C_0} = f(Pe) \qquad\qquad\qquad (E - 4.14)$$

where f is an unknown function. We know that $f(Pe) = \exp(-Pe)$ from the exact solution.

Dimensional analysis is such an important approach that it is worthwhile working through a second example.

Example 2

In this second example, we will work part way through the sequence of dimensional analysis and let you complete it.

A smooth wire (diameter d) is pulled at constant velocity v, through a reservoir containing a viscous liquid. The wetted length of the wire is L. Our task is to express the force F that has to be applied to the wire to keep it in motion in terms of relevant dimensionless numbers.

First we have to determine the quantities that determine F. These are in fact the velocity (v), the length of the wire (L), the diameter of the wire (d), the density of the liquid (ρ) and its viscosity η. The last two are properties of the liquid. We expect that the liquid in the reservoir will start to circulate due to the motion of the wire. Furthermore, the larger ρ and η the more difficult it is to keep the wire in motion.

We can write that the quantities that determine F are: v, L, d, ρ, η

Thus:

$$F \sim f(v, L, d, \rho, \eta) \qquad\qquad\qquad (E - 4.15)$$

We can now go to Stage 3 of dimensional analysis.

Thus:

$$F \sim v^{\alpha} L^{\beta} d^{\gamma} \rho^{\delta} \eta^{\varepsilon} \qquad\qquad\qquad (E - 4.16)$$

$\prod$ Now write in the units for F, v, L, d, ρ, η.

You should have written.

$$\left[\frac{kg.m}{s^2}\right] \sim \left[\frac{m}{s}\right]^{\alpha} . m^{\beta}, m^{\gamma}, \left[\frac{kg}{m^3}\right]^{\delta}, \left[\frac{kg}{m.s}\right]^{\varepsilon}$$

$\prod$ Now collect up each dimension as we did in the previous example. The kg first, then m, then s. (Write these down on a piece of paper before reading on).

Our calculations show that the following relationships hold:

kg $1 = \delta + \varepsilon$

m $1 = \alpha + \beta + \gamma - 3\delta - \varepsilon$

s $-2 = -\alpha - \varepsilon$

Thus now we have 3 equations for 5 unknowns. Consequently, we take two of them as independent ones and express the other three in terms of these two. For instance, take ε and β and work out what α, γ and δ are in terms of ε and β.

∏ See if you can write α, γ and δ in terms of ε and β. We will give you a start:

Since $1 = \delta + \varepsilon$ (E - 4.17)

then $\delta = 1 - \varepsilon$

Now try it with the others.

You should come to the conclusion that:

$\alpha = 2 - \varepsilon$

$\gamma = 2 - \beta - \varepsilon$ (E - 4.18)

$\delta = 1 - \varepsilon$

Since $-2 = -\alpha - \varepsilon$ then $\alpha = 2 - \varepsilon$

Also since:

$1 = \alpha + \beta + \gamma - 3\delta - \varepsilon$ then substituting in

$1 = 2 - \varepsilon + \beta + \gamma - 3(1 - \varepsilon) - \varepsilon = 2 - \varepsilon + \beta + \gamma - 3 + 3\varepsilon - \varepsilon$

$1 = -1 + \beta + \gamma + \varepsilon$

Thus $\gamma = 2 - \beta - \varepsilon$

Using these results in Equation 4.16 gives

$F \sim v^{2-\varepsilon} \, L^{\beta} \, d^{2-\beta-\varepsilon} \, \rho^{1-\varepsilon} \, \eta^{\varepsilon}$ (E - 4.19)

Now we come to the difficult part. We need to collect together like terms (ie all those with ε and all those with β).

Let us see if we can help.

$v^{2-\varepsilon}$ is the same as writing $v^2 . v^{-\varepsilon}$ or $\dfrac{v^2}{v^{\varepsilon}}$

∏ Now go down Equation 4.19 and split each of the quantities up in this way.

Your equation should now look like this:

$F \sim v^2 . v^{-\varepsilon} . L^{\beta} . d^2 . d^{-\beta} . d^{-\varepsilon} . \rho . \rho^{-\varepsilon} . \eta^{\varepsilon}$

∏ Now collect ε and β terms together, and identify the two sets of dimensionless numbers.

Your calculations should have resulted in the following arrangement.

$$F \sim v^2 d^2 \rho \left[\frac{\eta}{vd\rho}\right]^{\varepsilon} \left[\frac{L}{d}\right]^{\beta}$$

(E - 4.20)

or $F = k\, v^2 d^2 \rho \left[\dfrac{\eta}{vd\rho}\right]^{\varepsilon} \left[\dfrac{L}{d}\right]^{\beta}$ where k is a constant without dimensions

So the dimensionless groups are:

$$\left[\frac{\eta}{vd\rho}\right]^{\varepsilon} \quad \text{and} \quad \left[\frac{L}{d}\right]^{\beta}$$

If we rearrange 4.20 a little to give

$$\frac{F}{\rho v^2 d^2} = k \left[\frac{\rho vd}{\eta}\right]^{-\varepsilon} \left[\frac{L}{d}\right]^{\beta}$$

You may recognise a dimensionless number you have previously encountered, $\left[\dfrac{\rho vd}{\eta}\right]$ is of course Reynolds Number.

Thus, in the case we have just explored:

$$\frac{F}{\rho v^2 d^2} = k\, Re^{-\varepsilon} \left[\frac{L}{d}\right]^{\beta}$$

(E - 4.21)

<table>
<tr><td>

SAQ 4.2

</td><td>

1) Let us assume that the concentration of a reaction product is dependent upon the concentration of one of the reactants, the time, the length of the wire used to catalyse the reaction, and the depth of the liquid in which the reaction takes place:

C_p = concentration of product

C_r = concentration of reactant

x measures depth of liquid

t = time

L = length

From this information see if you can express the concentration of the product in terms of a relevant dimensionless number(s).

</td></tr>
</table>

The discussion so far has focussed onto the benefits of dimensional analysis, but it has also some significant drawbacks.

∏ Before reading on see if you can make a list of these.

The sorts of items we expect you to include are as follows:

as we mentioned earlier, dimensional analysis does not show how the groups are related.

For instance in example 1 we know that $\dfrac{C_L}{C_0} = f(Pe)$

but there is no information about f whatsoever. Likewise in SAQ 4.2 we calculated that:

$$\frac{C_p}{C_r} = f\left(\frac{x}{L}\right)$$

but had no information about f. Secondly, this one is more serious, the outcome of the analysis is completely determined by what we 'guessed' as the determining quantities B,...P. This seems a trivial remark, but it is not. It means that if the set B,...P contains one (or more) quantities that have no influence on the chosen one, the outcome of the analysis is nonsense. The same holds for the case when one has forgotten a relevant quantity. The following example illustrates the need for great care when using dimensional analysis.

Example 3

A polymerization takes place in a stirred vessel. The viscosity of the liquid changes during this reaction. But the reactor is coupled to a motor which drives the stirrer at a constant power P. How does the number of revolutions of the stirrer (N) change?

Solution:

determine P as a function of the relevant variables.

First try:

P depends on N and η

$P \sim N^{\alpha} . \eta^{\beta}$

$$\left[\frac{kg.m^2}{s^3}\right] \sim \left[\frac{1}{s}\right]^{\alpha} . \left[\frac{kg}{m.s}\right]^{\beta} \qquad\qquad (E - 4.22)$$

Equations:

kg	$1 = \beta$
m	$2 = -\beta$
s	$-3 = -\alpha - \beta$

This is a contradicting set, because β cannot equal both 1 and -2. Thus our first try is wrong. We have not selected the correct variables.

Second try:

P is determined by N, η and D, the diameter of the stirrer.

$$P \sim N^{\alpha}. \eta^{\beta}. D^{\gamma}$$

$$\left[\frac{kg.m^2}{s^3}\right] \sim \left[\frac{1}{s}\right]^{\alpha} \left[\frac{kg}{m.s}\right]^{\beta} \left[m\right]^{\gamma} \qquad \text{(E - 4.23)}$$

Equations:

$$\begin{array}{rcl} kg & 1 & = \beta \\ m & 2 & = -\beta + \gamma \\ s & -3 & = -\alpha - \beta \end{array} \right\} \quad \Rightarrow \quad \alpha = 2, \ \beta = 1, \ \gamma = 3 \qquad \text{(E - 4.24)}$$

Thus:

$P \sim N^2.\eta.D^3$ or in dimensionless numbers.

$$\frac{P}{N^2 \eta D^3} = \text{constant} \qquad \text{(E - 4.25)}$$

Thus, since the stirrer is driven at constant power P (and D is not varied in the tank), we find $\eta N^2 = \text{const.}$

In our second try we took N, η and D, but the combination $N * D$ is a velocity thus we could have taken v, η, D. Now we have three of the four quantities that form Re, so why not include the density ρ since we might expect that Re will somehow enter in the expression for P.

Third try:

$$P \sim N^{\alpha}. \eta^{\beta}. D^{\gamma}. \rho^{\delta} \qquad \text{(E - 4.26)}$$

$$\left[\frac{kg.m^2}{s^3}\right] \sim \left[\frac{1}{s}\right]^{\alpha} \left[\frac{kg}{m\,s}\right]^{\beta} \left[m\right]^{\gamma}. \left[\frac{kg}{m^3}\right]^{\delta}$$

Equations:

$$\begin{array}{rcl} kg & 1 & = \beta + \delta \\ m & 2 & = -\beta + \gamma - 3\delta \\ s & -3 & = -a - b \end{array} \right\} \quad \Rightarrow \quad \left\{ \begin{array}{l} \delta = 1 - \beta \\ \gamma = 5 - 2\beta \\ \alpha = 3 - \beta \end{array} \right. \qquad \text{(E - 4.27)}$$

Thus:

$$P \sim \rho\, N^3 D^5 \left(\frac{\rho ND^2}{\eta}\right)^{-\beta} \qquad \text{(E - 4.28)}$$

The combination $\dfrac{\rho ND^2}{\eta}$ is a Reynolds number. As a characteristic velocity v we can take the velocity of the tip of the stirrer to be $\pi D.N$, since per second a tip travels N revolutions and one revolution is equivalent to a distance πD. Hence:

$$\frac{\rho ND^2}{\eta} = \frac{1}{\pi} \frac{\rho(\pi ND)D}{\eta} \propto \frac{\rho vD}{\eta} = Re$$

At this stage you may not have recognised the Reynolds number but, with more experience, you will begin to identify such numbers. We will discuss Reynolds numbers in more detail later in this text.

Note that the factor $\frac{1}{\pi}$ is dropped for simplicity, obviously constants are of no interest in the dimensionless numbers. The dimensionless combination $\frac{P}{\rho N^3 D^5}$ is called the power number Po. The power number gives a ratio between the power input P and a combination of the stirrer dimensions, the rate of rotation and the density of the liquid. Expressed in another way, it is the ratio of the input of power to the amount of power transferred to kinetic energy in a fluid.

Our conclusion is:

$$Po \equiv \frac{P}{\rho N^3 D^5} \sim \left(\frac{\rho vD^2}{\eta}\right)^{-\beta} = Re^{-\beta} \tag{E - 4.29}$$

Notice that for $\beta = 1$ the second try is recovered.

Experimentally one finds:

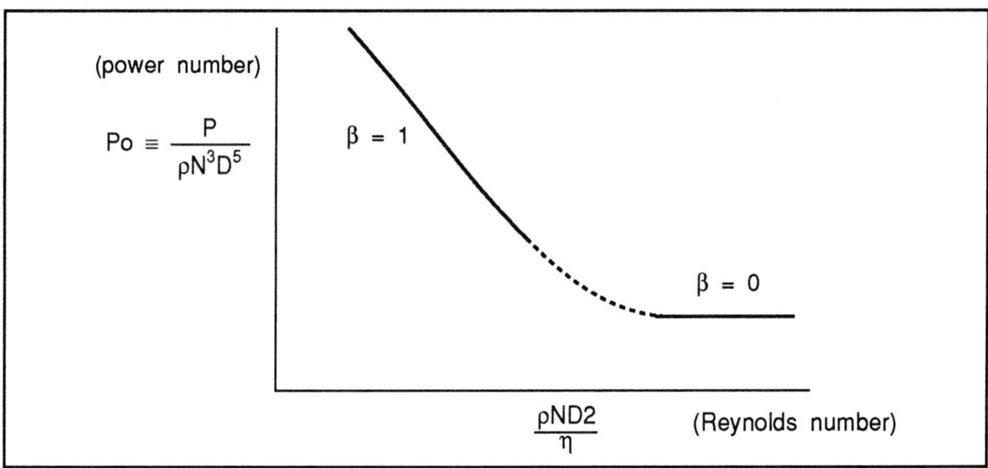

Figure 4.2 Relationship between power number and Reynolds number.

Thus over the lower range of Reynolds number, Reynolds number is inversely proportional to the power number. Notice that the original question of this example can only be answered if the value of β is known (ie depending on the Reynolds number).

Fourth try:

We could evenly well have tried $P = f(N, \eta, \rho)$, leaving out a length scale.

This results in:

$$P \sim \left(\frac{N\eta^5}{\rho^3}\right)^{1/2}$$

(E - 4.30)

∏ This result is wrong as we know from power input experiments. Why are the results wrong?

Your reasoning should have been that the power input is not independent of the size of the vessel (there is no length-scale in the expression). In other words a small or large vessel filled with the same liquid (thus the same ρ, η) requires the same power for a given N! The conclusion drawn from this example is that one has to be very careful with the dimensional analysis: starting with the wrong set of parameters renders wrong results.

Let us summarise what we have learnt from this example about the process of dimensional analysis. We have learnt that the key point is to ensure that we have included all the relevant variables.

Thus in our first attempt in example 3, we did not include all the variables that influence P. This became immediately obvious during the dimensional analysis as a contradicting set of dimension functions were produced.

In our second attempt we did not take the analysis as far as we could. In the third attempt, we ended up with a relationship in which two dimensionless groups (power number Po and Reynolds number Re) were directly related to each other. You will note however that this type of analysis has to be supported by experimental evidence. In this case experimentally the relationship $Po = Re^\beta$ was shown to be true but it turned out that β showed two extreme values in which $\beta = 1$ at low Reynolds numbers and $\beta = 0$ at high Reynolds numbers (see Figure 4.2).

∏ Now that we have met several dimensionless numbers, it might be a good idea to write them all down on a sheet of paper and to pin it up where you can see it regularly.

4.4 Buckingham - π theorem

In the examples discussed in the previous section we have seen that the number of dimensionless numbers necessary to describe the situation changes with the number of parameters we took into account. For instance, in the second try of example 3 we had four parameters and found one dimensionless group, whereas in the third try we had five parameters yielding 2 dimensionless groups. There is a general rule, called the Buckingham - π theorem, that enables us to predict in advance the number of dimensionless groups (p) from the number of parameters (n) and the number of basic units (m) that occur. It says:

$$p = n - m$$

(E - 4.31)

In words, this equation says that the number of dimensional groups equals the number of parameters minus the number of basic units.

Example 1:

Consider again example 3 from the previous section:

Our first try was:

$$P \sim N^{\alpha} . \eta^{\beta}$$

$$\left[\frac{kg.m^2}{s^3}\right] \sim \left[\frac{1}{s}\right]^{\alpha} \left[\frac{kg}{m.s}\right]^{\beta}$$

thus 3 parameters: P, N, η; 3 basic units kg, m, s. Hence the number of dimensionless groups p = 3 - 3 = 0, this means that in general this is a contradicting set.

Our second try was:

$$P \sim N^{\alpha} . \eta^{\beta} . D^{\gamma}$$

thus 4 parameters: P, N, η, D and 3 basic units: kg, m, s result in p = 4 - 3 = 1 dimensionless group.

Our third try was:

$$P \sim N^{\alpha} . \eta^{\beta} . D^{\gamma} . \rho^{\delta}$$

now 5 parameters and 3 basic units, thus p = 5 - 3 = 2 dimensionless groups.

SAQ 4.3

1) In carrying out a dimensional analysis, we try the relationship that Mass (M) is dependent upon time and viscosity.

ie $M \sim t^\alpha \eta^\beta$

In units:

$$[kg]^1 \sim [s]^\alpha \left[\frac{kg}{m.s}\right]^\beta$$

Use the Buckingham π theorem to determine:

    a) whether the set of variables which influence M we have chosen is complete and possibly correct;

    b) the likely number of dimensionless groups.

2) In a similar analysis to that we described in 1, the following relationship is proposed:

$C_x \sim C_0^\alpha . D^\beta . v^\gamma . L^\delta$, where v is velocity, C = concentration, L = length, D = diffusion coefficient.

In units:

$$\left[\frac{kg}{m^3}\right]^1 \sim \left[\frac{kg}{m^3}\right]^\alpha . \left[\frac{m^2}{s}\right]^\beta . \left[\frac{m}{s}\right]^\gamma . [m]^\delta$$

Use the Buckingham π theorem to determine.

    a) whether the variables we have chosen are possibly correct.

    b) the likely number of dimensionless groups.

4.5 Conclusion

Dimensional analysis is a useful tool in that it shows the possible ways in which the variables involved can be grouped. It is a dangerous tool in that it can give incorrect results if the physical nature of the problem is not understood and it can lead to false conclusions if a significant variable is omitted from the problem. It is therefore clear that it must be used with caution.

Summary and objectives

In this chapter we have explained the technique of dimensional analysis and practised the processes that take place during this type of analysis. We also described the dangers that inappropriate use of this type of analysis carries. In the final part of the chapter we dealt with the Buckingham π theorem which provides a quick means of testing whether a proposed set of variables could be interlinked in the manner envisaged and also as a quick means of determining how many dimensionless groups might be achieved.

Now that you have completed this chapter you should be able to:

- carry out dimensional analysis on a wide variety of situations;

- check for inconsistancies in proposed inter-relationships between variables;

- apply the Buckingham π theorem.

Flow phenomena and flow regimes

Flow phenomena and flow regimes

5.1 Introduction

The growth of the micro-organisms in a bio-reactor is strongly dependent on the local conditions in terms of the temperature and the concentrations of substrates, oxygen and metabolites. These local conditions are brought about as the resultant of the transport and the transfer of heat, mass and momentum. Both molecular and convective transport mechanisms play an important role in bringing the nutrients from the point of supply down to the level of the micro-organisms anywhere in the broth. Convective currents throughout the reactor and diffusion of individual molecules are therefore both effecting the local concentrations which in the end determine the growth of micro-organisms. After the section in Chapter 3 on molecular processes it is now appropriate to increase our knowledge about flows, flow phenomena, and flow regimes particularly in stirred vessels. In addition, forces exerted by flows on immersed bodies and pressure drop over flow lines are treated in some detail. This chapter provides an introduction to fluid mechanics that may serve as a basis for understanding flow properties.

5.1.1 Laminar flow

laminar flow

The most simple, regular and orderly type of flow is laminar flow: where the flow (of a gas or a liquid) takes place in more or less parallel, plane or concentric, layers (or 'laminae', from Latin). In the concept of laminar flow, adjacent layers are allowed to diverge or converge slightly, but the flow paths of adjacent volume elements never cross. Adjacent layers may have different velocities and, hence, may mutually exert

shear forces

shear forces. Whether and to what degree adjacent layers may differ in velocity (as a result of forces exerted) depends on the fluid's viscosity. Note once more that the unit of the dynamic viscosity coefficient is Nsm^{-2}, indicating the amount of momentum that can be transferred per unit area. A high value of the viscosity coefficient represents a strong internal coherence of the fluid due to an intense interaction of the molecules. Actually, these shear (or friction) forces are exactly the momentum flows referred to in Chapter 3. Adjacent layers (see Figure 5.1) have different velocities (momentum concentrations) and consequently momentum is transferred.

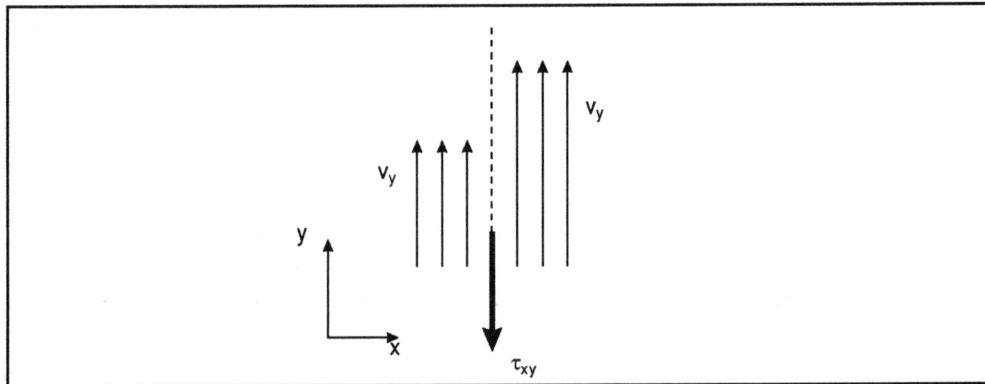

Figure 5.1 Laminar flow showing two layers flowing at different velocities (see text for details).

The molecules of the faster moving layer give up part of their y-momentum to the molecules of the slower moving layer. This re-distribution of y-momentum takes place perpendicularly to the direction of flow, and depends on the local gradient in the y-velocity (or momentum concentration). This transport of momentum perpendicularly to the direction of flow is also often described in terms of shear or friction: a layer of fluid may drag an adjacent sluggish layer along while slowing down itself (conservation of momentum). This shear stress is denoted by τ_{xy} and depends on the prevailing velocity gradient at least for constant density.

shear stress

A fluid for which this is valid is called a Newtonian liquid. More specifically we can write:

$$\tau_{xy} = -\phi''_{P_{y},x} = -\eta\frac{dv_y}{dx} = -v\frac{d(\rho v_y)}{dx} \qquad \text{(E - 5.1)}$$

Newtonian liquid

where $\phi''_{P_{y},x}$ is the momentum flux, η is the coefficient of dynamic viscosity (see Section 3.3.6), v is the coefficient of kinematic viscosity and ρ is the density. A Newtonian liquid is therefore a liquid in which τ_{xy} is directly proportional to the velocity gradient $\left(\dfrac{dv_y}{dx}\right)$ and η and for which is constant.

drag force

Adjacent layers are therefore able to mutually exert a viscous drag force. While this viscous drag is exerted, the kinetic energy (being a form of mechanical energy) of the molecules associated with the macroscopic flow rate is converted, or dissipated, into the more chaotic mode of heat (thermal energy).

We can perhaps describe this in the following way. When 'slow' (low kinetic energy ie low temperature) molecules get into a layer with 'fast' molecules, the latter will collide with the 'slow' ones, first over one, then over the next resulting in turmoil and chaos! In other words the mechanical energy of the macroscopic flow rate is converted into the chaotic mode of heat.

friction heat

5.1.2 Reynolds number

In any layer, the product of velocity (v) (or: flow rate per unit cross-sectional area, in ms^{-1}) times momentum concentration ($\rho.v$) (in Nsm^{-3}) denotes the convective flux of momentum. This convective flux may be compared to the viscous shear stress experienced from and exerted on adjacent layers. As expressed by Equation 5.1 this stress essentially is a molecular flux of momentum in a direction perpendicular to the direction of the convective momentum flux. The ratio of the convective momentum flux in one direction to the molecular momentum flux in the other has become known as the Reynolds number Re:

Reynolds number

$$Re = \frac{\rho v^2}{\eta \dfrac{v}{d}} = \frac{\rho v d}{\eta} = \frac{v d}{v} \qquad \text{(E - 5.2)}$$

in which d represents a characteristic length scale in the flow field that is suited for estimating the velocity gradient of Equation 5.1. This Reynolds number is dimensionless and may be viewed as the ratio of the inertia of the flow to the action of the viscous shear in the flow.

inertia

5.1.3 Turbulent flow

As long as the Reynolds number does not exceed a certain critical value, volume elements in the flow are not able to acquire momentum (due to the action of forces) to such an extent that shear stress (or viscosity) could not cope with it. As soon as the Reynolds number is sufficiently high, however, the inertia of the flow makes that volume elements start skidding and topple over. The velocity or, preferably, the momentum differences are then so large that viscosity is no longer able to smooth them out. The flow is said to become turbulent: eddies, or vortices, arise which are much more effective in re-distributing momentum (or velocity) differences. We can represent the

differences between laminar and turbulent flow by a simple diagram.

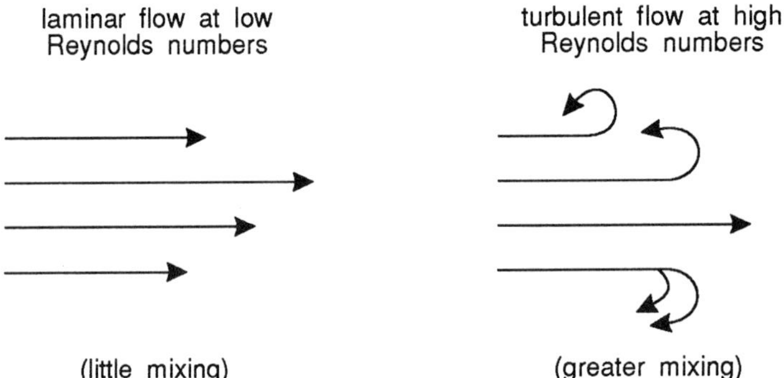

Eddies are coherent elements of fluid which are in a whirling mode of flow, which get their energy from the main flow, and which in their turn induce new, smaller eddies. In a really turbulent flow, eddies of a wide variety of sizes are present which form a complex albeit coherent three-dimensional system of whirls. Their interactions are completely governed by physical laws, but to the human eye the complexity looks like chaos. In this respect there is a close analogy with the motions of molecules in a gas. The eddies are able to transport mass, energy and momentum, just like molecules do. An important difference is that eddies are able to transport these over much longer distances than molecules can do between collisions.

In turbulent flows the eddies derive their kinetic energy from the main flow. This amount of kinetic energy is communicated throughout the whole spectrum of eddy sizes (conservation of energy) until finally it is dissipated in the tiniest eddies due to the action of viscosity. At the scale of these tiniest eddies the transport of momentum by the individual molecules can no longer be ignored. In other words, the momentum concentrations and fluxes have fallen down to a level that viscosity can cope with them.

The size of these so-called Kolmogorov eddies in which the energy dissipation due to internal friction takes place, depends on both the kinematic viscosity and the amount of kinetic energy involved in the main flow. The kinetic energy of the main flow is related to the power supplied to the flow system, in bio-reactors often via the impeller. More power supplied may be guessed to result in tinier Kolmogorov eddies, while a larger viscosity may imply larger ones. Within the Kolmogorov eddies, molecular transport is the only way of life!

5.2 Flows around obstacles

⊓ In thinking about fluids, what do you think is meant by the term drag? (Think what happens when you first throw a log into a fast flowing stream. At first it does not move, but then it begins to move in the direction of the stream, first slowly. Gradually it may reach the velocity of the stream).

From such common-day experiences, it is known that flows exert drag forces on bodies immersed in a fluid. Generally, such forces consist of two quite different components. These are:

- form drag;

- friction drag.

These two components will now be discussed separately.

form drag The form drag is due to the pressure distribution around the body which arises from the interaction of the flow with the body. At the front of the body, the flow impinges upon the body (see Figure 5.2). The kinetic-energy concentration $\frac{1}{2}\rho\,v^2$ of the flow is

stagnation pressure completely converted into the so-called stagnation pressure (this conversion follows from, and obeys, the mechanical energy balance, in a simplified form; it also explains why the kinetic-energy concentration is often denoted as the velocity head!) and the flow is continuous over the surface of the body. At the rear, in the wake of the body, the flow has separated from the surface, creating a dead zone of low pressure. The pressure difference over the body imparts a force on the body in the direction of the flow.

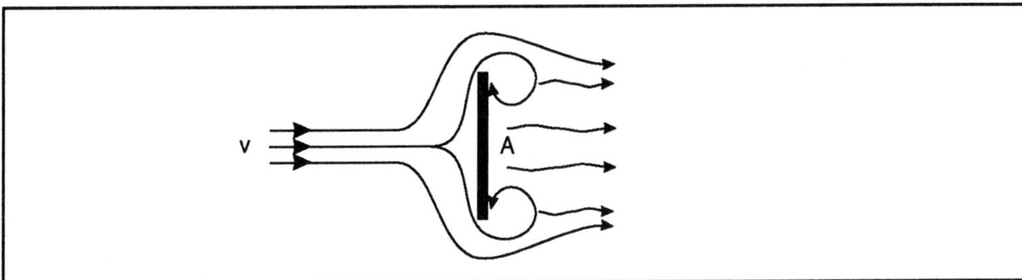

Figure 5.2 Flow patterns round a body (see text for description).

friction drag The flow along the surface of the body exerts a shear or friction force on the body: momentum directed alongside the body is transferred radially inwards, towards the body, as a result of which the body is dragged along. This shear upon the body is denoted as the friction drag. Over most of the surface of the body, the direction of the shear stress is such that the friction drag acts in the direction of the main flow, while in the wake, shear may be exerted in the opposite direction.

Alternatively when a body moves through a fluid, it exerts a drag force on the fluid. Again, both form drag and friction drag can be distinguished. With the high pressure at the nose of the body, the low pressure in the wake, and friction along the boundaries of the body, the drag again depends strongly on the body's speed. In general terms: the interaction force between body and fluid depends on the velocity difference between body and fluid. Note that, for example, in a vessel the impeller and the baffles are obstacles to the flow. This is also true for a gas bubble dispersed in a liquid.

Reynolds number

Calculating this interaction force on the basis of the above definitions of form drag and friction drag would require the flow field around the body to be known in detail. From this flow field, the pressure distribution and the shear stress profile could be calculated. In many instances of practical interest such a calculation is tedious and/or only marginally possible (in view of the present state of the art in computational fluid flow techniques). In engineering practice it is therefore useful to pursue a more practical procedure based on empiricism. The total interaction force (ie both form drag and friction drag) turns out to be first of all a function of the velocity head (the kinetic-energy concentration) of the flow. By dimensional analysis (see Chapter 4) it is found that the ratio of force to velocity head times d_p^2 is some function of the particle Reynolds number (Re_p) where:

$$Re_p = \frac{\rho \, v \, d_p}{\eta}$$

(E - 5.3)

in which v denotes the fluid's velocity with respect to the particle, ρ and η relate to the fluid, and d_p is the particle diameter. This Reynolds number dependence on the interaction force reflects the effect of the details of the flow field around the particle. This flow field is typified by the ratio of convective momentum flux (velocity head) to molecular momentum flux (friction). Hence:

$$F = C_w \, A \, \frac{1}{2} \, \rho \, v^2$$

(E - 5.4)

C_w = drag coefficient. C_w is a function of the particle Reynolds number (see Figure 5.3), A is the projection of the surface perpendicular to the direction of flow. For highly turbulent flow conditions around a body, this C_w is constant and F only depends on the velocity head, viscosity (ie molecular transport) playing no role at all. For spherical

creep flow

particles with particle Reynolds numbers smaller than 1 (generally denoted as creep flow), C_w is inversely proportional to Re_p:

$$C_w = \frac{24}{Re_p}$$

(E - 5.5)

Substituting this result in the earlier expression for F (for a spherical particle $A = \frac{\pi}{4} d_p^2$)

Stokes' law

results in Stokes' law for form drag and friction drag together:

$$F = 3 \, \pi \, \eta \, d_p \, v$$

(E - 5.6)

Figure 5.3 shows some examples of the relationship between C_w and Reynolds number for particles of various shapes.

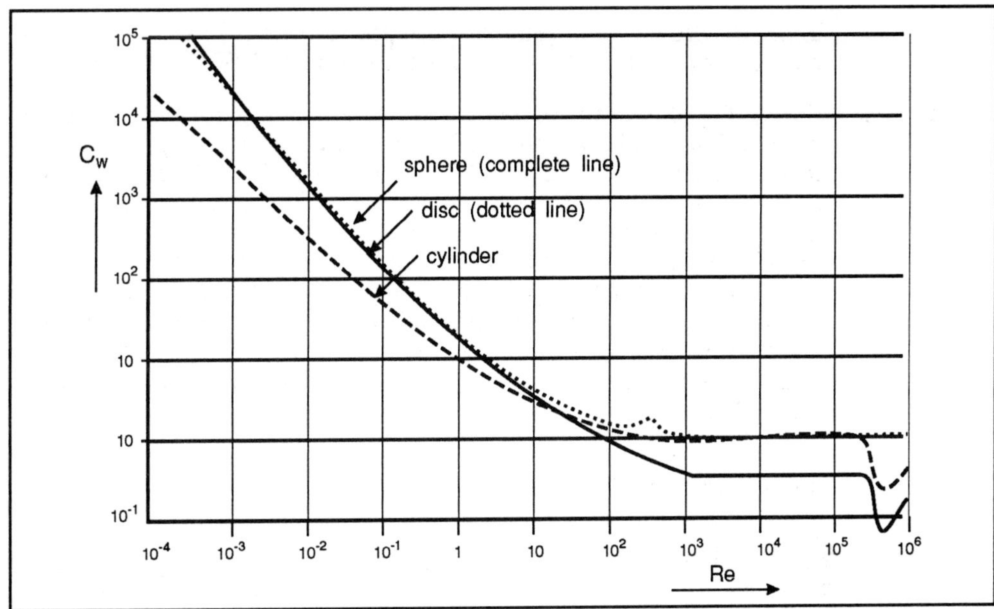

Figure 5.3 Relationship between drag coefficient C_w and Reynolds number (Re). Redrawn from Janssen, LPBM and Warmoeskeken, MMC. (1987) Transport Phenomena Data Companion. Published by Edward Arnold Ltd. Sevenoaks, Kent.

Ⅱ Let us work through an example.

A spherical hail-stone (diameter d = 3 mm, density ρ_{hail} = 915 k m^{-3}) falls steadily through stagnant air. The density of air is 1.2 kg m^{-3}. Calculate the free fall velocity of the cell mass in terms of the drag coefficient (C_w). g = 9.81 m s^{-2} and η_{air} = 1.8 x 10^{-5} Nsm^{-2}.

Solution: The hail-stone mass falls steadily, thus the form and friction drag forces exerted by the air on the hail-stone just compensate the forces of gravity and buoyancy.

The form and friction drag forces exerted by the air on the hail-stone = $C_w A \dfrac{1}{2} \rho_{air} v^2$ (see Equation 5.4).

The forces of gravity and buoyancy are given by the differences in the density of the hail-stone and air multiplied by the volume and the gravitational acceleration constant.

Thus the force of gravity and buoyancy

$$= \frac{\pi}{6} d^3 \ (\rho_{hail} - \rho_{air}) \ g$$

Since the hail-stone is falling steadily

$$C_w A \frac{1}{2} \rho_{air} v^2 = \frac{\pi}{6} d^3 \ (\rho_{hail} - \rho_{air}) \ g \qquad\qquad (E - 5.7)$$

Since the area A for a spherical particle is given by:

$$A = \frac{\pi}{4} d^2$$

we thus arrive at:

$$C_w v^2 = \frac{4}{3} d \frac{(\rho_{hail} - \rho_{air})}{\rho_{air}} g = \frac{4}{3} d \frac{(913.8)}{1.2} g \qquad (E - 5.8)$$

$$\text{Thus } v = \frac{\sqrt{29.88}}{C_w} \qquad (E - 5.8a)$$

Our problem is that $C_w = C_w$ (v), that is frictional drag at velocity v. Using Figure 5.3, we have to solve iteratively for v. Let our first guess be that Re > 10^4, thence $C_w = 0.43$. Substituting into Equation 5.8a, we find v = 8.3 ms^{-1}. Now, using $\rho_{air} = 1.2$ kg m^{-3} and $\eta_{air} = 1.8 \times 10^{-5}$ Nsm^{-2} we calculate Re = 1660. Thus, the second value for C_w from Figure 5.3 is still $C_w = 0.43$. Hence v = 8.3 ms^{-1}.

The example we have just worked through is for a cell mass falling through air. We could use an entirely analogous calculation to determine the rate of sedimentation of particles in a settling tank or to determine the rate at which cells will sediment in an unstirred vessel.

5.3 Pressure drop in pipe systems

The pressure drop over horizontal and straight pipes is due to the shear stress or friction exerted on (or experienced by) the pipe walls. Mechanical energy is dissipated into heat (we referred to this in an earlier section). Again, this pressure drop due to friction is first of all related to the velocity head of the flow, on the basis of dimensional analysis, the pressure drop is given by Fanning's equation:

$$\Delta p = 4f \frac{L}{D} \frac{1}{2} \rho v^2 \qquad (E - 5.9)$$

where D = diameter of the pipe, L = length, v = velocity, ρ = density.

The so called friction factor f is a function of the Reynolds number Re based on the pipe diameter (Re = $\rho v D / \eta$). This dependence is shown in Figure 5.4. Note that for laminar flow through pipes (ie for Reynolds numbers up to about 2 x 10^3), the frictional force is proportional to the Reynolds number and can be represented by the relationship:

$$4f = \frac{64}{Re} \qquad (E - 5.10)$$

For turbulent flow (Re value above about 3 - 4 x 10^3), the relationship between friction factor and Reynolds number is weaker.

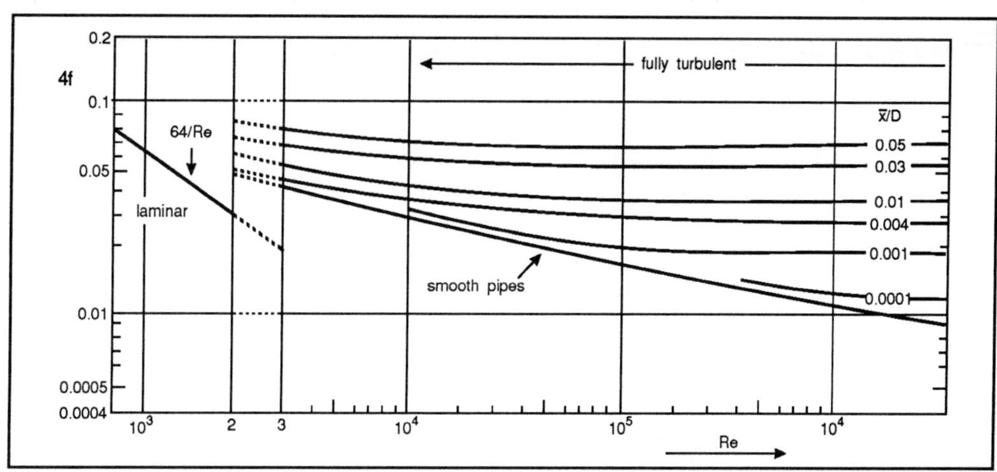

Figure 5.4 Relationship between the value of 4f and Reynolds Number. x̄/D is a measure of the roughness of the pipe (see text for details). Redrawn from Smith, J.M., Stammers, E. and Janssen, LPBM (1986) Fysische transport verschÿnselen 1, Delftse Üitgevers Maatschappÿ bv, Delft, The Netherlands.

So far our discussion on the pressure drop in pipe system has only focussed onto horizontal, straight and smooth pipes. Life is of course never as simple as that!

If the pipe is not horizontal such as shown below

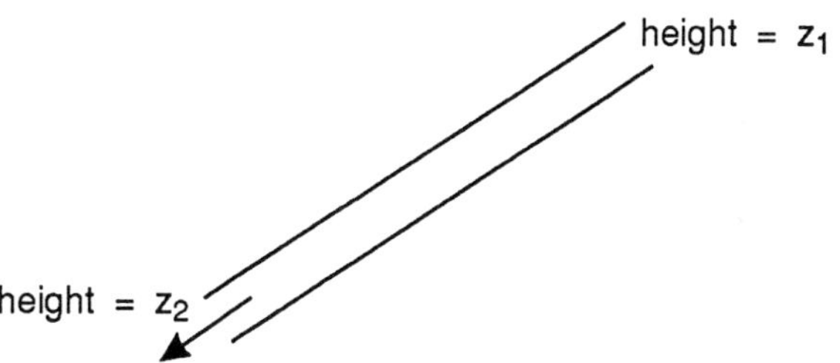

then Equation 5.9 becomes:

$$\Delta p = 4f \frac{L}{D} \ \frac{1}{2} \ \rho v^2 + \rho.g \ (z_2 - z_1)$$

(E - 5.10a)

in accordance with Equation 3.9 (the mechanical energy balance).

If the pipe has turns in it then this will also have an impact on the pressure drop (we will not deal with this here).

We must also take into account the effects of the roughness of the pipe. This too will have an effect on the value of f. In Figure 5.4 we show several relationships between 4f and Re for pipes of different roughness. Look at the right hand side of the figure: x̄/D it is a measure of relative roughness in which x̄ is the (absolute) roughness of the pipe representing the mean size of scars and pits and D is the diameter of the pipe. Thus the greater the value of x̄/D, the less smooth the pipe. Note that the frictional factor increases with the relative roughness of the pipe.

<table>
<tr><td>

SAQ 5.1

</td><td>

Water is transported through a horizontal, smooth tube of length L (=4 km). The mass flow rate, ϕ_M, is 1 kgs^{-1}. The diameter (D) of the tube is 5 cm. Calculate the pressure drop across the tube. The density and viscosity of water are 10^3kg m^{-3} and 10^{-3} Ns m^{-2} respectively. Report your answer in bars, (1 bar = 10^5Nm^{-2}). To help you we suggest you first calculate the velocity of the water. This can be done from using the mass flow rate, the density of water and the cross-sectional area of the tube.

Once you have calculated the velocity, it is then possible to calculate Reynolds number from:

$$Re = \frac{v}{\nu} D \text{ or } \frac{\rho \, v \, D}{\eta}$$

The Reynolds number can then be used to determine the friction factor from Figure 5.4.

</td></tr>
</table>

5.4 Flow patterns in stirred vessels

A full derivation and description of flow patterns in a variety of (bio-)process equipment is far beyond the scope and level of this chapter. This would require a proper treatment of momentum balances (the so-called Navier-Stokes equations) as well as a detailed discussion on turbulence and turbulence models. Computer codes are available for calculating, on this basis, the flow field in process equipment in quite some detail. Even the most recent advances of computational fluid flow techniques, however, render such calculations by no means simple or standard. Fortunately a lot of empirical data are available, in particular on flow patterns and associated aspects in stirred vessels. These data enable us to make some description of the flow patterns and processes within (bio-)process equipment.

Π A common mixing device used in bioreactors is an impeller. Before reading on, see if you can draw at least two types of impeller. Do all impellers act in the same manner? You will be able to check your answer in the following discussion.

Most impellers induce large scale circulatory motions in the stirred vessels and in that respect are similar. But, however, the pressure distribution around the impeller depends upon the geometry of the impeller. We can divide impellers into two classes:

* those which generate currents parallel with the axis of the shaft (ie axial flow type of propellers);

* those which generate currents radially (ie turbines throw fluid radially outwards).

Some examples of impellers are shown in Figure 5.5.

Did you include these types in your list?

axial and radial flow

Also shown in Figure 5.5 are the two types of flow (axial and radial) generated by these impellers.

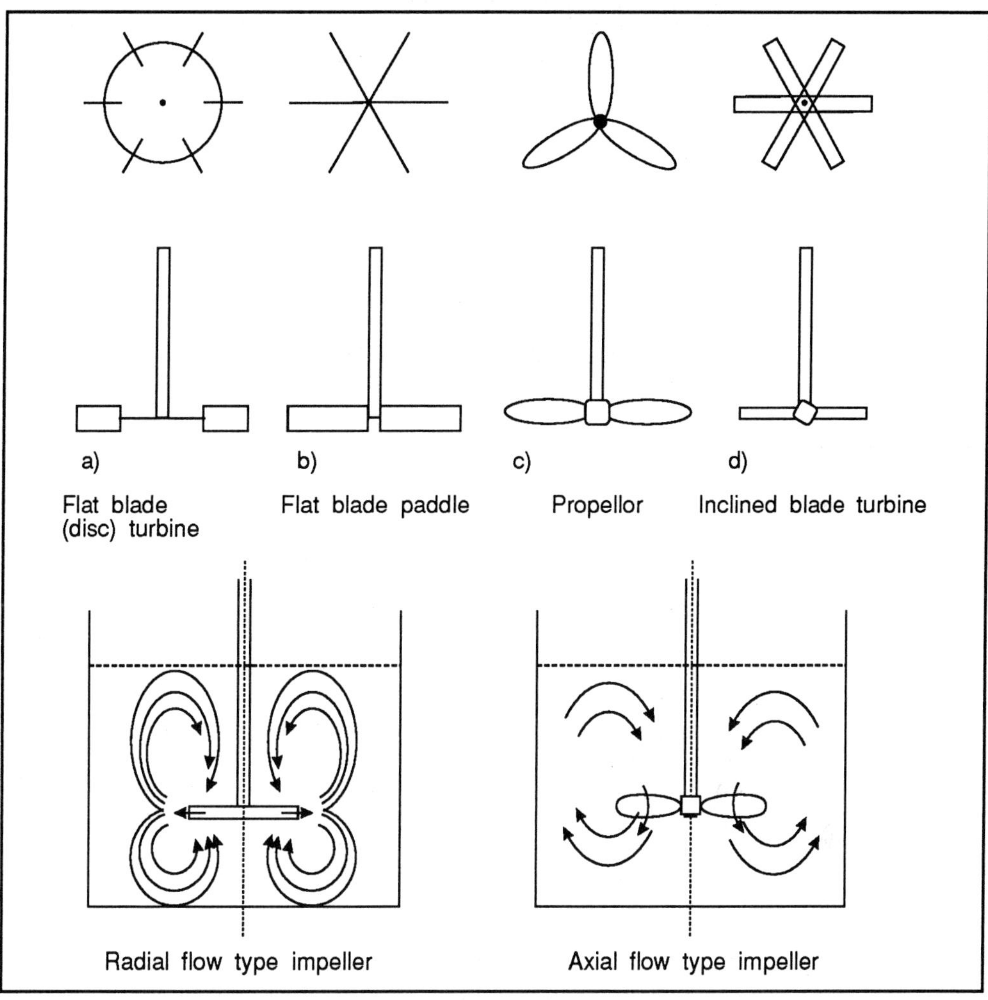

Figure 5.5 Impeller types and the flow patterns they generate. Redrawn from Beck, WJ & Muttzall. (1975) Transport Phenomena. John Wiley and Sons Ltd. Chichester, UK.

In these stirred vessels, the impeller brings the liquid contents in a revolving motion by supplying momentum to the liquid. The blades of the impeller continuously exert a force on the liquid which, of course, is given by Equation 5.4. The resistance the liquid offers to the revolving motion of the impeller must be surmounted. All energy supplied to the liquid via the impeller is ultimately dissipated by the action of viscosity. This energy supply rate aims at getting sufficiently small eddies to ensure fast mixing. Again a Reynolds number typifies the mode of swirling flow by comparing velocity head (in the swirling motion) and internal friction (or radical exchange of momentum):

$$Re = \frac{\rho(ND)^2}{\eta(ND)/D} = \frac{\rho \, N \, D^2}{\eta}$$

$$(E - 5.11)$$

in which N denotes the number of revolutions and D the diameter of the impeller, and apart from a factor π, ND represents the circumferential velocity of the impeller (circumferential velocity = $N\pi D$).

<table>
<tr><td>SAQ 5.2</td><td>A culture vessel is stirred by an impeller of diameter 1m. If the density of the culture medium is 1000 kg m^{-3} and its η = 10 Ns m^{-2}, what will the Reynolds number be when:

1) the number of revolutions of the impeller is 100 revolutions per second.

2) the number of revolutions of the impeller is 1 revolution per second?</td></tr>
</table>

For Reynolds numbers exceeding about 10,000, the flow in a stirred vessel is very turbulent. Such large-scale fluid motions quickly distribute mass, momentum and/or heat over the whole vessel, while turbulent eddies of all scales, along with molecular mechanisms, take care of homogenization on a smaller scale. To what degree the power supplied to the impeller is used for inducing a circulatory flow throughout the vessel may be expressed in terms of pumping efficiency that relates the product of flow rate times pressure drop over the impeller to the power input. Turbines, produce a lot of eddies in their immediate vicinity, while at more remote positions the turbulence is less intense. Eventually, all power supplied is converted into the turbulent energy within the tiny Kolmogorov eddies. In many bioreactor (fermentation) systems a compromise has to be made. On the one hand there is a need to generate homogeneity within the system by ensuring that localised areas of nutrient deficiency or toxin build up do not develop. This demands good mixing. This of course can be achieved with high impeller speeds and thus high Reynolds numbers. But this turbulent flow sets up stress forces across cellular materials and can cause mechanical damage to the cells.

stress forces may damage cells

Which impeller is used or recommended in a particular case, strongly depends on the viscosity of the liquid, on the scale of the mixing vessel, and on the purpose of the mixing operation (blending miscible liquids, dispersing gas, suspending solids, carrying out chemical reactions or growing cells). A proper selection of impeller type and size requires a large degree of experience. Usually, vertical baffles along the vessel wall prohibit the liquid contents from forming a vortex with a core consisting of air. Vortex mixers have been used for small scale operations. For a fuller description of impellers and bioreactor design and operation see the Biotol texts 'Operational Modes of Bioreactors' and 'Bioreactor Design and Product Yield'.

5.5 Non-Newtonian liquids

Thus far we have been dealing with liquids which obey Newton's law. This is not true for all liquids. A key feature of so-called Newtonian liquid is that there is a linear relationship been momentum flux (or shear stress τ_{xy}) and the velocity gradient: $(\frac{dv_y}{dx})$. In some liquids, this linearity does not hold.

The breakdown of the Newtonian relationship is always an indication of strong molecular interactions. We will now examine some examples of some non-Newtonian liquids.

5.5.1 Power law liquids

For power law liquids the shear stress τ_{xy} and the gradient $\dfrac{dv_y}{dx}$ are related via the empirical relation

$$\left|\tau_{xy}\right| = k \left|\frac{dv_y}{dx}\right|^n \qquad\qquad\qquad\text{(E - 5.12)}$$

consistency   Here k ('consistency') and n ('index') are constants. Note that always $n > 0$. For values of n between 0 and 1 a liquid is denoted as being pseudo-plastic, while for $n > 1$ a liquid is called dilatant. For $n = 1$ we recover the relation for Newtonian fluids. Examples of liquids with non-Newtonian rheology are paint, ink and cement. Some microbial suspensions change from Newtonian to power-law liquids if the microbial concentration increases.

Thus (see Figure 5.6):

$0 < n < 1$: pseudo-plastic (example ink, paint)
$n = 1$: Newtonian (many)
$n > 1$: dilatant (example cement)

5.5.2 Bingham liquids

The rheology of Bingham liquids is given by:

$$\left\{ \begin{array}{ll} \left|\tau_{xy}\right| - \tau_0 = \eta \left|\dfrac{dv_y}{dx}\right| & \text{for } \left|\tau_{xy}\right| > \tau_0 \\[3mm] \left|\dfrac{dv_y}{dx}\right| = 0 & \text{for } \left|\tau_{xy}\right| \leq \tau_0 \end{array} \right. \qquad\qquad\text{(E - 5.13)}$$

in which τ_0 is called the yield stress and represents the strength of molecular interactions which must be overcome by the velocity gradients in the flow field (see also Figure 5.6).

From Equation 5.13 we see that as long as the shear stress does not exceed the yield stress, the gradient $\left|\dfrac{dv_y}{dx}\right|$ is zero, or in other words the velocity is constant (but not necessarily zero!). Examples of Bingham liquids are clay and toothpaste. They are also referred to as Bingham plastics.

5.5.3 Visco-elastic liquids

The rheology of visco-elastic liquids is quite complicated. These liquids behave partially like viscous liquids and partially like solids, ie they possess elastic properties. Thus if the flow of the liquid is stopped because you have removed the driving force, the liquid will 'flow' a little in the direction opposite to the initial flow. A nice example of a visco-elastic liquid is shark-fin soup. Stir the soup with your spoon so that it rotates in your cup. If you stop stirring (take out the spoon), you will see that the rotation slows down due to viscous effects. But the flow does not stop, the flow direction changes sign and the soup rotates a little backwards.

The relation between the shear stress and the gradient in the velocity now contains the derivative of the shear stress with respect to time. This term accounts for the elastic properties of the visco-elastic liquid. The relation reads as:

$$\tau_{xy} + \lambda \frac{d\tau_{xy}}{dt} = - \eta \frac{dv_y}{dx}$$

$$(E - 5.14)$$

Many biological fluids behave like visco-elastic liquids, examples are found in fibrinogen solutions and fungal suspensions. The dissolved molecules or suspended structures have three-dimensional structures that can be distorted in a flow, but which take their original shape as soon as the flow stops.

5.5.4 Suspensions

For simple suspensions several proposals for the viscosity η_s of the suspension are found in literature. One of them is given by Einstein:

$$\eta_s = \eta_c (1 + 2.5 \ (1 - \varepsilon))$$

$$(E - 5.15)$$

in which η_c represents the viscosity of the mother fluid and ε denotes the (volume) fraction of the dispersed particles. The term in brackets accounts for the increase in the apparent viscosity due to the enhanced exchange of momentum resulting from the presence and motion of the dispersed particles.

Figure 5.6 shows the relationship between τ_{xy} (shear stress) and $\frac{dv_y}{dx}$ (velocity gradient) for both Newtonian and non-Newtonian fluids.

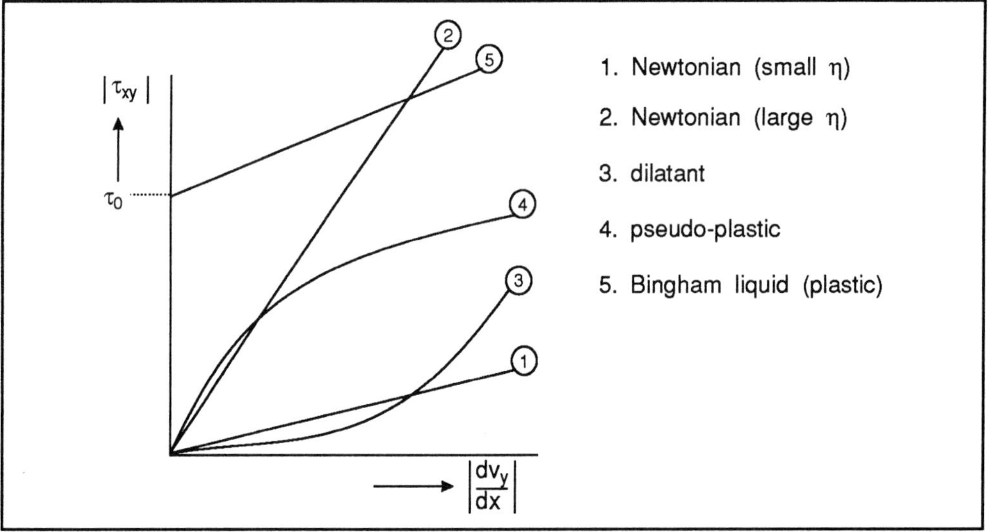

Figure 5.6 Relationships between shear stress (τ_{xy}) and velocity gradient ($\frac{dv_y}{dx}$) for Newtonian and non-Newtonian fluids (see text for details).

SAQ 5.3

Complete the following sentences using the words provided below.

1) When a body is placed in a liquid flow, it is subject to two types of drag forces called [   ] and [   ] drag.

2) When a liquid flows such that the layers of the liquid are moving in parellel, the liquid is said to be showing [   ] flow.

3) Flat bladed paddle types of impellers produce [   ] flow in stirred tank.

4) Inclined blade turbine impellers produce [   ] flow in stirred tanks.

5) When the Reynolds number exceeds 10,000 in a stirred tank, the liquid in the tank will show [   ] flow characteristics.

6) The smallest eddies in the liquid in a stirred tank are called [   ] eddies.

7) Liquids in which the shear stress τ_{xy} is proportional to the velocity gradient $(\frac{dv_y}{dx})$ are said to be [   ] liquids.

8) Liquids in which the velocity gradient $\frac{dv_y}{dx} = 0$ providing the shear stress (τ_{xy}) is smaller than the stress yield (τ_0) are called [   ].

Word Selection: Newtonian, form, laminar, axial, turbulent, Bingham plastics, Newtonian, radial, Kolmogorov, frictional.

SAQ 5.4

A hollow steal, smooth sphere (diameter 5 mm, mass 0.07 g) is released in a liquid. The sphere reaches a steady fall velocity of $5 \ 10^3$ ms^{-1}. The density of the liquid is 900 kg m^{-3}. The distance from the sphere to the walls of the vessel, containing the liquid, is very large (g = 9.81 m s^{-2}).

1) Calculate the friction coefficient C_w of the sphere.

2) Calculate the viscosity of the liquid.

SAQ 5.5

In a closed vessel, completely filled with liquid, an impeller mixes the liquid. We slowly increase the number of revolutions per minute of the impeller. It is found that at 400 revolutions per minute the flow field in the vessel changes from laminar to turbulent.

In a geometrically similar vessel with four times larger dimensions, completely filled with a different liquid (with a viscosity twice as high as the viscosity of the liquid mentioned above, but all other properties identical) one performs a similar experiment. Determine the number of revolutions per minute at which the flow field in this vessel will change from laminar to turbulent.

Summary and objectives

In this chapter we have discussed a variety of phenomena associated with the flow of liquids and gases. We have distinguished between laminar and turbulent flow and described the flow around obstacles in a fluid stream. Much of the discussion focussed on so called Newtonian liquids (ie these in which $\tau_{xy} = \eta \dfrac{dv_y}{dx}$) in which we established a variety of relationships especially relating to the flow and pressure drop in pipe systems. We also described flow patterns in stirred vessels and discussed flow patterns produced by different types of impellers. Finally we briefly described liquids which do not conform to Newtonian behaviour.

Now that you have completed this chapter you should be able to:

* describe the difference between laminar and turbulent flow;

* calculate Reynolds number from provided data;

* use data relating to pipe dimensions, fluid velocity and density to calculate the pressure drop in pipe systems;

* describe the flow patterns in stirred vessels generated by a variety of impellers;

* describe what is meant by power-law liquids, Bingham plastics and visco-elastic liquids.

The transport of heat

The transport of heat

6.1 Introduction

In earlier chapters, we introduced the concepts of energy balances and flow phenomena. By now, you should be aware that the transport of heat is predominantly accomplished by two mechanisms; conduction and convection. In stagnant (non-mobile) systems, heat is transported by conduction. In non-stagnant systems (systems in which populations of molecules move, for example in a stirred tank), heat is transported by convection and conduction. In circumstances where convection is taking place, the transport of heat by conduction is often of minor importance.

In considering heat transport we must determine from the outset whether or not both conduction and convection processes are taking place. If both processes are taking place then a description of energy balances and heat transport must take into account both processes.

In this chapter, we are going to extend your knowledge of heat transport since heat transport is of great importance in a wide range of biotechnological processes. The chapter is divided into two phases. In the first phase we will consider the transport of heat by conduction. This will include discussion of heat transport in a range of systems with different geometries and lead to a consideration of heat transfer coefficients. In the second phase, we will examine the transport of heat by convection.

6.2 Conduction of heat

6.2.1 Steady-state heat conduction

In this section, we will begin by examining heat transport by conduction in systems in steady-state.

SAQ 6.1

What is conduction?

In Chapter 3, we dealt with Fourier's law (Section 3.3.2).

Can you remember Fourier's law? If so, write down the mathematical form of this law.

Fourier's law describes the relationship between heat flow (ϕ_q'') by conduction and the temperature gradient. Essentially the transport of heat by conduction is driven by a temperature gradient and the rate of heat transported is proportional to the temperature gradient.

Mathematically we expected you to write:

$$\phi_q'' = -\lambda \frac{dT}{dx}$$

(E - 6.1)

where $\frac{dT}{dx}$ is the temperature gradient; λ is a proportionality constant, usually known as the thermal conductivity coefficient.

The minus-sign is needed to show that heat flows from a region of high temperature to a region of low temperature. The heat flow is proportional to the local gradient in the temperature $\left(\frac{dT}{dx}\right)$ and thus in general ϕ_q'' changes from place to place, even if we restrict ourselves to considering steady-state.

Let us examine the steady-state of a few simple geometries more closely. We will do this by examining some examples.

Example 1: Flat geometry in a single layer

Consider the layer of (solid) material with conductivity λ in Figure 6.1.

In a steady-state, there is no accumulation of heat in the slab. Therefore heat in = heat out. This is valid for every slab (x_1 - x_2) within the layer.

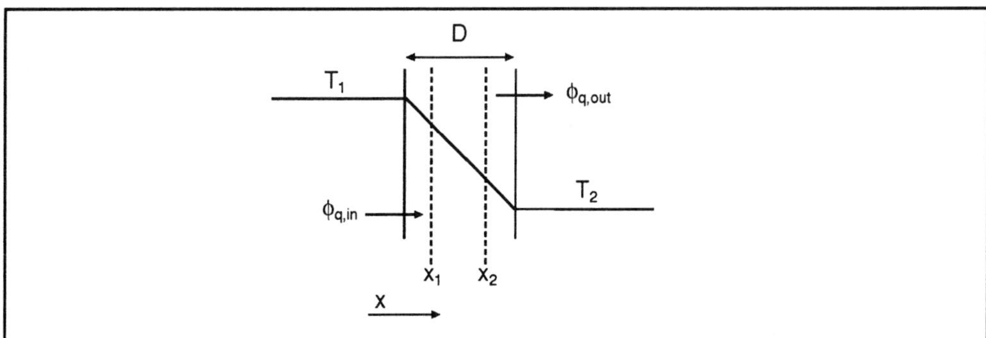

Figure 6.1 Heat transport through flat slabs (see text for details).

In the steady-state the heat balance over the slab x_1 - x_2 reads as:

$$0 = \phi_{q,in} - \phi_{q,out} = \phi_{q,x_1} - \phi_{q,x_2}$$

(E - 6.2)

Hence

$$\phi_{q,x_1} = \phi_{q,x_2}$$

(E - 6.3)

As this result holds for any x_1 and x_2, the heat flow is independent of x and is a constant. This also holds for the heat flux ϕ_q'' (x) since this is just ϕ_q / A (A is the area of the plane $x = x_1$). Remember we used the notation '' to denote per unit area. Using Equation 6.1 and 6.3 we therefore conclude that:

$$\frac{dT}{dx} = -\frac{\phi_q''}{\lambda} = \text{constant}$$

(E - 6.4)

Thus the temperature gradient across the layer is constant.

Integration of Equation 6.4 together with the boundary conditions $T(x=0) = T_1$, $T(x=D) = T_2$ and the definition $\Delta T = T_1 - T_2$ yields:

$$T(x) = -\frac{\phi_q''}{\lambda} \cdot x + T_1 \qquad\qquad (E - 6.5)$$

or

$$\Delta T = \phi_q'' \cdot \frac{D}{\lambda} \qquad\qquad (E - 6.6)$$

These two equations are quite important so remember them.

Equation 6.5 shows that the temperature profile in the layer is linear, while Equation 6.6 expresses the dependence of the flux ϕ_q'' on the driving force ΔT.

If we compare Equation 6.6 with Ohm's law

$$\Delta V = I \cdot R \qquad\qquad (E - 6.7)$$

where ΔV is potential difference, I = current, R = resistance, we obtain a somewhat different view on conduction. Ohm's law says that if we apply a voltage difference ΔV over a resistance R, this driving force ΔV generates a current (=flow) I through the resistance which is simply proportional to ΔV (see Figure 6.2). The proportionality constant is just R. R characterises the resistance of the piece of material to the transport of electricity through it: the higher R the smaller the current for a given ΔV.

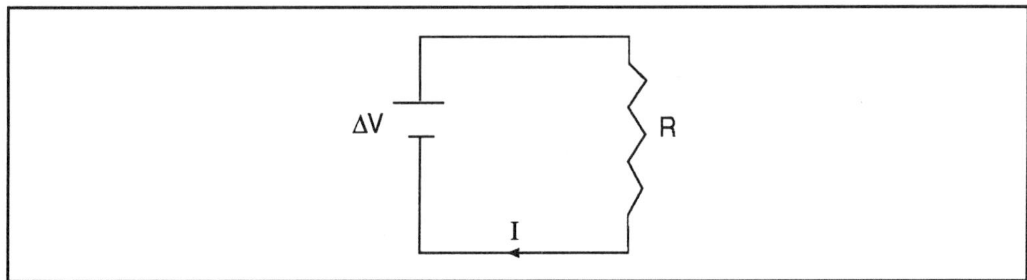

Figure 6.2 A simple electric circuit ΔV = potential difference, I = current, R = resistance.

Now let's look at Equation 6.6 in a similar way. Here the driving force is ΔT and current is the heat flow (per unit area) ϕ_q''. Thus we can regard $\frac{D}{\lambda}$ as a resistance, now to heat flow: the higher $\frac{D}{\lambda}$, the smaller the current (ϕ_q'') for a given driving force ΔT!

Notice that in steady-state heat conduction in a flat layer (slab) everywhere in the layer the following simple relations hold:

$$\phi_q'' = \text{constant}; \frac{dT}{dx} = \text{constant and that } \Delta T = \phi_q'' \frac{D}{\lambda}$$

<table>
<tr><td>SAQ 6.2</td><td>If a piece of wood allows the passage of 50 J m^{-2} s^{-1} when the temperature differences between its two faces is 10 K, how much heat will it pass when the temperature difference is 20 K?</td></tr>
</table>

Example 2: Flat geometry with two layers

Now let us consider two layers with different coefficients of conductivity in a steady-state (see Figure 6.3).

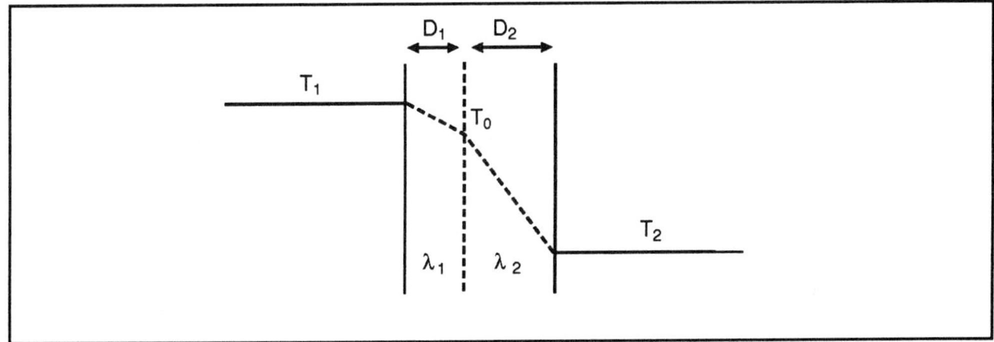

Figure 6.3 Two layers with different coefficients of conductivity (see text for further details).

We can solve this problem in terms of the driving force $\Delta T = T_1 - T_2$, the coefficients of thermal conductivity λ_1, λ_2 and the thickness of the layers D_1, D_2. It is convenient to introduce an unknown temperature T_0 at the interface between the two materials. Since it is in a steady-state, T_0 is constant. Thus we can solve the problem by regarding it as two layers each with its own driving force $T_1 - T_0$ and $T_0 - T_2$ respectively. Then all we have to do is to copy the results from Example 1 and eliminate T_0.

$$\text{layer 1: } T_1 - T_0 = \phi_{q,1}'' \cdot \frac{D_1}{\lambda_1} \tag{E - 6.8}$$

$$\text{layer 2: } T_0 - T_2 = \phi_{q,2}'' \cdot \frac{D_2}{\lambda_2} \tag{E - 6.9}$$

Now in this geometry $\phi_{q,1}''$ obviously equals $\phi_{q,2}''$ since Equation 6.2 still holds for any x_1 and x_2 in the layers. Hence combination of the Equations 6.8 and 6.9 renders:

$$\Delta T \equiv T_1 - T_2 = \phi_q'' \cdot \left(\frac{D_1}{\lambda_1} + \frac{D_2}{\lambda_2} \right) \tag{E - 6.10}$$

If we compare Equation 6.10 with the case of two electrical resistances in series (Figure 6.4) we see that Example 2 is quite analogous. In the case of the electric circuit we can write: $\Delta V = I(R_1 + R_2)$.

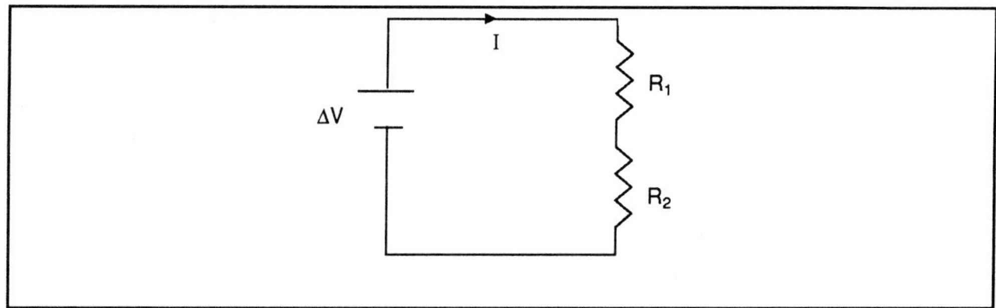

Figure 6.4 Simple electric circuit in which two resistances (R_1 and R_2) are connected in series.

<table>
<tr><td>

SAQ 6.3

</td><td>

Two sheets of different materials (material A and B) of equal thickness are fastened together. The materials have different conductivities such that $\lambda_A = 2\lambda_B$. If the temperature difference $(T_1 - T_2)$ is 10K. (See Figure below) what is the temperature at the interface of A and B (ie T_0) if $T_1 = 300$ K.

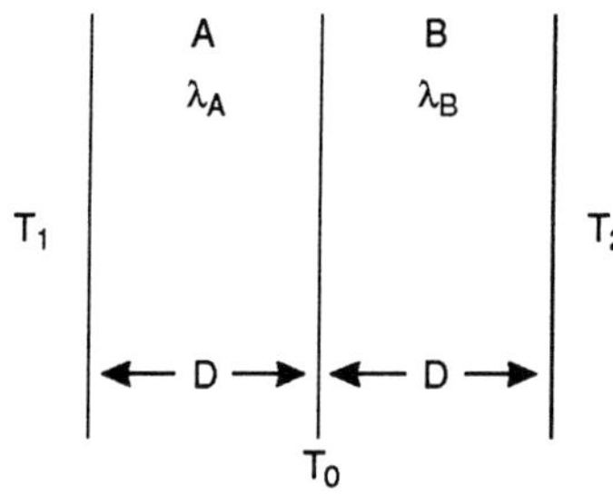

</td></tr>
</table>

Example 3: Cylindrical geometry

In both Example 1 and Example 2 we could show that in the steady-state ϕ_q'' is independent of the position in the material. This is an exception rather than a rule. This is readily illustrated with the geometry shown in Figure 6.5.

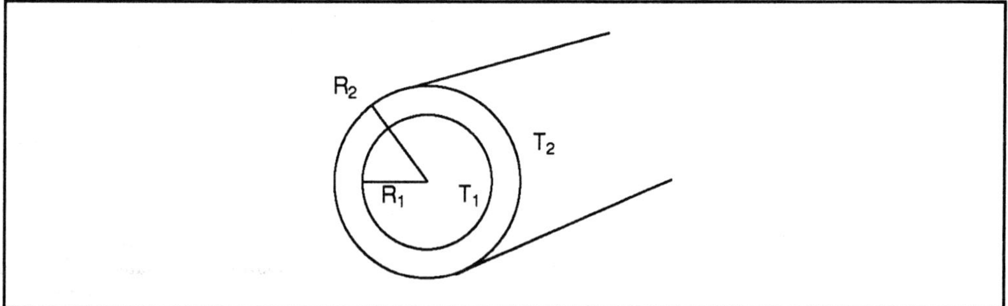

Figure 6.5 Heat conduction through curved surface (see text for details).

The inside of a (very long) hollow cylinder (inner radius R_1) is kept at constant temperature T_1, while the outside (outer diameter R_2) is at fixed temperature T_2. Thus the driving force for the heat flow can be written as $\Delta T = T_1 - T_2$. You can probably think of many examples where this type of situation might arise. The one that springs readily to mind is a pipe carrying hot water in a heating system.

What is now the steady-state temperature profile between $r = R_1$ and $r = R_2$ and how does ϕ_q'' depend on the r-coordinate (the radial distance from the axis of the cylinder)?

Let us first construct the heat balance over the 'ring' with length L and radii $r, r + \Delta r$ (see Figure 6.6).

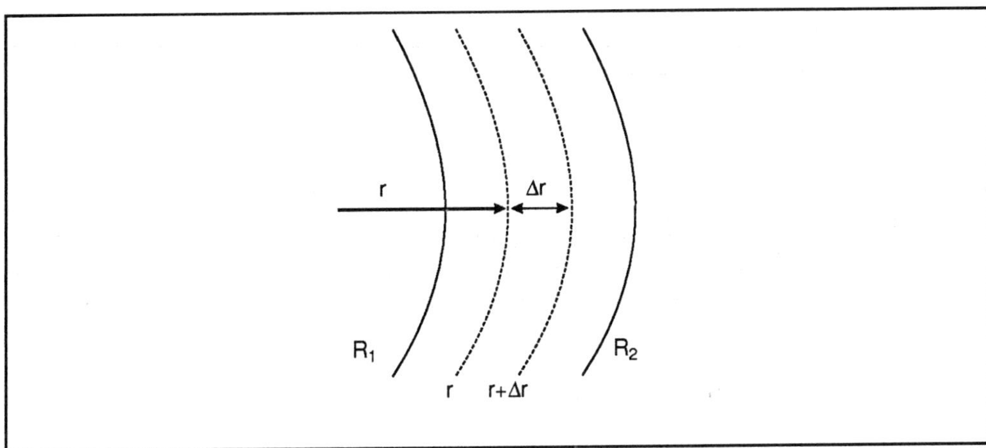

Figure 6.6 Model used to construct a heat balance over a curved surface (see text for details).

In a steady-state

$$\phi_{q,in} - \phi_{q,out} = 0 \qquad (E - 6.11)$$

Thus $\phi_{q,in} = \phi_{q,out}$ is constant. But what is ϕ_q'' in such a curved geometry? From symmetry we see that ϕ_q'' is parallel to the radius. Hence:

$$\phi_q'' = -\lambda \frac{dT}{dr} \qquad (E - 6.12)$$

and since

$$\phi_q = A \cdot \phi_q'' = 2\pi rL \cdot \phi_q'' \quad \text{(since the area of a cylinder of radius r and length L} = 2\pi r\, L\,) \qquad (E - 6.13)$$

we can write, combining Equations 6.11 to 6.13:

$$-\lambda \cdot 2\pi rL \cdot \frac{dT}{dr} = \text{constant} = C_1 \qquad (E - 6.14)$$

(since from Equation 6.11 $\phi_{q,in} = \phi_{q,out}$ as the heat flow over any ring is constant)

∏ Is ϕ_q'' constant over any ring?

The answer is no. The area of a cylinder (A) depends on r (since $A = 2\pi rL$). Since $\phi_q = A.\phi_q''$ and ϕ_q is constant, then ϕ_q'' must also vary with r.

To express this in another way, the heat flow per unit area (ϕ_q'') is greater for the inner radius of the cylinder than it is for the outer layer.

$\prod$ Is the temperature gradient across the layers of the cylinder constant?

Again the answer is no. The heat flow per unit area is given by $\phi_q'' = -\lambda \dfrac{dT}{dr}$ (see Equation 6.12).

Since ϕ_q'' changes with values of r, so must the temperature gradient $\left(\dfrac{dT}{dr}\right)$.

We would of course like to be able to determine the temperature profile and heat flux in such systems.

The next few steps are mathematically a little bit more complex. Try to follow them. If your mathematical knowledge is limited then at least try to remember the results (Equations 6.16, 6.17 and 6.18).

Equation 6.14 has the general solution.

$$T(r) = -\frac{C_1}{2\pi\lambda L}\ln r + C_2$$

(E - 6.15)

Thus the temperature at radius r is related to ln r.

The two constant C_1 and C_2 are found using the boundary conditions $T(R_1) = T_1$, $T(R_2) = T_2$.

It turns out that the temperature profile is given by:

$$\frac{T(r) - T_2}{T_1 - T_2} = \frac{\ln (r/R_2)}{\ln (R_1/R_2)}$$

(E - 6.16)

This equation shows us that the temperature profile at radius r can be calculated if we know the temperature inside (T_1) and outside (T_2) of the cylinder and the inside and outside radii (R_1 and R_2) of the cylinder. We can see from the relationship that the temperatures of layers within the cylinder are related to ln r. We can represent the temperature profile within the cylinder walls thus:

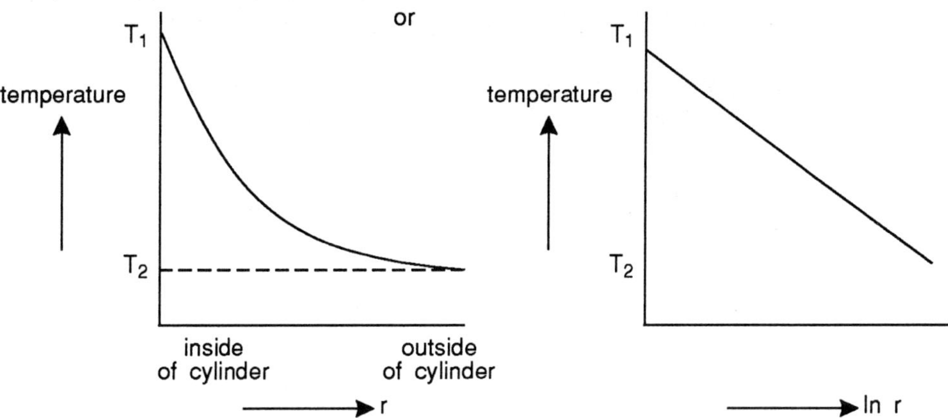

Note that a plot of temperature against ln r gives a straight line.

The temperature gradient within the cylinder walls is therefore not linear.

What about the heat flux through the wall? We know that at steady-state $\phi_{q,in} = \phi_{q,out}$ but, since A increases as r increases.

$\phi_{q,in}''$ does not $= \phi_{q,out}''$.

We have to take into account both the changing area as r increases and also the change in the driving force of heat flux, the temperature gradient. We remind you that the flux per unit area is given by:

$$\phi_q'' = -\lambda \frac{dT}{dr}$$

Solving for the temperature gradient $\left(\dfrac{dT}{dr}\right)$ is mathematically quite complex but it can be shown from Equation 6.16 that the temperature profile at radius r is given by:

$$\frac{dT}{dr} = \frac{(T_1 - T_2)}{\ln (R_1/R_2)} \frac{1}{r}$$

Thus since $\phi_q'' = -\lambda \dfrac{dT}{dr}$

$$\phi_q'' = -\lambda \frac{(T_1 - T_2)}{\ln R_1/R_2} \frac{1}{r} \tag{E - 6.17}$$

Since $\phi_q = 2\pi r L \phi_q''$ (E - 6.13) we could also write:

$$\Delta T = \frac{\ln R_2/R_1}{2\pi L\lambda} \phi_q \tag{E - 6.18}$$

Compare this again with Ohm's law ($V = IR$). We can see the current (ϕ_q) is proportional to the driving force (ΔT) but the resistance is not easily guessed. The resistance is in effect $\dfrac{\ln R_2/R_1}{2\pi L\lambda}$.

Cylinder geometry does not alter the strategy for determining temperature profiles and heat flux. What it does do is to make the mathematics much more complex. You may have had some difficulty with these mathematical derivations. Nevertheless we would expect you to be able to use the derivations produced in Equation 6.16 - 6.18.

Let us try a numerical example.

The water in a pipe is maintained at a temperature of 4°C. This pipe runs through an enclosed box containing air at 30°C. What will be the rate of heat transfer from the air into the water if the thermal conductivity coefficient of the material of the pipe is 10^2 J s^{-1} m^{-1} K^{-1}, the inside radius of the pipe is 5mm and the walls of the pipe are 5mm thick and the length of the pipe is 2m? Assume steady-state conditions apply and assume that the temperatures of the two surfaces of the pipe are 4° and 30°C. In real life of course this would not be so because of the effects of a stagnant film of water and air at the two surfaces of the pipe. We will deal with these effects later. (Attempt this before reading our solution).

It is always a good idea to make a little drawing representing the problem.

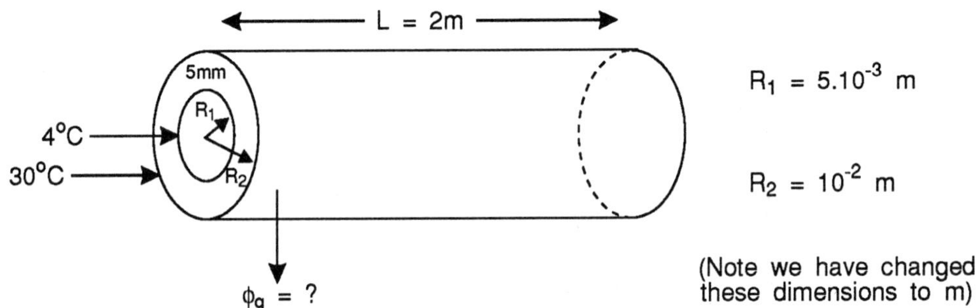

We can use the relationship.

$$\Delta T = \frac{\ln R_2/R_1}{2\pi L\lambda}\ \phi_q\ \text{(see Equation 6.18)}$$

$$4 - 30 = \frac{\ln 2}{2\pi \cdot 2 \cdot 10^2}\ \phi_q$$

$$-\frac{26 \cdot 400\pi}{\ln 2} = \phi_q\ \text{(in J s}^{-1})$$

Thus $\phi_q = -47160$ J s^{-1} $= -4.7 \times 10^4$ J s^{-1} The minus sign indicates that the heat flows radially inwards from the outside air towards the water in the pipe.

Thus heat will flow into the water at a rate of 4.7×10^4 J s^{-1}.

We could extend this problem by asking what would be the temperature at a point midway through the walls of the pipe (ie at a radius of 7.5mm). To solve this we would need to use Equation 6.16.

$$\frac{T(r) - T_2}{T_1 - T_2} = \frac{\ln (r/R_2)}{\ln (R_1/R_2)}$$

On the other hand if we wanted to calculate the heat flux per unit area (ϕ_q'') at this radius we would need to use Equation 6.17.

$$\phi_q'' = -\lambda \frac{(T_1 - T_2)}{\ln R_1/R_2} \frac{1}{r}$$

Example 4: Spheres

Consider a metal sphere (of radius R) kept at temperature T_1 surrounded by a stagnant medium (with temperature T_∞ at large distance from the sphere). In the steady-state the heat flow through any spherical 'plane' r is constant:

$$\phi_q = 4\pi r^2 \cdot \left(-\lambda \frac{dT}{dr}\right) = \text{constant} \tag{E - 6.19}$$

From Equation 6.19 we solve the temperature profile and subsequently calculate the heat flow. The result is:

$$\phi_q = 4\pi R \lambda (T_1 - T_0) \tag{E - 6.20}$$

From this description of the transport of heat through simple geometric shapes it must be obvious that except for flat plates, the relationship between heat flow, temperature gradient and thermal conductivity is mathematically quite complex. Consider for example a curved hollow, cooling coil used to cool a bioreactor. In such a circumstance we are not only having to deal with cylindrical geometry, but also with the curves within the length of the cylinder. Obviously the relationship between ϕ_q, temperature gradient and λ is difficult to determine from a simple mathematical approach. In these circumstances it is usual to make approximations or to determine the relationship between these parameters using data derived from experiments or models. We will deal with such systems a little later.

By rewriting Equation 6.20, we can find again an analogy with Ohm's law; the resistance to heat transport in the case of conduction in a spherical geometry is $\dfrac{1}{4\pi R\lambda}$ and thus depends upon the radial position.

<table>
<tr><td>

SAQ 6.4

</td><td>

A furnace wall is composed of a layer of fire proof stones ($\lambda_{fp} = 1.21$ J m^{-1} K^{-1} s^{-1}), a layer of isulation stones ($\lambda_i = 0.080$ J/mKs and a layer of bricks ($\lambda_b = 0.69$ J m^{-1} K^{-1} s^{-1}). Each layer has a thickness of 10 cm. The temperature at the inner side of the furnace is 872°C, at the outside 32°C. The furnace is operated in the steady-state.

Assume that the wall is flat.

1) Draw a sketch of the situation.

2) Write down, for every layer, the relation between the heat flux and the driving force across the layer in question.

3) Calculate the temperature at the interfaces of the layers.

4) In calculating the temperature of the interface, what have you assumed?

</td></tr>
</table>

6.2.2 Non steady-state heat conduction

Penetration theory and Fourier numbers

In the previous section we concentrated on steady-state heat conduction. Now we will deal with the more general (but also more complex) case of non steady-state heat conduction. Hence, the energy balance now also contains the change of the amount of thermal energy in our control volume with respect to time. Because of the complexity of the problem let us consider the simple geometry shown in Figure 6.7.

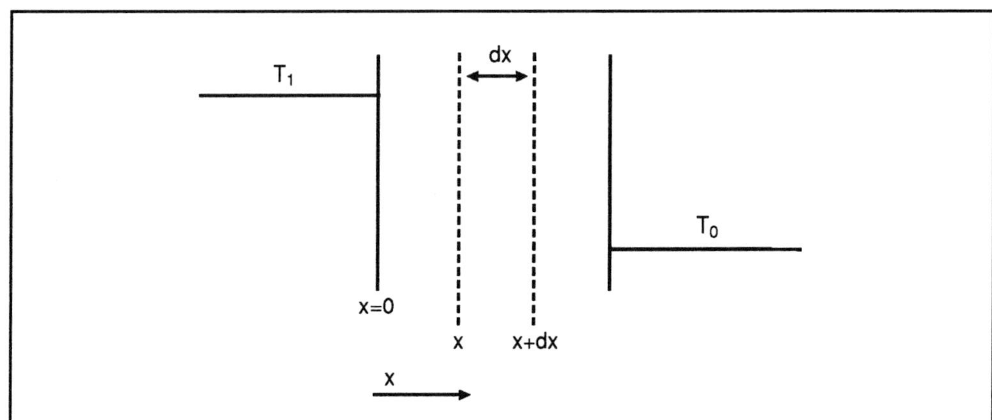

Figure 6.7 Representation of a cube initially at T$_0$ when one of the faces (x = 0) is brought to a higher temperature (see text for more details).

A very large (cubic shaped) piece of material is initially at temperature T_0. Suddenly, at $t = 0$, one of the boundaries, $x = 0$, is brought to a higher temperature T_1 and kept at T_1 for the rest of the experiment. The question is, how does this temperature T_1 penetrate into the material?

Clearly, immediately after the temperature at $x = 0$ is raised to T_1 a heat flux runs into the material since there is a driving force, initially over a very small distance. Thus a heat flow starts due to this sharp temperature gradient. This flow will gradually warm up the parts of the cube close to $x = 0$ thus breaking down the steep gradient. Hence we expect the heat flow to decrease in time.

Let us look somewhat more precisely at this and construct a non-steady thermal energy (U) balance for a layer bounded between x and $x + dx$. Our control volume is $V = L.W.dx$ (L is the length in the y-direction, W in the z direction), thus:

$$\frac{\partial U}{\partial t} = \frac{\partial(\rho V c_p T)}{\partial t} = LW . dx . \rho c_p \frac{\partial T}{\partial t} = \phi_{q,in} - \phi_{q,out} \qquad (E - 6.21)$$

(Note the symbol ∂ is used in the expressions for the derivatives, rather than d, to stress T is a function of both t and x. Remember that c_p = specific heat, ρ = density).

$$\phi_{q,in} = - \lambda LW \left[\frac{\partial T}{\partial x}\right]_x \qquad (E - 6.22)$$

$$\phi_{q,out} = - \lambda LW \left[\frac{\partial T}{\partial x}\right]_{x + dx} \qquad (E - 6.23)$$

Hence:

$$\frac{\partial T}{\partial t} = \frac{\lambda}{\rho c_p} \frac{\left[\frac{\partial T}{\partial x}\right]_{x + dx} - \left[\frac{\partial T}{\partial x}\right]_x}{dx} = \frac{\lambda}{\rho c_p} \frac{\partial^2 T}{\partial x^2} \qquad (E - 6.24)$$

Thus we derived the differential equation that governs the problem:

$$\frac{\partial T}{\partial t} = \alpha \frac{\partial^2 T}{\partial x^2} \qquad (E - 6.25)$$

where α = thermal diffusion coefficient. We remind you that $\alpha = \dfrac{\lambda}{\rho c_p}$ (see Equation 3.3)

The boundary conditions are:

$T(x = 0) = T_1$ for $t \geq 0$

$T(t = 0) = T_0$ for $x > 0$

$T = T_0$ for $x \rightarrow \infty$ for every t $\qquad (E - 6.26)$

The last condition suggests that we regard the cube as so large that the right boundary is not influenced by the changed condition at the left. In other words the solution we obtain by solving Equation 6.25 with the conditions described in 6.26 is only valid for times too short for the heat flow to reach the boundary opposite to the heated boundary. It is for this reason that the result is called penetration theory. It tells how a temperature difference penetrates into the bulk of the material.

Before we look at the exact solution let us first find how $T(x,t) - T_0$ depends on (x,t) and the properties of the material. We use for this purpose the dimensional analysis we discussed in Chapter 4. From Equation 6.25 we know that $T(x,t) - T_0$ will depend on t, x and α. But we have already seen that the driving force $T_1 - T_0$ enters even in the steady-state solution. Thus we obtain:

$(T_1 - T_0)$, t, x, α

Π It would be good practice to attempt to carry out the dimensional analysis for this yourself. Do this on a piece of paper before reading on.

Here is our solution.

The solution may take the form $(T - T_0) \sim (T_1 - T_0)^\alpha \ t^\beta \ x^\gamma \ \alpha^\delta$

The units are therefore:

$K \sim K^\alpha \ s^\beta \ m^\gamma \ (m^2 \ s^{-1})^\delta$

$K \sim K^\alpha \ s^\beta \ m^\gamma \ m^{2\delta} \ s^{-\delta}$

Collecting together units.

K: $1 = \alpha$
s: $0 = \beta + (-\delta)$ thus $\beta = \delta$
m: $0 = \gamma + 2\delta$ thus $-\dfrac{\gamma}{2} = \delta = \beta$

Thus:

$(T - T_0)^1 \sim (T_1 - T_0)^1 \ t^{-\gamma/2} . x^\gamma . \alpha^{-\gamma/2}$

Thus $\dfrac{(T - T_0)}{(T_1 - T_0)} \sim \dfrac{x^\gamma}{t^{\gamma/2} \alpha^{\gamma/2}} = \left(\dfrac{x}{\sqrt{\alpha t}}\right)^\gamma$

$$(E - 6.27)$$

If you manage to derive Equation 6.27 for yourself, well done.

From Equation 6.27 we see what we intuitively might have expected: the temperature profile penetrates into the material. The longer we wait the further a particular value of the temperature propagates into the material. Thus the value of $T - T_0$ is related to $\dfrac{1}{\sqrt{t}}$ providing $T_1 - T_0$ remains constant.

Thus the penetration proceeds at a decreasing rate with time. Equation 6.27 is quite useful but of course we still do not know the actual relationship between $\dfrac{(T - T_0)}{(T_1 - T_0)}$ and $\dfrac{x}{\sqrt{\alpha t}}$.

To obtain this we need the exact solution of Equation 6.27. We do not intend to go through the mathematical derivation, but the exact solution shows us that the temperature at a position x and at time t $(T(x_1,t))$ is related to α, x and t in the following way.

$$\frac{T(x,t) - T_0}{T_1 - T_0} = 1 - \frac{2}{\sqrt{\pi}} \int_0^{\frac{x}{2\sqrt{\alpha t}}} \exp(-s^2)\, ds \qquad\qquad \text{(E - 6.28)}$$

where the final term in this equation is called the error function.

This type of temperature profile is shown in Figure 6.8.

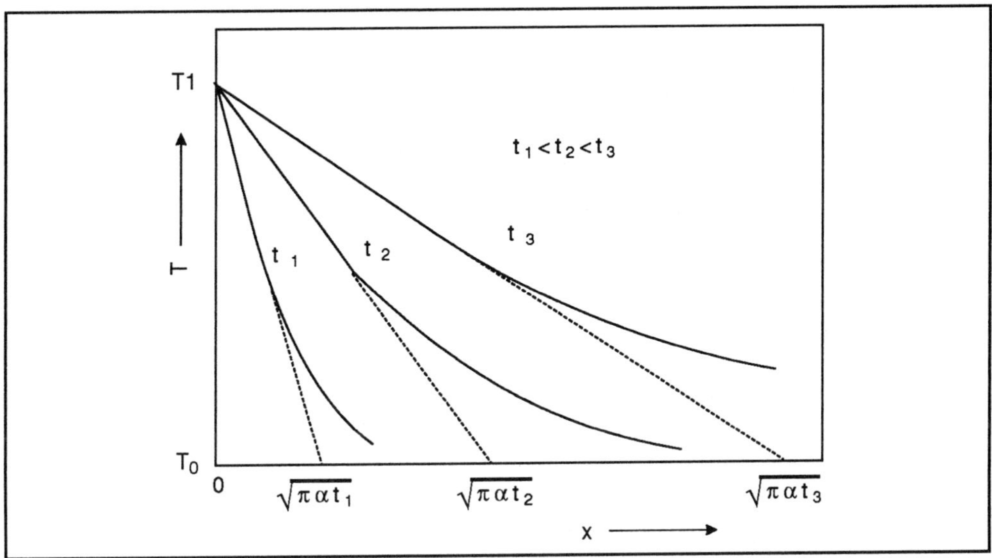

Figure 6.8 Plots of temperature (T) against position (x) in a layer being heated from one side. t_1, t_2 and t_3 represent three time intervals after heating has been commenced.

If we calculate from Equation 6.28 the heat flux through the plane $x = 0$, we find

$$\phi_q'' = -\lambda\left[\frac{dT}{dx}\right]_{x=0} = -\lambda\,\frac{1}{\sqrt{\pi\alpha t}}\,(T_1 - T_0) \qquad\qquad \text{(E - 6.29)}$$

Thus the heat flux through the plane $x = 0$ is dependent upon the difference between T_1 and T_0 and is inversely proportional to $\sqrt{\pi\alpha t}$.

Hence, as we expected the heat flux decreases in time. The driving force ΔT is a constant but the resistance $\frac{1}{\lambda}\sqrt{\pi\alpha t}$ increases with time. The heat flux is proportional to the gradient in T. The second equality of Equation 6.29 shows that the tangent to the temperature curve at $x = 0$ goes through the point $T = T_0$, $x = \sqrt{\pi\alpha t}$ (see Figure 6.8). The distance $x = \sqrt{\pi\alpha t}$ is called the penetration depth. It is very useful to estimate how far the heat has penetrated into the material.

penetration depth

Ⅱ In Figure 6.8, what is the penetration depth at time = t_1? What is the penetration depth at time = t_3?

The answers we expect to find were that at t_1, the penetration depth $= \sqrt{\pi\alpha t_1}$ and, at t_3, $= \sqrt{\pi\alpha t_3}$

Actually at the penetration depth the temperature has changed by only about 20% of the difference $(T_1 - T_0)$.

This is a figure well worth remembering.

Obviously, the group $\dfrac{x}{\sqrt{\pi\alpha t}}$ is the dimensionless group that governs penetration. If D is the length scale in the x-direction of the object, then clearly we may only use penetration theory if the penetration depth is considerably smaller than D, since in that case the flow has not reached the end of the object and our boundary condition $T = T_0$ for large values of x still holds. Thus we obtain a criterion: if, say,

$\dfrac{\sqrt{\pi\alpha t}}{D} < \dfrac{1}{2}$, or $\dfrac{\alpha t}{D^2} < 0.1$ then penetration theory holds; otherwise we have to consider the

case when the driving force is no longer constant. The group $\dfrac{\alpha t}{D^2}$ is known as the Fourier

Fourier number

number.

$$Fo = \frac{\alpha t}{D^2}$$

(E - 6.30)

On the one hand Fo denotes time made non-dimensional by use of the penetration time (D^2/α). On the other hand $Fo^{1/2}$ is the ratio of penetration depth to D.

It is worthwhile working through an example of applying penetration theory.

A large plate separates a chamber from a flow channel. In the channel flows a liquid at a constant temperature (20°C). The plate and the gas in the chamber are also at 20°C. Suddenly the temperature of the incoming liquid is increased to 40°C (the flow velocity is so large that this means in practice that the lower side of the plate is suddenly at 40°C (see Figure 6.9). We would like to estimate the time that it takes for the gas in the chamber to begin to change its temperature.

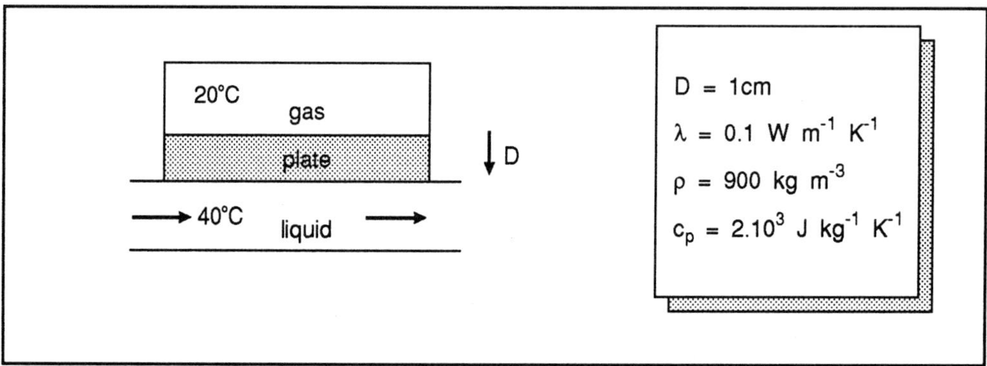

Figure 6.9 A plate (shaded strip) bounded by a gas chamber at 20°C and a flow of water at 40°C (see text for additional details).

A solution to this problem is as follows:

The temperature of the gas will change when heat flows into it. This happens when the heat has penetrated through the plate. Thus our first guess is to calculate the time it takes for the penetration depth $x = \sqrt{\pi \alpha t}$ to become equal to D:

$$\sqrt{\pi \alpha t} = D \Rightarrow$$

$$t = \frac{D^2}{\pi \alpha} = \frac{\rho c_p D^2}{\pi \lambda} = 570s \text{ (approximate)}.$$

We have used the data from Figure 6.9 for this calculation.

Of course this overestimates the actual time since we know that at the penetration depth the temperature rise is 20% of ΔT. Hence a second guess could be:

$$\sqrt{\pi \alpha t} = \frac{D}{2} \Rightarrow$$

$$t = \frac{1}{4} \frac{D^2}{\pi \alpha} = 143s$$

Obviously the truth will be somewhere in between.

As mentioned above, when Fo > 0.1 the situation is quite different. The heat flow has reached the opposite end of the cube, plate etc. Temperature may change everywhere in the material in question. Now it is found that the temperature profiles for successive moments in time become geometrically similar. The driving force ΔT is decreasing in time but the resistance is found to be a constant depending on the geometry of the object under consideration. Figures 6.10 and 6.11 show the mean temperature and the temperature of the centre of the object, respectively, as functions of the Fo-number. As you can see from these Figure 6.10 the relationship between $\dfrac{T_1 - <T>}{T_1 - T_0}$ (where $<T>$ is the mean temperature) and Fo depends on the geometry of an object. Likewise the temperature of the centre (T_c) as a function of Fo also depends on the geometry of the object (Figure 6.11).

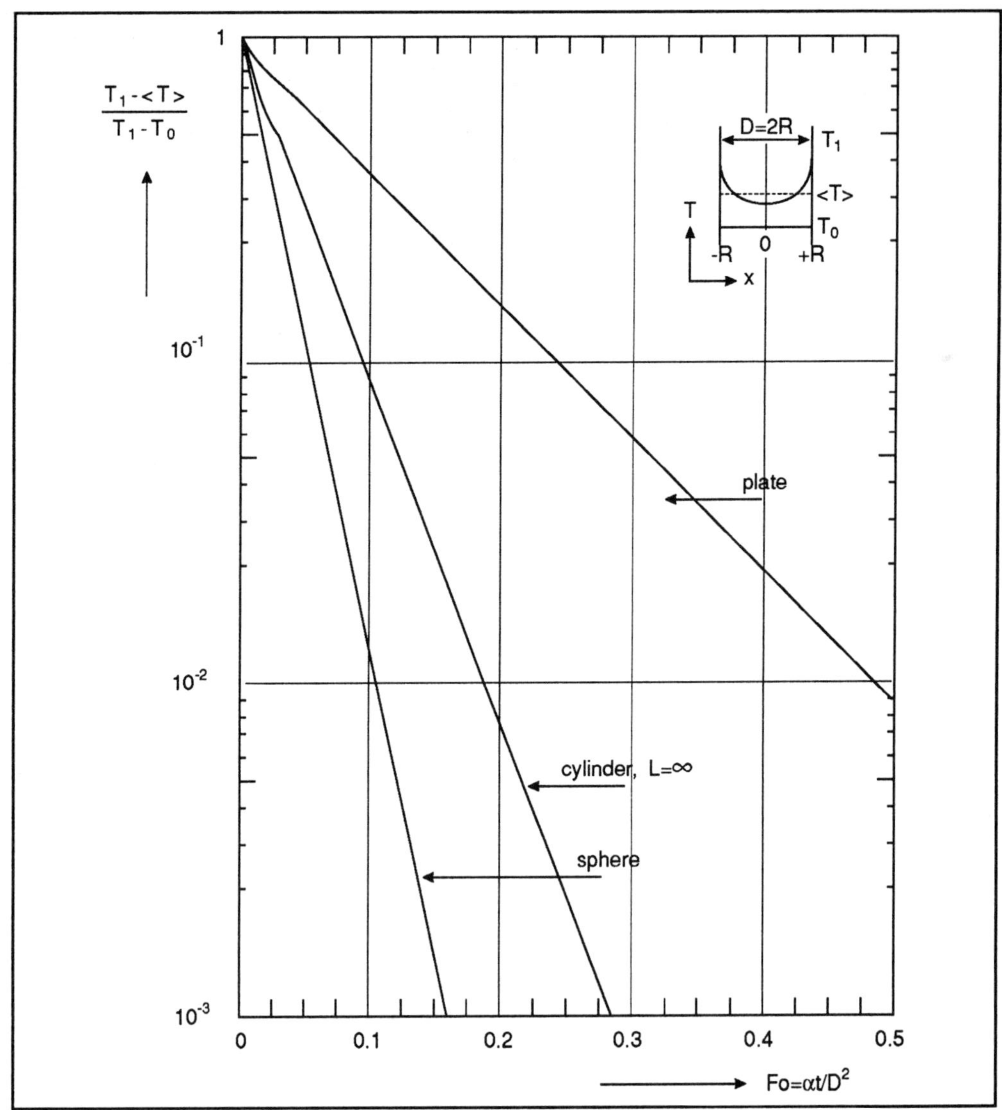

Figure 6.10 Relationships between mean temperature <T> and Fo for differently shaped objects. T_1 = fixed temperature of the outside of this object (for $t \geq 0$), T_0 = initial temperature of the object, D = diameter = 2 x radius = 2R (see text for further details).

From the data displayed in Figure 6.10 and 6.11, which objects warm up quicker, spheres or cylinders? Take the diameter of the sphere and cylinder as equal.

The quicker the object warms up the faster the values of $\dfrac{T_1 - T_c}{T_1 - T_0}$ and $\dfrac{T_1 - \langle T\rangle}{T_1 - T_0}$ will become smaller. Since Fo is proportional to time (t), then the steepest line represents the fastest warming object. Thus you should have concluded that spheres warm up faster than cylinders.

You will need the data represented in these figures to solve the next SAQ.

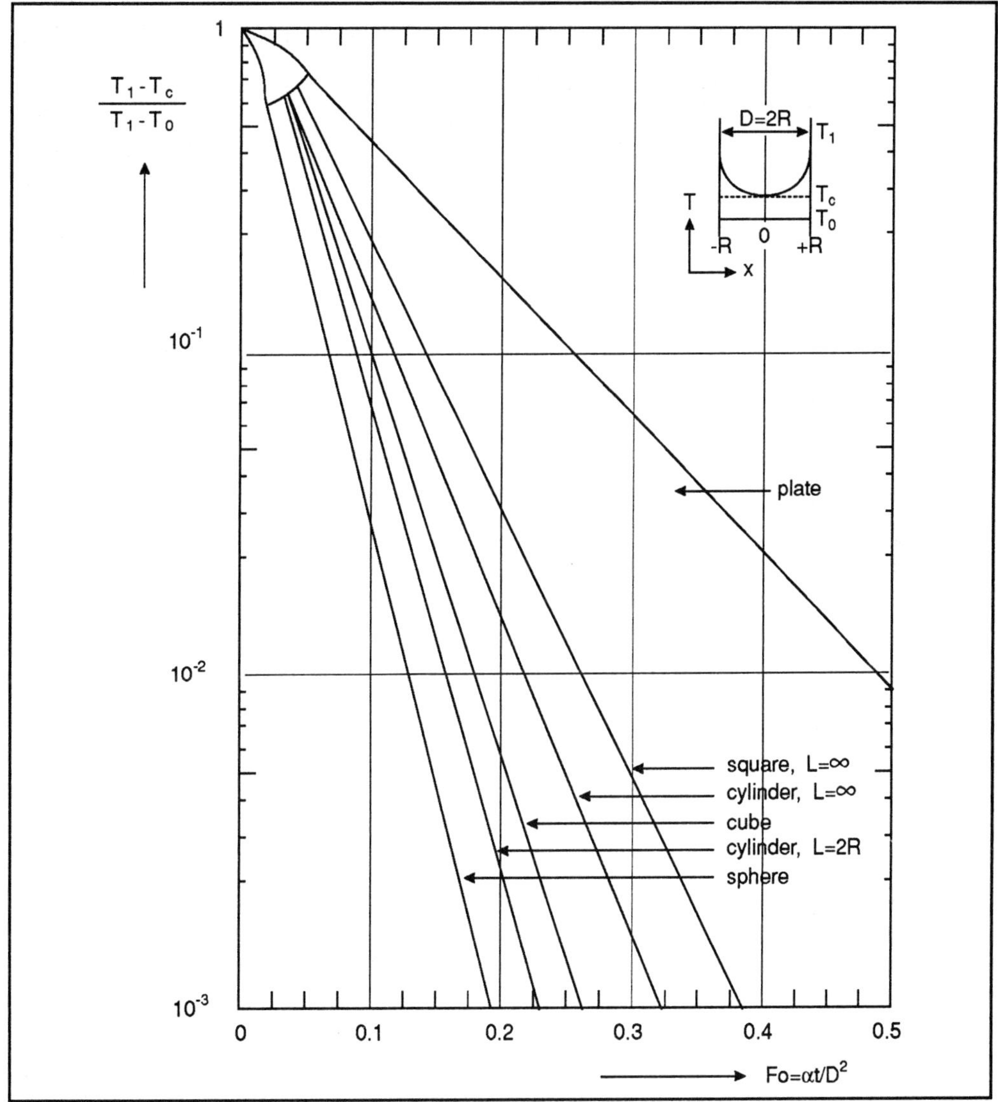

Figure 6.11 Relationship between the temperature at the centre (T_c) and Fo for differently shaped objects (see Figure 6.10 and text for further details).

SAQ 6.5

A cylinder is transferred from ice at 0°C into boiling water at 100°C. 10 minutes after the transfer, the temperature of the centre of the cylinder has risen to 60°C. Use the data presented in Figure 6.11, to determine what the temperature at the centre of the cylinder will be

1) 20 minutes after the transfer into the boiling water.

2) 30 minutes after the transfer into the boiling water.

6.3 Total and partial heat transfer coefficient

Having seen the analogy between transport of heat and electricity, we could formulate the dependence of the heat flow ϕ_q on the driving force ΔT according to:

$$\phi_q = h\, A\, \Delta T \qquad\qquad (E-6.31)$$

heat transfer coefficient

where A is the area of the object, ΔT the driving force and h the so-called heat transfer coefficient. Equation 6.31 is known as Newton's law of cooling, although it can equally well be used for describing how an object increases in temperature.

This is quite a pragmatic approach, since all the difficulties met in describing and calculating heat transport are put into the unknown h. Nevertheless the idea is attractive. It stresses the coupling between driving force (ΔT) and flow rate (ϕ_q). It is also a phenomenological approach. In many instances a detailed description of the heat transport is too complicated and h may be obtained from experiments and semi-theoretical calculations. This is particularly the case when convection plays a part in the heat transfer. In situations in which the transport of heat takes place by conduction only, h can often directly be related to the thermal conductivity. Of course Equation 6.31 is sometimes written as a relation between flux and driving force.

$$\phi_q'' = h\, \Delta T \qquad\qquad (E-6.32)$$

where now we are referring to heat transfer over a unit area. Notice that the reistance to heat transport is now $\dfrac{1}{h}$

In the remainder of this section all heat conduction problems as well as some related cases will be re-interpreted in terms of the heat transfer coefficient. In the following section heat transfer cases will be treated in which convective transport plays a major role.

Example 1:

Remember Example 1 of Section 6.2.1 (see Figure 6.1). We obtained for the flat plate in a steady-state:

$$\phi_q'' = \frac{\lambda}{D}\, \Delta T \qquad\qquad (E-6.33)$$

(see Equation 6.6)

Hence for a flat plate

$$h = \frac{\lambda}{D} \qquad\qquad (E-6.34)$$

Example 2:

In Example 4 of Section 6.2.1 we obtained for the heat flow from the surface of a sphere into a stagnant medium

$$\phi_q = 4\pi R\lambda\, (T_1 - T_0) \qquad\qquad (E-6.35)$$

(see Equation 6.20)

If we write this problem in terms of the heat transfer coefficient

$$\phi_q = h \, 4\pi R^2 \, (T_1 - T_0) \tag{E - 6.36}$$

We obtain from Equation 6.35 and 6.36

$$h = \frac{\lambda}{R} = \frac{2\lambda}{D} \tag{E - 6.37}$$

with $D = 2R$ (the diameter of the sphere).

Example 3:

For non steady-state; the result from penetration theory (for $Fo < 0.1$) can be written in the form of Newton's law. From Equation 6.29

$$\phi_q'' = -\frac{\lambda}{\sqrt{\pi \alpha t}} \, (T_1 - T_0) \tag{E - 6.38}$$

we see that:

$$h(t) = \frac{\lambda}{\sqrt{\pi \alpha t}} = \sqrt{(\lambda \rho c_p)/(\pi t_e)} \tag{E - 6.39}$$

Clearly the heat transfer coefficient in this case is time dependent. If the penetration process continues already t_e seconds, than the mean of the heat transfer coefficient ($<h>$) during this time t_e equals

$$<h> = \frac{1}{t_e} \int_0^{t_e} h\,(t)\, dt = 2\,\sqrt{(\lambda \rho c_p)/(\pi t_e)} = 2\, h(t_e) \tag{E - 6.40}$$

∏ It might be useful here to draw up a kind of revision chart of these relationships using the following format.

Geometry of object	heat transfer relationship	heat transfer coefficient
flat plate	$\phi_q'' = \dfrac{\lambda}{D}\,\Delta T$	$h = \dfrac{\lambda}{D}$

We do not expect you to remember all of these relationships for the various geometries and for non steady-state conditions. But you should know that these relationships exist and be able to apply them.

In most cases we have considered only one medium through which the heat transport takes place. Then obviously only one resistance ie one heat transfer coefficient has to be taken into account. However, in general, there will be more than one medium through which heat is transported. Thus we expect several resistances. In Example 2 of Section

6.2.1 we discussed steady heat transport through two solid layers due to a driving force (temperature difference). The result can be written as:

$$\phi_q = \left(\frac{D_1}{\lambda_1} + \frac{D_2}{\lambda_2}\right)^{-1} . A . \Delta T$$

(E - 6.41)

In terms of our pragmatic concept we write for this problem:

$$\phi_q = U. A . \Delta T$$

(E - 6.42)

total heat transfer coefficient Notice that we have written U instead of h anticipating that here there are two resistances. U is called the total heat transfer coefficient. From Equations 6.57 and 6.58 we obtain:

$$\frac{1}{U} = \frac{D_1}{\lambda_1} + \frac{D_2}{\lambda_2} = \frac{1}{h_1} + \frac{1}{h_2}$$

(E - 6.43)

partial heat transfer coefficient Here h_1 and h_2 are the partial heat transfer coefficients.

The concept discussed above is quite general. For instance, if we consider the case of two fluid media (at different bulk temperatures T_1 and T_2) separated by a solid wall (of thickness D and with thermal conductivity λ) intuitively we expect the temperature profile (in the steady-state) to be something like that shown in Figure 6.12. That the wall temperature T_{w1} is different from the bulk temperature T_1 is due to the fact that the convective heat transport from the bulk towards the near wall region is such that it cannot maintain T_1 at the wall. Similarly, there is a hotter 'film' at the other side of the wall (in medium 2).

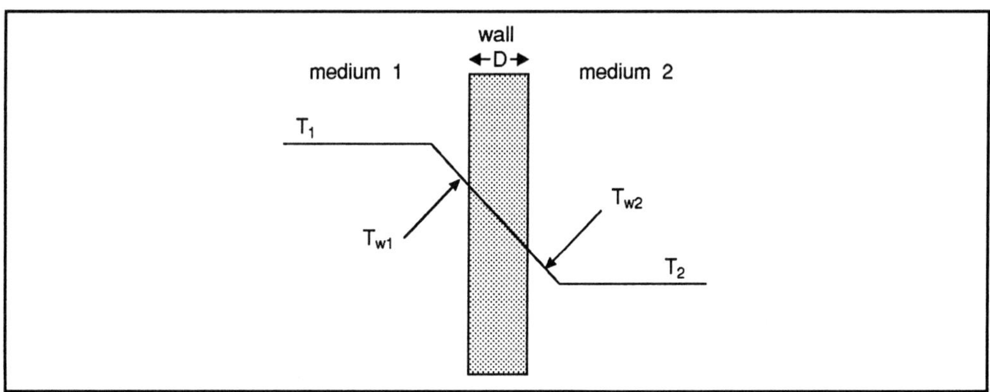

Figure 6.12 Diagramatic representation of the temperature profile through two media separated by a solid wall (see text for further details).

In terms of Newton's law we write:

$$\phi_q = h_1 A (T_1 - T_{w1})$$

$$\phi_q = \frac{\lambda}{D} A (T_{w1} - T_{w2})$$

$$\phi_q = h_2 A (T_{w2} - T_2)$$

(E - 6.44)

Using Equation 6.44 to eliminate T_{w1} and T_{w2} gives:

$$\phi_q = \left(\frac{1}{h_1} + \frac{D}{\lambda} + \frac{1}{h_2} \right)^{-1} A \, (T_1 - T_2) \qquad \text{(E - 6.45)}$$

Hence for the total heat transfer coefficient U we find:

$$\frac{1}{U} = \frac{1}{h_1} + \frac{D}{\lambda} + \frac{1}{h_2} \qquad \text{(E - 6.46)}$$

Of course to proceed with this example we need expressions for h_1 and h_2. Both are determined by the specific conditions in the experiment: is there forced convection, free convection, what is the geometry etc?

SAQ 6.6

A wall of a cubic vessel is made up of three layers of different materials each with different coefficients of conductivity.

Layer A is 5 cm thick and its coefficient of conductivity = $0.6 \text{ kJ m}^{-1} \text{ K}^{-1} \text{ s}^{-1}$

Layer B is 5 cm thick and its coefficient of conductivity = $0.4 \text{ kJ m}^{-1} \text{ K}^{-1} \text{ s}^{-1}$

Layer C is 10 cm thick and its coefficient of conductivity = $0.2 \text{ kJ m}^{-1} \text{ K}^{-1} \text{ s}^{-1}$

1) Calculate the heat transfer coefficient for each layer.

2) Calculate the total heat transfer coefficient for the wall.

3) If the temperature difference between the two sides of the wall is 10°C and the total area of the wall is 40 m^2, calculate the heat flow rate through the wall.

6.4 Nusselt number

In the previous section we introduced the heat transfer coefficient h as an unknown proportionality factor between heat flux and driving force ΔT. Now we need numerical values, correlations etc for h. It is more convenient to give these for a dimensionless number derived from h, since based on the findings of dimensional analysis we might anticipate relationships between dimensionless groups like Re and Pr.

Nusselt number   If we think of the analogy with Ohm's law, we can easily construct a dimensionless number. In steady-state conduction through a wall (thickness D) the resistance is $\frac{D}{\lambda}$, in Newton's law the resistance is $\frac{1}{h}$. Thus our dimensionless group called the Nusselt number Nu can be defined as :

$$\text{Nu} = \frac{\text{resistance in steady–state conduction}}{\text{actual resistance}} = \frac{D/\lambda}{1/h} = \frac{hD}{\lambda} \qquad \text{(E - 6.47)}$$

Thus for a wall since $h = \dfrac{\lambda}{D}$ (Equation 6.34) then $Nu = \dfrac{D/\lambda}{1/h} = 1$.

$\prod$ Using this approach and Equation 6.37 see if you can calculate the Nusselt number for conduction from a sphere into an infinite medium.

Your answer should come to $Nu = 2$. (Since $h = \dfrac{2\lambda}{D}$)

Below we have listed some examples of Nusselt numbers for other geometric shapes:

- penetration of heat in a 'half infinite' object (penetration theory discussed in Section 6.2.2, see Equation 6.39): $Nu = 0.57\ Fo^{-1/2}$;

- conduction inside a sphere (of radius R) for $Fo > 0.1$: $Nu = 6.6$ (in Figures 6.10 and 6.11).

$\prod$ Draw a table of Nu values for the different objects. We will be meeting more Nu values later.

6.5 Heat transfer by forced convection

Thus far we have only considered heat flow in stagnant media, ie conduction. Let us now focus onto heat transfer due to forced convection. The adverb 'forced' indicates that the convective flow is a consequence of external forces and is only to a minor extent influenced by heat transport itself. As an example we could think of the flow of water in a tube, where heat is transferred from the wall of the tube into the water. What will in this case, determine Nu? Well first of all we recognise two transport mechanisms for heat:

- conduction perpendicular to the wall into the fluid;

- convection more or less parallel to the wall.

If we restrict ourselves to fluids with constant density, gravity effects can be ignored. Furthermore we ignore changes of the viscosity with temperature. Now we can perform a dimensional analysis to find h and thus Nu. We consider the case of tube flow where the fluid enters at T_0 and the wall of the tube is at temperature T_w, then we expect h to depend on ρ, $<v>$, D, η, λ, c_p.

where:

ρ = density
$<v>$ = mean velocity
D = diameter
η = coefficient of dynamic viscosity
λ = thermal conductivity coefficient
c_p = specific heat

But it will also be dependent on x, the distance from the entrance, since both the flow and the heat transfer may need some distance to adjust to the new pipe conditions. Hence:

$$h = h(\rho, <v>, D, \eta, \lambda, c_p, x) \qquad \text{(E - 6.48)}$$

Using a dimensional analysis approach we obtain:

$$Nu = f(Re, Pr, Gz) \qquad \text{(E - 6.49)}$$

$$\text{with } Re = \frac{\rho<v>D}{\eta} \text{ (Reynolds), } Pr = \frac{v}{\alpha} \text{ (Prandtl), } Gz = \frac{\alpha x}{<v> D^2} \text{ (Graetz)} \qquad \text{(E - 6.50)}$$

Graetz number We have now introduced you to another dimensionless number, the Graetz number, which we will discuss below.

We point out that the Fourier number $(Fo) = \frac{\alpha t}{D^2}$ (6.30) is related to the Graetz number.

In fact $Fo = Gz$ when $t = \frac{x}{<v>}$.

Ⅱ Examine the definition of the Graetz number given in Equation 6.50 and then describe in words what the Graetz number measures.

Since:

$$Gz = \frac{\alpha x}{<v> D^2}$$

We can describe the Graetz number as a ratio between the thermal diffusion coefficient (α) multiplied by the distance (x) from the tube entrance and the mean velocity multiplied by the square of the diameter. Thus we can see that the Graetz number is derived from the two basic mechanisms:

- convection ($<v>$ and x are related to convection parallel to the wall);

- diffusion in a radial direction (D^2/α is penetration time - see Equation 6.30).

Graetz (or Fourier) takes the entrance effect into account.

The appearance of Re is easy to understand, since Re determines whether the flow is laminar or turbulent. Obviously turbulent flow and laminar flow react different upon the different wall conditions. Laminar flow has a layered structure and heat has to pass from one layer to another whereas in turbulent flow the eddies take liquid from close to the wall and transport it to the inner parts and vice versa. Thus in turbulent flow the radial temperature profile is rather flat; only close to the wall, does the temperature change in a small film from the mean to the wall temperature (see Figure 6.13).

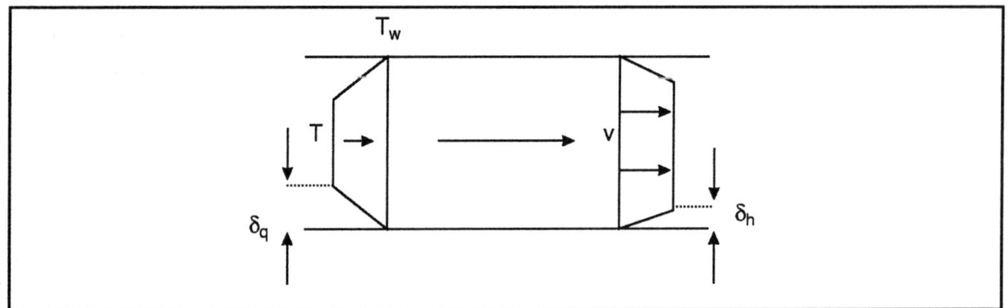

Figure 6.13 Diagramatic representation of a turbulent flow through a pipe showing temperature and velocity profile. T = temperature of the flow. T_w = temperature of the wall of the vessel. v = velocity of the flow. Note that the velocity of the flow is slowest close to the wall and that there is a temperature gradient over a short distance close to the wall of the vessel.

Finally, the Prandtl number takes into account that the hydraulic boundary layer (δ_h) for the velocity profile is not necessarily the same as the thermal boundary layer (δ_q). The thickness of these boundary layers are of course dependent on the molecular transport properties ν and α, and are further affected by the intensity of the turbulent flow.

From experiments and complex calculations several reliable correlations of Nu are available.

laminar flow

for laminar tube flow (Re < 2000) it has been calculated that:

$$Nu = 3.66 \text{ for Gz} > 0.1 \tag{E - 6.51}$$

and if the velocity profile has established itself

$$Nu = 1.08 \, Gz^{-1/3} \text{ for Gz} < 0.05$$

or averaged over the length of the pipe

$$<Nu> = 1.62 \, Re^{1/3} \, Pr^{1/3} \left(\frac{x}{D}\right)^{-1/3} \tag{E - 6.52}$$

turbulent tube flow:

$$Nu = 0.027 \, Re^{0.8} \, Pr^{0.33} \text{ for Re} > 10^4 \text{ and Pr} \geq 0.7 \tag{E - 6.53}$$

forced flow around a sphere:

$$Nu = 2.0 + 0.66 \, Re^{1/2} \, Pr^{1/3} \tag{E - 6.54}$$

∏ You should not attempt to learn these relationships by heart but you should know that they exist and know where to find them and be able to use them.

Again you might find it useful to draw yourself a kind of revision chart of Nu numbers in the form of a table. For example:

Type of flow	conditions	Nusselt number
Conduction	steady-state, wall	Nu = 1
Conduction	sphere, to infinite medium	Nu = 2
Convection	laminar tube flow	
	Re <2000 G_z > 0.1	Nu = 3.66

Let us tackle an important type of problem. We would like for example to determine how much heat can be removed from a bioreactor using a cooling system composed of a hollow tube through which cold water is pumped. The following example will show you the type of calculation that needs to be done.

Water flows through a tube (with inner diameter D = 3 cm) at a rate of 37.2 l min^{-1}. The temperature of the incoming water is 20°C, while the wall temperature equals 40°C. The length L of the tube is 5.2 m. Determine, in the steady state, the temperature of the water as it leaves the pipe. Data of water: ρ = 1000 kg m^{-3}, η = 9.10^{-4} Ns m^{-2}, λ = 0.61 W m^{-1}K^{-1} and c_p = 4.18 10^3 Jkg^{-1} K^{-1}.

What will the temperature of the water be at the outlet?

The situation described in this problem is illustrated in Figure 6.14.

To help you, first calculate the mean velocity of the water. From this you should be able to calculate the Reynolds number and determine whether or not the flow is laminar or turbulent. Then construct a heat balance over a section of the tube from which the temperature at the exit can be determined.

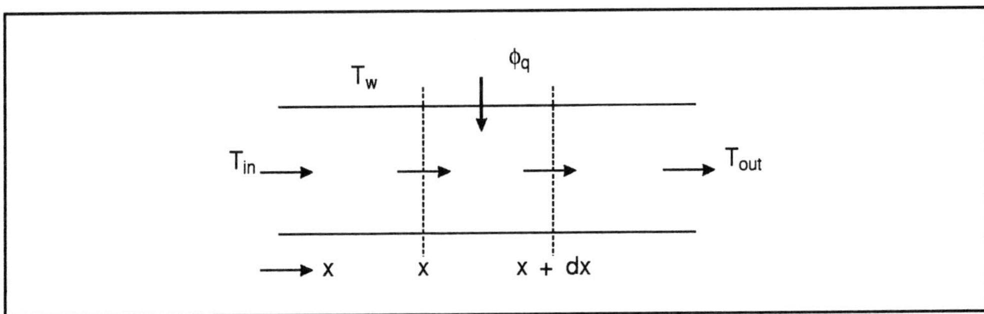

Figure 6.14 Illustration of the problem posed by the in-text activity.

Here is our solution. For simplicity we have not shown all of the stages in the arithmetic steps.

First we determine whether the flow is laminar or turbulent:

$$<v> = \frac{\phi_v}{\frac{\pi}{4} D^2} = 0.88 \text{ ms}^{-1} \Rightarrow$$

$$Re = \frac{\rho <v> D}{\eta} = 2.9 \times 10^4 \Rightarrow \text{turbulent}$$

Thus for the transfer of heat from the wall to the water we have $Nu = 0.027\,Re^{0.8}\,Pr^{0.33}$ for $Re > 10^4$ and $Pr \geq 0.7$ (see E - 6.53).

Note that $Pr = 6.2$

Next we construct (for the steady-state) a heat balance over a slab from x to $x + dx$ (see Figure 6.14) (because the flow is turbulent we can work with the mean temperature of the water at position x).

$$0 = \boxed{\begin{array}{c}\text{heat in the}\\ \text{inflow}\end{array}} - \boxed{\begin{array}{c}\text{heat in the}\\ \text{outflow}\end{array}} + \boxed{\begin{array}{c}\text{heat gained}\\ \text{through the}\\ \text{walls of}\end{array}}$$

$$0 = \phi_v\,\rho c_p\,[T]_x - \phi_v \rho c_p\,[T]_{x+dx} + h.\,\pi D dx.\,(T_w - T) \qquad\qquad\text{(E - 6.55)}$$

Hence:

$$\phi_v \rho c_p \frac{dT}{dx} = h.\,\pi D.\,(T_w - T) \qquad\qquad\text{(E - 6.56)}$$

This equation can be integrated since h is a constant and we obtain

$$\frac{T_w - T}{T_w - T_{in}} = \exp\left(-\frac{h\pi D}{\phi_v \rho c_p}\,x\right) \qquad\qquad\text{(E - 6.57)}$$

Thus we calculate for the exit temperature $T_{out} = 30°C$.

SAQ 6.7

It is proposed to use a large hall to grow plants at elevated temperatures. The hall, with a volume of 12 000 m^3 has to be provided with a new heating installation. The heat losses from the hall are mainly caused by:

- heat transfer via the windows (area 400 m^2; total heat transfer coefficient $U = 5\ W\ m^{-2}\ K^{-1}$);

- a flow of cold air from outside into the hall ($\phi_v = 3\ m^3\ s^{-1}$).

Data: density of air $\rho = 1.2\ kg\ m^{-3}$; specific heat of air $c_p = 1000\ J\ kg^{-1}\ K^{-1}$.

1) Draw a sketch of the situation.

2) Set up a heat balance over the growth hall.

3) How large is the power of the heater to be if a minimum temperature increase with respect to the open air of 30°C has to be reached?

4) How long does it take to heat up the hall to a temperature of 15°C if the temperature of the open air is 5°C and half the power calculated in 3) is used?

SAQ 6.8	A liquid flows through a tube with an inner diameter $D = 5$ cm at a rate of 40 l min^{-1}. The temperature of the incoming liquid (T_{in}) is lower than that temperature of the walls (T_w). The length of the tube is 2.5 m. Given that for the liquid $\rho = 1000$ kg m^{-3}; $\eta = 1 \times 10^{-3}$ Ns m^{-2}.

1) Calculate the mean velocity of the liquid in the tube.

2) Determine whether the flow is laminar or turbulent.

3) Derive the relationship $\dfrac{T_w - T}{T_w - T_{in}} = \exp\left(-\dfrac{h\pi D}{\phi_v \rho c_p} x\right)$ for this system,

in which T_w is the temperature of the tube wall. T_{in} is the temperature of the fluid entering the tube, T is the temperature of the flow leaving the tube. D is the diameter of the tube, ρ = density of the liquid, c_p = specific heat of the liquid, h is the heat transfer coefficient, ϕ_v is the flow rate, x = co-ordinate along of the tube. (Hint - use a heat balance over a slab of water as described in Figure 6.14).

4) In order to calculate the actual temperature of the liquid leaving the tube (at $x = L$), what further information other than that given would be needed?

Summary and objectives

In this chapter we have examined both convective and conductive heat transport phenomena. We began by examining heat conduction in which we described heat transport being driven by a temperature gradient (difference) and resisted by the heat conductivity of the system. We have learnt that heat transport is influenced by the geometry of the system. We have examined how heat penetrates an object (penetration theory) and derived a dimensionless unit known as the Fourier (Fo) number. We have also introduced the concepts of partial (h) and total (U) heat transfer coefficients and Nusselt numbers.

In the final part of the chapter we dealt with heat transfer by forced convection. We established that in these cases the relationships for Nusselt number are quite different for laminar and turbulent flows.

Now that you have complete this chapter you should be able to:

- calculate heat flow by conduction from supplied data;

- calculate temperature profiles in simple and multilayered systems from supplied data;

- use data supplied to calculate the temperature within an object using penetration theory and Fourier's numbers;

- calculate and use partial and total heat transfer coefficients from supplied data;

- define and use a whole range of terms which are employed in describing flow phenomena and heat transport;

- apply balance equations to calculate heat transfer and temperatures.

Mass Transport

Mass Transport

7.1 Introduction

In the previous chapter we examined the transport of heat. In this chapter we will examine the transport of mass. From what we have already learnt, we might anticipate that when a concentration gradient exists within a fluid of two or more components, there is a tendency for each constitutent to flow in such a direction as to reduce the concentration gradient. This process is mass transfer.

We remind you that in Chapters 2 and 3, we said that molecular transport of mass (eg of component A in a binary mixture) and molecular transport of thermal energy (ie heat) take place in a completely analogous way, viz. due to the chaotic motions and collisions of individual molecules. The two processes can therefore be described in quite similar terms. Mass fluxes and heat fluxes are directly proportional to gradients of mass concentrations and thermal energy concentrations, the proportionality constants having the same dimension, viz. m^2s^{-1}. The proportionality constants, viz. the diffusion coefficient D and the thermal diffusivity α, are physical properties and reflect the mobility of individual molecules (at a given state of aggregation, composition, pressure and temperature) and their ability to transport or pass thermal energy.

In Chapter 3, we also saw that convective transport of some component 'A' does not essentially differ from convective transport of heat. In both cases fluid elements featuring some concentration (of component A and/or of heat) are taken along by the flow. Thus transfer of mass and heat at a boundary or phase interface into or out of a flow alongside this boundary or phase interface takes place by analogous processes. In both cases there is an intimate interaction between convection (in a direction more-or-less parallel to the boundary or phase interface) and molecular exchanges (in a direction perpendicular to the flow). In other words, there is a competition between molecules and fluid elements which deliver or pick up mass or heat.

The treatment of mass transport in this chapter will be strongly based on the analogy described above and will draw on the experiences you have gained from the preceding chapter. There are however some differences between mass and heat transfer. These differences will be highlighted. In the first part of this chapter we will examine diffusion and the transfer of mass across interphases. In the final part of the chapter, we will examine convective mass transfer. This is quite a long chapter, so do not attempt to study it all at one sitting.

7.2 Mutual diffusion of molecules

∏ As an aid to refreshing your memory, write down a brief description of diffusion and check it with the following one provided.

Diffusion is the movement, under the influence of a physical stimulus, of an individual component through a mixture. The most common cause of diffusion is a concentration gradient. There are however different causes of diffusion, many of which are given special names. Thus, molecular diffusion induced by a pressure gradient is called pressure diffusion, whereas that induced by temperature is called thermal diffusion. Molecular diffusion from the exterior into a system is sometimes called forced diffusion.

Let us turn our attention to the quantitative aspects of diffusion.

Consider a plane stagnant layer (thickness D) containing either two gaseous or two liquid components A and B. The concentrations C_A and C_B (in mole m^{-3}) in this layer are dependent on the x-coordinate that is perpendicular to the layer. Note that everywhere $C_A + C_B = C$ = constant, where C denotes the total number of moles per m^3. Imagine a concentration difference ΔC_A over the layer with simultaneously a concentration difference ΔC_B that equals $-\Delta C_A$. You might find it helpful to draw a small figure of this, so that you can visualise what is going on. Two oppositely directed, but equally sized mole fluxes ϕ_A and ϕ_B are the result. The molecules of A and B are supposed to be able to move freely together and to exchange positions. A mass balance for component A over a slab within the layer, analogous to the heat balance in section 6.1 shows that in a steady state   the mass flow and thus the concentration gradient are constant, ie do not depend on x.

Hence, the concentration profile of A is given by:

$$C_A(x) = - \frac{\phi''_{M,A}}{D} \cdot x + C_{A,1} \qquad\qquad (E - 7.1)$$

which resembles Equation 6.5. NB. D = diffusion coefficient.

Similarly:

$$\Delta C_A = \phi''_{M,A} \cdot \frac{D}{D} \qquad\qquad (E - 7.2)$$

where D = thickness or depth of the solution.

This has similarities to Ohm's law. The resistance to mass transport is therefore equal to D/D. Strictly speaking the above equations pertain to a situation of mutual binary diffusion in which molecules of both types move freely and effectively in opposite direction due to oppositely directed concentration gradients. We will see later that Fick's law may also be used when a single component diffuses at low concentration through an abundance of inert gas.

Let us work through some examples to gain experience in applying the concepts and relationships described above.

Example 1

In a stagnant medium, a fast first order chemical reaction takes place. A component 'A' is converted in this reaction, hence there is a negative production of A. The rate of this production is for a first order reaction:

$$r_A = - k_r \cdot C_A$$

first order
reaction rate
constant

r_A = reaction rate, C_A = concentration of A, k_r is called the first order reaction rate constant. The medium is bounded at one side (x=0) by a wall, where the concentration C_{A0} is maintained, whereas far from this wall the concentration of A is zero. This system is illustrated in Figure 7.1.

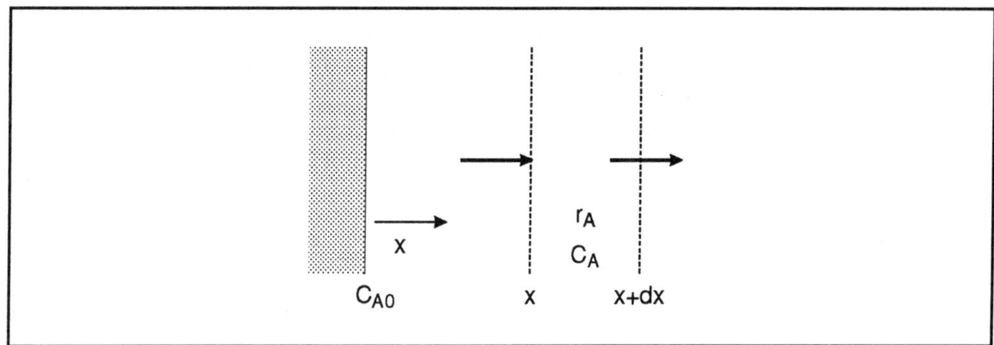

Figure 7.1 Diffusion of substance A in a system in which the concentration of A is maintained at C_{A0} at the wall (shaded area) and $C_A = 0$ at a long distance from the wall. r_A = reaction rate for production/loss of A (see text for further details)

We want to calculate the concentration profile of A and the mass flux of A from the wall into the medium, both in a steady state. For this purpose we set up a mass balance for A over the thin slab between x and x + dx (see Figure 7.1).

$$0 = \boxed{\text{diffusion in}} - \boxed{\text{diffusion out}} - \boxed{\begin{array}{c}\text{amount of A}\\\text{chemically converted}\end{array}}$$

$$0 = -D\left[\frac{dC_A}{dx}\right]_x + D\left[\frac{dC_A}{dx}\right]_{x+dx} - k_r . C_A . dx \qquad \text{(E - 7.4)}$$

Hence:

$$\frac{d^2C_A}{dx^2} - \frac{k_r}{D} C_A = 0 \qquad \text{(E - 7.5)}$$

The general solution of Equation 7.5 is:

$$C_A(x) = C_1 \exp\left(\sqrt{(k_r)/(D)} . x\right) + C_2 \exp\left(-\sqrt{(k_r)/(D)} . x\right) \qquad \text{(E - 7.6)}$$

The integration constants C_1 and C_2 are found using the two boundary conditions: $C_A (x = 0) = C_{A0}$ and $C_A (x \rightarrow \infty) = 0$. This renders as a result for the concentration profile:

$$C_A(x) = C_{A0} \exp\left(-\sqrt{(k_r)/(D)} . x\right) \qquad \text{(E - 7.7)}$$

Now the mass flux of A from the wall into the medium can be equated:

$$\phi''_{M,A}(x = 0) = -D\left[\frac{dC_A}{dx}\right]_{x=0} = \sqrt{k_r D} . C_{A0} \qquad \text{(E - 7.8)}$$

Equation 7.8 is quite important.

It means that we can calculate the flux ($\phi''_{M,A}$) of A from the wall, if we know k_r, D and C_{A0}.

This derivation for determining a concentration profile is quite straight forward. As you might anticipate the situation for non-linear geometry becomes more complex.

Example 2: Cylindrical geometry

Consider a diffusion process (in a stagnant fluid) in a cylindrical geometry, eg in the annular space between two concentric cylinders, as effected by radial concentration differences. A mass balance for A over a shell for a steady state shows that the mass flow of A in the radial direction is independent of r,

$$\phi''_{M,A} = -D \cdot 2\pi rL \cdot \frac{dC_A}{dr} = C_1 \qquad\qquad (E\text{-}7.9)$$

which resembles Equation 6.14. (You might like to repeat the derivation yourself.). You similarly should be able to derive that, due to a concentration difference ΔC_A between r $= R_1$ and r $= R_2$, the mass flow is given by:

$$\Delta C_A = \frac{\ln R_2/R_1}{2\pi LD}\ \phi_{M,A} \qquad\qquad (E\text{-}7.10)$$

which on the analogy of Equation 6.18 expresses that the mass flow depends on the concentration difference and on $\ln(R_2/R_1)$ rather than on the difference $(R_2\text{-}R_1)$! This is related to the size of the cylindrical areas through which the transport takes place.

Example 3: Spherical geometry

Just as we did for heat transport, we could also derive the mass of A in a mutual diffusion process in a stagnant fluid. By analogy with Equation 6.19 we should conclude that:

$$\phi''_{M,A} = 4\pi r^2 \left(-D\ \frac{dC_A}{dr}\right)$$

Let us consider a shell around a sphere for which the internal radius is R_1 and the external radius is R_2.

Thus:

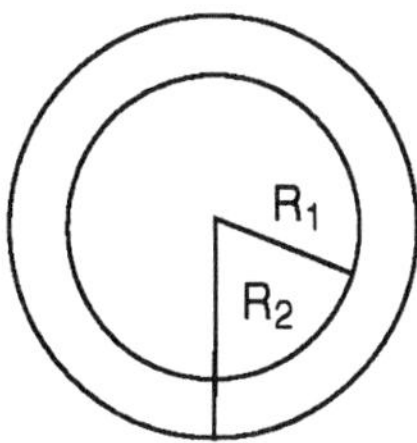

We would like to be able to calculate the rate of mass transfer across this 'shell'.

The concentration of A at position r is given by $C_A(r) = \dfrac{\phi_{M,A}}{4\pi Dr}$

Thus by analogy, the concentration of C_A at radius R_1 is given by:

$$C_A\,(r = R_1) = \frac{\phi_{M,A}}{4\pi D R_1}$$

and at radius R_2

$$C_A\,(r = R_2) = \frac{\phi_{M,A}}{4\pi D R_2}$$

For simplicity we will write $C_{A1} = C_A\,(r = R_1)$ and $C_{A2} = C_A\,(r = R_2)$

$$\text{Thus } C_{A1} - C_{A2} = \frac{1}{4\pi D}\left(\frac{1}{R_1} - \frac{1}{R_2}\right) \cdot \phi_{M,A} \tag{E - 7.11}$$

(If you would like to derive this yourself we provide the following hint. Put up a mass balance over a spherical shell, and find that the mass flow is independent of r).

∏ It would be a helpful form of revision to make a list of the relationships described above and to write the comparative list for heat transfer. We would suggest you use the following format. We have begun the table for you.

Geometry	Mass transport	Heat transport
Slab (no net product of mass or heat)	$\Delta C_A = \phi''_{M,A}\,\dfrac{D}{D}$	$\Delta T = \phi''_q\,\dfrac{D}{\lambda}$
Cylindrical geometry	$\Delta C_A = \dfrac{\ln R_2/R_1}{2\pi L D}\,\phi_{M,A}$	$\Delta T = \dfrac{\ln R_2/R_1}{2\pi L \lambda}\,\phi_q$

In these examples, we have used the fact that mass transfer can be treated analogously to heat transfer. The differences in the equations are:

$\phi_{M,A}$ (mass flow of A) replaces ϕ_q (heat flow);

D (diffusion coefficient) replaces λ (coefficient of conductivity).

mass transfer coefficient

We can take this analogy a stage further by using the term mass transfer coefficient k in place of heat transfer coefficient h. Thus, by analogy to Equation 6.31 the mass flow $\phi_{M,A}$ can be expressed as the product of k times transfer area A times driving force ΔC_A:

$$\phi_{M,A} = k \,.\, A \,.\, \Delta C_A \tag{E - 7.12}$$

(compare with $\phi_q = h \,.\, A \,.\, \Delta T$)

This result is the mass transfer analogue of Newton's law of cooling. Again, the coefficient k is the proportionality constant between mass flux and driving force. The reciprocal of k is the resistance to mass transfer. Using the concept of a mass transfer coefficient k and condensing all uncertainty about the mass transfer mechanism and mass transfer rate into this k has become common engineering practice. Check whether or not you have understood this analogy by doing SAQ 7.1.

<table><tr><td>SAQ 7.1</td><td>If the difference in concentration of substance A at two points in a system is 0.5 mol.m^{-3}, the mass transfer coefficient k for the system = 0.1 ms^{-1} and the transfer area = 1 m^2, what is the mass flow rate?</td></tr></table>

Let us now attempt to solve mass transfer problems associated with process engineering. In the example described below we will be examining the adsorption (ie mass transfer) of SO_3 from an SO_3 - air mixture into concentrated H_2SO_4.

Practical example

A SO_3-air mixture, containing 7 mol % SO_3, is fed to the bottom of an absorption tower. A flow of concentrated H_2SO_4 flows down the tower. At the liquid-mixture interface the SO_3 is absorbed: the concentration of SO_3 at the interface is negligible. The interfacial area equals a = 100 m^2/m^3 of tower volume. The mixture velocity, calculated with respect to the empty tower, is v_0 = 2ms^{-1} (we call this the superficial velocity). The mass transfer coefficient in the gas phase equals k = 0.02 ms^{-1}. Now the question is: what length L of column is required, if 98% of the SO_3 has to be absorbed? The column is to be operated in a steady state. This may appear quite a complex problem but there is a fairly straight forward solution.

To find the answer, we have to calculate the concentration profile of SO_3 and thus set up a mass balance over a slab of the tower between x and x + dx. Since the molar concentration of SO_3 is relatively low we may take the volume flow through the tower to be (roughly) constant, thus the mass balance for a steady state reads as:

amount entering the slab - amount leaving the slab - the amount removed in the gas phase in the slab

$$Av_0.C(x) - Av_0.C(x + dx) - k.(a.Adx) . C(x) = 0 \qquad \text{(E - 7.13)}$$

The first two terms are the convective flows of SO_3 through the plane x and x + dx respectively, whereas the third term accounts for the SO_3 flow from the mixture inside the control volume to the absorbing H_2SO_4. In this last term, the fact that the concentration at the interphase equals zero is already taken into account. Equation 7.13 gives the differential equation.

We can manipulate Equation 7.13 further. Thus $v_0 C (x) - v_0 C (x + dx) = ka\, dx\, C(x)$

Thus:

$$\frac{dC}{dx} - \frac{ka}{v_0} C(x) = 0 \qquad \text{(E- 7.14)}$$

If we call the concentration of SO_3 of the incoming mixture flow C_0 we obtain from Equation 7.14 the required concentration profile

$$\ln\left[\frac{C(x)}{C_0}\right] = -\frac{ka}{v_0} x \qquad \text{(E - 7.15)}$$

Thus, if the concentration at the outlet must be $C(x) = 0.02\, C_0$, (since 98% of C_0 has been absorbed), we can calculate that the length of the tower (x = L) = 3.9 m.

The problem we have just solved has many similarities to the problem we tackled in the previous chapter in which we examined the transfer of heat into water flowing through a cylinder (tube). This sort of problem is encountered when, for example, we want to design and operate an adsorption tower.

It is only in the case of mass transport by diffusion in simple geometries that the concept of k can be related to the notion of the diffusion coefficient. By considering diffusive transport of mass (at low concentration) from a sphere into a large immobile medium (eg evaporation of a droplet in a stagnant atmosphere) it can be found that

$$k = \frac{2D}{D}$$

(E - 7.16)

where D = diffusion coefficient, D = diameter of the sphere.

This resembles Equation 6.37 $\left(h = \frac{2\lambda}{D} \right)$

[Note the derivations of Equation 7.16 and 6.37 are completely analogous.]

7.3 Sherwood number

Sherwood number

Using the analogy of the Nusselt number for heat transfer, the dimensionless Sherwood number Sh expresses the ratio of the resistance to mass transport due to diffusion only (being D/D) to the actual resistance as made up by convection and diffusion:

$$Sh = \frac{D/D}{1/k} = \frac{kD}{D}$$

(E - 7.17)

(compare this with Equation 6.47)

For steady-state diffusion around a sphere in a stagnant medium Equation 7.16 can be formulated as

$$Sh = 2.$$

(E - 7.18)

∏ From what you learnt about Nusselt numbers in chapter 6 and using the analogy described above and in Equation 7.17 what is the Sherwood number for mass transfer through a flat plane?

You should have come to the conclusion that Sh = 1 (look back to Section 6.4 if you had difficulty remembering this).

SAQ 7.2

The diffusion coefficient of A in a semi-liquid membrane = $2 \times 10^{-3} m^2 s^{-1}$ and the membrane is 1 mm thick. If the membrane is held in a flat plane, what is the mass transfer coefficient of this system?

Let us work through a more sophisticated example.

Example 4: Moth-ball

A moth-ball (diameter $D_0 = 1$ cm) is suspended in stagnant air. Moth-balls are made of pure naphthalene. The vapour pressure p_v of naphthalene in air at room temperature is 0.05 mm Hg. As a consequence the concentration of the naphthalene vapour at the surface of the moth-ball is a constant and equals (using the ideal gas law);

$$C^* = \frac{p_v}{RT} \text{ (in mol m}^{-3})$$

How long does it take before the diameter of the moth-ball is reduced to 5 mm?

Data: diffusion coefficient of naphthalene in air: $D = 7.0 \times 10^{-6} \text{ m}^2 \text{ s}^{-1}$

density of (solid) naphthalene $\rho = 1150 \text{ kgm}^{-3}$

molar mass of naphthalene $M = 106 \text{ kg kmol}^{-1}$

gas constant $R = 8.31 \text{ J.mol}^{-1}\text{K}^{-1}$

Assuming room temperature $= 20°C$

Our solution to this problem is as follows:

In this case there is mass transport from a sphere into stagnant air. Therefore the Sherwood number equals 2. The mass flow from the sphere to the air is given by:

$$\phi_M = k \cdot \pi D^2 \cdot M \cdot (C^* - C_{air}) \tag{E - 7.19}$$

with

$$k = Sh \frac{D}{D} = \frac{2D}{D} \text{ and } C_{air} = 0 \tag{E - 7.20}$$

The mass balance for the moth-ball reads as:

$$\frac{d}{dt}(\rho V) = \frac{d}{dt}\left(\rho \frac{\pi}{6} D^3\right) = -k \cdot \pi D^2 \cdot M \cdot (C^* - C_{air}) \tag{E - 7.21}$$

combining with the relation for k and C^* and remembering that $C_{air} = 0$ yields:

$$D \frac{dD}{dt} = -4 \frac{DM}{\rho} \frac{p_v}{RT} \tag{E - 7.22}$$

hence, solving this differential equation using the inital condition: $t = 0 \rightarrow D = D_0$ gives:

$$\left(D^2 - D_0^2\right) = -8 \frac{DM}{\rho} \frac{p_v}{RT} \cdot t \tag{E - 7.23}$$

But $D = D_0/2$ since the moth-ball had an original diameter of 1 cm and has been reduced to 5mm. Thus $D = D_o/2$ at $t = 61.4$ days.

We will have, thus far, considered systems that are stagnant. What happens when the system is mixed.

7.4 Film theory

film theory   Consider the transfer of some component A from a (plane or curved) wall (or interface) into a liquid phase that is mechanically stirred due to the input of a certain amount of power. Due to the action of the stirrer, a wide gamut of eddies of varying size and strength arises. While these eddies mix the bulk of the liquid phase, they get dragged when they approach the wall. Close to the wall the viscosity of the liquid simply precludes large values of the velocity components as well as excessive values of velocity gradients. Some type of wall layer can be visualised that is free from eddies and in which the transport of momentum, heat and components occurs only by the molecular mechanism. Eddies are simply not capable of periodically refreshing this wall layer. As a result, this wall layer or boundary layer or film is stagnant, and mass transport through it must take place by the action of diffusion. The resistance to mass transfer is thought to be concentrated in this thin film, since outside this film the transport of mass, heat and momentum is quite effectively accounted for by the eddies, ie by the convective mechanism. When the time required for establishing the concentration profile in the film is small compared to the full transfer process time, the transport may be described in terms of a (quasi-) steady state. From Equation 7.2 you can see that the resistance to mass transport is directly proportional to the thickness of the film and furthermore depends on the diffusion coefficient D. The thickness δ_c of the film strongly depends on a number of parameters associated with the flow in the bulk of the liquid. The above concept of a film or wall layer that visualizes the idea of resistance to mass transport, was originally prosed by Lewis and Whitman in 1924.

$\prod$ Make a list of what you think are the essential requirements that enable us to apply film theory to mass transfer in mixed systems.

From the description given above, we would have expected you to have included in your list the following:

- the film must be very thin;

- the concentration gradient must be set up quickly to establish a steady state;

- the quantity of solute within the film must be relatively small to the amount passing through it.

One-dimensional mass transport by diffusion in a plane geometry under transient conditions is governed by the partial differential equation.

$$\frac{\partial C}{\partial t} = D \, \frac{\partial^2 C}{\partial x^2}$$

(E - 7.24)

Equation 7.24 is equivalent to Equation 6.25 we derived in Chapter 6 when we were discussing heat transport. For boundary conditions similar to those mentioned as when we derived Equation 6.25 the mass flux at the boundary is:

$$\phi''_{M,A} = -D\left[\frac{dC}{dx}\right]_{x=0} = D\,\frac{C_1 - C_0}{\sqrt{\pi D t}} = \sqrt{(D)/(\pi\,t)}\,(C_1 - C_0) = k\,(C_1 - C_0)$$

$$(E\text{-}7.25)$$

which looks like Equation 6.29. It expresses the time-dependent mass flux as the product of a constant driving force ($C_1 - C_0$) times a time-dependent mass transfer coefficient k $= \sqrt{(D)/(\pi\,t)}$.

penetration
theory

The distance $\sqrt{\pi D t}$ is again called the 'penetration depth' for diffusion. Every time a fluid experiences a new concentration at its boundary that differs from the bulk concentration, molecules start penetrating into (or reversely, start leaving) the fluid. Contact time determines over which 'penetration depth' the concentration field can be adjusted to the new boundary condition. The situation, however, gets really complicated when the new boundary condition is not constant in time, in particular when the boundary concentration changes as a result of the transport (or transfer) process itself.

Suppose some component A must be removed from a gas. To this end, the gas is passed in the form of bubbles through some liquid in which A is easily soluble. In other words, A is to be absorbed by the liquid from the gas. When a bubble rises through the stagnant liquid, it continuously comes into contact with fresh volume elements of liquid which flow alongside such a bubble. During the period t_p of this passage each volume element experiences at its boundary the high concentration of A in the gas bubble: the time interval t_p is of the order of the bubble diameter divided by the rise velocity of the bubble since if the bubble has a diameter D and a velocity v_b, it will take $\dfrac{D}{v_b}$ to enter a new volume element of liquid. The volume element of liquid adjusts the concentration of A over a depth equal to the penetration depth that corresponds to the time interval t_p. During the time interval t_p the momentary transfer rate of component A into such a volume element of liquid decreases with time according to Equation 7.25. The total mass of A transferred into the volume element of liquid during t_p follows the integration of this expression over t_p. In other words, the mean mass transfer coefficient ($<k>$) in the time interval t_p is equal to twice the mass transfer coefficient at the moment t_p.

$$<k> = 2\sqrt{(D)/(\pi t_p)} \qquad\qquad (E\;7.26)$$

Let us try a numerical example.

Example: Diffusion of a pollutant

Ten years ago, on a dumping-group, some polluting material had been dumped. During the intervening years this substance has slowly (especially via rainwater) penetrated into the ground. This process can be described like a diffusion process with an effective diffusion coefficient D equal to 2.10^{-10} m^2 s^{-1}. The layer of dumped material is so thick that during all these years the concentration at the ground surface just below the layer has a constant value C^* equal to 2 kg m^{-3}. This situation is illustrated in Figure 7.2.

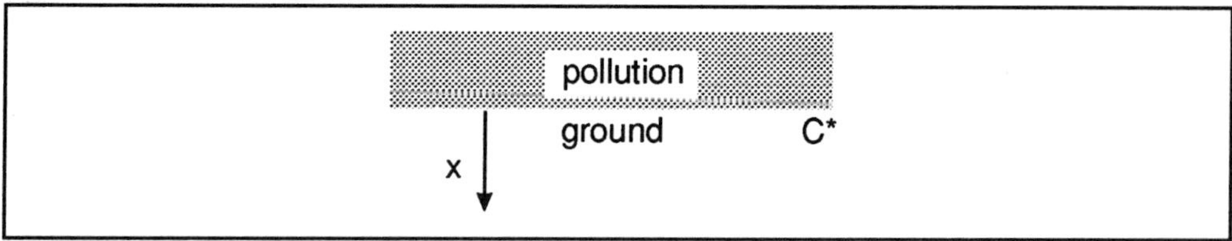

Figure 7.2 The sepage of pollution from a dump (shaded area) into the ground beneath. (C* concentration of pollutant in the ground just below the polluted layer. See text for further details).

The question we would like to answer are:

1)   over what distance has the polluting substance diffused into the ground during these 10 years?

2)   how much of this substance has diffused into the ground?

3)   suppose one removes, not only the polluted layer, but also 30 cm of the ground. What is the mean concentration of the pollutant in this 30 cm layer?

Our solutions to these problems are as follows:

The diffusion process into the ground is described by penetration theory, hence the substance is diffused into the ground over the penetration depth. Thus the answer to 1) is:

$$x_e = \sqrt{\pi D\, t_e} \qquad\qquad\qquad\qquad\qquad (E - 7.27)$$

Since $D = 2.10^{-10}\ m^2\ s^{-1}$ and $t_e = 10$ years $= 3.155 \times 10^8 s$.

$$x_e = \sqrt{0.198}\,m\ =\ 0.445m\ =\ 45cm$$

The mean flux into the ground during the 10 years is easily calculated using the mean value for the mass transfer coefficient k given in Equation 7.26.

$$\langle\phi_M''\rangle = \langle k\rangle \cdot (C* - 0) = 2\sqrt{\frac{D}{\pi t_e}}\ C* \quad (\text{note} < \ > \text{which denotes 'mean'}).$$

$$\qquad\qquad\qquad\qquad\qquad\qquad\qquad\qquad (E - 7.28)$$

Hence, the mass diffused into the ground per unit area in these years is given by

$$M" = \langle\phi_M''\rangle \cdot t_e\ =\ \frac{2}{\pi}\ \sqrt{\pi D\, t_e}\cdot C*\ =\ 0.57\ kg\ m^{-2}$$

$$\qquad\qquad\qquad\qquad\qquad\qquad\qquad\qquad (E - 7.29)$$

As we have seen in Chapter 6, the profile can be approximated by a straight line through the points $(x = 0, C = C*)$ and $(x = x_e, C = 0)$. This is illustrated in Figure 7.3.

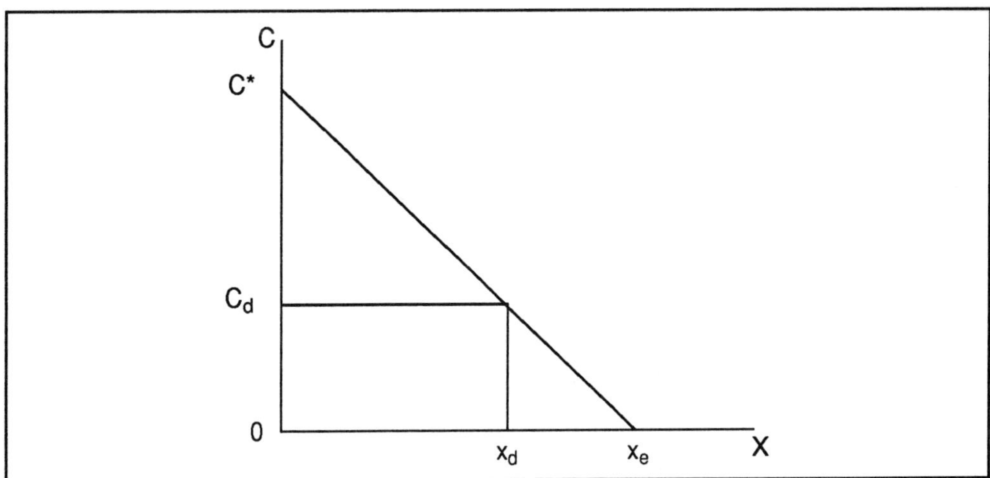

Figure 7.3 Concentration gradient of a pollutant in the ground beneath a dump (see text for details).

Thus the answer to 3) is (writing $x_d = 30$ cm):

$$<C> = \frac{C* + C_d}{2} = C*\left(1 - \frac{x_d}{2x_e}\right) = 1.33 \text{ kg m}^{-3}$$

(E - 7.30)

You can probably think of many situations where we would employ such relationships. For example we might envisage this type of calculations to be useful in calculating the penetration of compounds into absorbents used in the purification of products from bioreactors. Let us consider another example.

Example: Adsorption into water droplets

Consider a spherical water drop (diameter $D = 5$ cm) that falls at a constant vertical velocity in stagnant air. Above an agriculture area it moves through a smog-cloud containing a high ammonia concentration. The thickness of the cloud is 250 m. The NH_3 concentration in equilibrium with the vapour pressure p_o of NH_3 in the cloud is $C*$ and equals 0.34 kg m^{-3}. During the fall of the drop its concentration at the surface remains $C*$. The diffusion coefficient of NH_3 in water is $D = 1.46 \ 10^{-9}$ m^2 s^{-1}.

What is the mean NH_3-concentration in the drop when it leaves the cloud?

This problem is quite complex but it can be solved if tackled logically. First, the steady fall velocity is calculated. We will not repeat the calculation here since we dealt with this expect in an earlier chapter. Using the approach indicated in Chapter 5 $v_s = 12$ m s^{-1}. Thus the drop travels for a time interval:

$$t = \frac{\text{thickness of the cloud}}{\text{drop velocity}} = h/v_s = 20.8s \text{ through the cloud.}$$

The penetration depth of ammonia into the drop at the moment the drop leaves the cloud is therefore $x_e = \sqrt{\pi D \ t} = 3.09 \ 10^{-4}$ m (approx). Thus $x_e << D/2$ and we may approximate the sphere surface as a flat plate (with area πD^2). The total mass that enters the drop during this time interval is calculated by equating the mean mass flow during this interval t. This renders according to Equation 7.25 and 7.26.

$$\Delta M = \pi D^2 . 2 \sqrt{(D)/(\pi t)} \, (C* - 0) . t \tag{E - 7.31}$$

$$= \pi D^2 . 2 \sqrt{(D)/(\pi t)} . C* . t$$

$$= 2 D^2 . C* \sqrt{(\pi^2 . D . t^2)/(\pi t)}$$

$$= 2 D^2 . C* \sqrt{\pi . D . t}$$

$$= 2 D^2 . C* . x_e$$

The mean NH_3 concentration in the drop when it leaves the cloud is then easily calculated:

$$<C> = \frac{\Delta M}{\frac{\pi}{6} D^3} = \frac{12 \, C^* x_e}{\pi D} = 0.008 \; kg \; m^{-3} \tag{E - 7.32}$$

Use the experience you have gained from these two worked examples to answer the following SAQ's.

SAQ 7.3	A drop of dye solution falls onto a surface. The dye immediately begins to penetrate the surface. The dye has an effective diffusion coefficient $D = 6.31 \times 10^{-5} m^2 \, s^{-1}$ within the surface it penetrates; the drop of dye solution is sufficiently large such that the concentration of dye remains constant at a point just below the drop for at least a minute. The concentration of dye at the point just below the drop is $10 \, kg \, m^{-3}$.
	
	Calculate:
	
	    the depth the dye solution will have penetrated after 10s;
	
	    the mass of dye that will have diffused into the absorbing layer in these 10s.

SAQ 7.4	A droplet of water is falling down a column of inert gas containing SO_2. The column is 100 m long and the droplet has a diameter (D) of 4 cm and falls at a velocity of $10 \, ms^{-1}$. The concentration of SO_2 in the column is $0.4 \, kg \, m^{-3}$. The diffusion coefficient of SO_2 in water under the conditions which the column operates $= 2 \times 10^9 \, m^2 s^{-1}$. Calculate the mean SO_2 concentration in the drop when it reaches the bottom of the column.

7.5 Unilateral diffusion of molecules

In many instances of diffusion in binary mixtures the boundary conditions are such that, in spite of concentration gradients for both components, there is a net transport of molecules of only one of the two components. this can be illustrated by means of the Winkelman experiment (see Figure 7.4) in which molecules of some volatile substance A diffuse through air. Effectively a large number of them travel from the surface of the liquid reserviour towards the rim of the tube from where these are taken away by a current of air. At a first glance, this is in complete agreement with the concept of Fick's law: namely a flux as a result of some concentration difference over some distance. The concentration difference in question is the difference between the concentration just above the liquid level and the concentration at the rim of the tube. The latter concentration equals zero (all molecules of A arriving at the rim are immediately taken along by the air current). The concentration just above the liquid level is determined by the temperature of the liquid and, in the case when the ideal gas law applies, equals the partial vapour pressure p_A divided by RT. Yet, Fick's law does not apply in cases like this. This may best be explained by considering what is happening to the air molecules in the Winkelman experiment.

Winkelman experiment

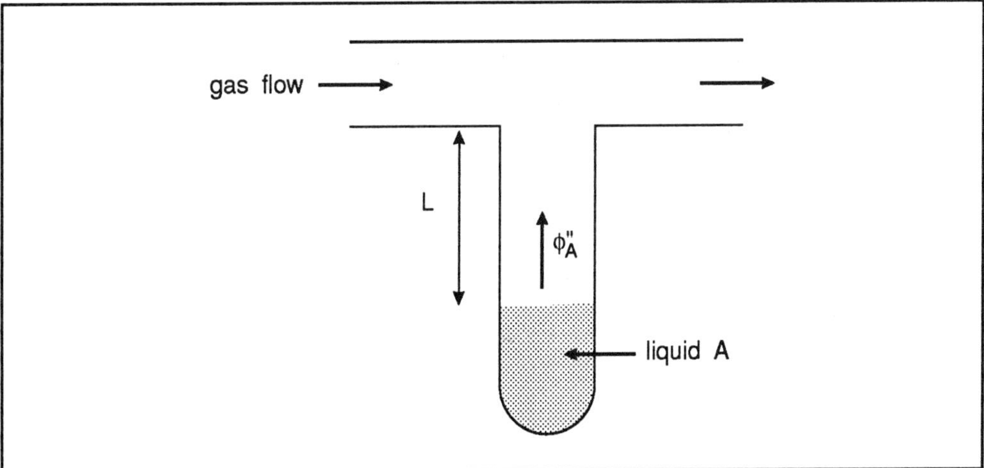

Figure 7.4 Diagrammatic representation of the Winkelman experiment (see text for details).

The air molecules are also subjected to a concentration gradient that is connected to the concentration gradient of component A. The concentration of air molecules (denoted by B) decreases from the value p/RT at the rim (where no A is present) down to the low value $(p-p_A)/RT$ just above the liquid level (the formulae apply to ideal gases only). In spite of this concentration gradient there is no net flux of air from rim towards liquid level. Obviously, some flux ϕ'' is fully compensating Fick's diffusive flux rendering a zero net transport of air. This so-called drift flux ϕ'' is of the convective type and is related to the evaporative flux from the liquid reservoir that tries to increase the local (vapour) pressure.

drift flux

This is demonstrated in the next section.

7.5.1 Stefan's Law

First, we write the net mole fluxes of component A and of air (component B) as the sum of a diffusive flux according to Fick's law and a drift flux contribution:

$$\phi_A'' = -D \cdot \frac{dC_A}{dx} + \phi'' \cdot \frac{C_A}{C}$$

$$(E - 7.33)$$

$$\phi_B'' = -D \cdot \frac{dC_B}{dx} + \phi'' \cdot \frac{C_B}{C}$$

$$(E - 7.34)$$

Note $\phi'' \dfrac{C_A}{C}$ and $\phi'' \dfrac{C_B}{C}$ drift flux contributions. One $\left(\phi'' \dfrac{C_A}{C}\right)$ describes the convective type transport related to the evaporative flux of the liquid. The other is for the compensatory gas in this case air.

Inspection and further reflection on the forms of (7.33) and (7.34) reveals that the drift flux ϕ'' must comprise both component A and air (B). As the two concentration gradients are equal in size, but opposite in sign (direction), and because of $C_A + C_B = C$, adding 7.33 and 7.34 yields:

$$\phi'' = \phi_A'' + \phi_B''$$

$$(E - 7.35)$$

In the case of mutual diffusion (section 7.2) the fluxes ϕ_A'' and ϕ_B'' are equal in size, but opposite in sign, rendering the drift flux ϕ'' zero. The Equations 7.33 and 7.34 then turn into Fick's law expressions.

In the case of the Winkelman experiment, there is no transport of B(air); thus ϕ_B'' equals zero.

Thus ϕ'' equals ϕ_A'' .

Substituting this into Equation 7.33 gives:

$$\phi_A'' = -D\frac{dC_A}{dx} + \phi_A'' \frac{C_A}{C}$$

Thus $\phi_A'' - \phi_A'' \dfrac{C_A}{C} = -D\dfrac{dC_A}{dx}$

and $\phi_A'' \left(\dfrac{C - C_A}{C}\right) = -D\dfrac{dC_A}{dx}$

Therefore $\phi_A'' = -D\dfrac{C}{C - C_A}\dfrac{dC_A}{dx}$

$$(E - 7.36)$$

and

$$\phi_B'' = 0$$

$$(E - 7.37)$$

Equation 7.36 is Stefan's law.

For the Winkelman experiment a mole balance for any slice of thickness dx between liquid level and rim of the tube shows that in a steady state:

$$0 = \left[\phi_A''\right]_x - \left[\phi_A''\right]_{x+dx} \tag{E-7.38}$$

or, in other words,

$$\frac{C}{C-C_A} \cdot \frac{dC_A}{dx} = k = \text{constant} \tag{E-7.39}$$

Solving this differential equation for C_A along with the boundary conditions $C_A = C_{A0}$ at $x = 0$ and $C_A = C_{AL} = 0$ at $x = L$, yields

$$C_A(x) = C - (C - C_A) \cdot \left(\frac{C - C_{AL}}{C - C_{A0}}\right)^{x/L} \tag{E-7.40}$$

Using this result for elaboring the derivative in Equation 7.36 produces the following expression for the mole flux of A:

$$\phi_A'' = \frac{DC}{L} \cdot \ln\left[\frac{C - C_{AL}}{C - C_{A0}}\right] \tag{E-7.41}$$

This result, based on Stefan's law, differs from the erroneous Fick's law result

$$\phi_A'' = -D \frac{C_{AL} - C_{A0}}{L} \tag{E-7.42}$$

by Stefan's correction factor (f_D)

$$f_D = \frac{C}{C_{A0} - C_{AL}} \cdot \ln\left[\frac{C - C_{AL}}{C - C_{A0}}\right] \tag{E-7.43}$$

Expressions for mass fluxes derived on the basis of Fick's law, such as Equation 7.10 and 7.11 become valid for unilateral mass transport when their right-hand terms are multiplied by Stefan's correction factor f_D. You might like to verify this. Similarly, it can be shown that the diffusive mass transport around a sphere in a stagnant medium under steady-state conditions is governed by:

$$Sh = 2 f_D$$

Stefan's law Stefan's law applies to all problems of unilateral mass transfer which are typified by the phenomenon that only a specific compound A is transported through some space from one boundary to another due to the boundary conditions. The molecules of A diffuse through the molecules of some other substance which do not make up some net transport. Usually the molecules of A are removed by a convective flow or by absorption, or consumed by some chemical reaction, as soon as they arrive at the boundary of their destination. Note that for low concentrations of A Stefan's law simplifies to Fick's law again and Stefan's correction factor f_D approaches unity.

In any mass transfer problem the value of Stefan's correction factor should therefore be verified in order to ascertain whether Fick's law or Stefan's law must form the basis for the solution of the problem.

Let us now work through a numerical example of the Winkelman experiment.

Example: The Winkelman experiment

In order to determine the diffusion coefficient of 'A', the following experiment is carried out. Liquid A, contained in the vertical tube, vapourizes due to a gas flow through the horizontal tube. The transport of the 'A' vapour from the liquid to the gas flow is governed by diffusion (see Figure 7.4).

The incoming gas flow contains no A and is so large, that the A concentration at the outlet of the gas flow is negligible. The rate of change of the A liquid level is so small that this has no effect on the concentration profile of the vapour in the vertical tube.

When the distance from the liquid level to the horizontal tube is 0.1 m it is found that the rate of change of the liquid level: $v = 1.75 \ 10^7 \ \text{ms}^{-1}$.

Can we calculate the diffusion coefficient of A in air from this experiment?

Additional data: temperature $T = 48°C$, pressure $p = 1$ bar;

density of liquid A is $\rho = 1540 \ \text{kg m}^{-3}$

vapour pressure of A at 48°C is $p_v = 282$ mm Hg;

molar mass of A is $M = 154 \ 10^3 \ \text{kg mol}^{-1}$

Our solution is as follows:

Since the gas does not penetrate into the A liquid, this experiment is clearly described by Stefan's equations. Thus the mole flux of the vapour (denoted by A) through the vertical tube is given by Equation 7.41 that is:

$$\phi_A'' = \frac{D\,C}{L} \cdot \ln\left[\frac{C - C_{AL}}{C - C_{A0}}\right] \tag{E - 7.45}$$

The total molar gas concentration C, (air and vapour) is calculated from the ideal gas law equated at the top of the vertical tube: $C = \frac{p}{RT} = 37.5 \ \text{mol m}^{-3}$. Similarly, the concentration C_{A0} at the liquid-vapour interface is obtained:

$C_{A0} = \frac{p_v}{RT} = 13.9 \ \text{mol m}^{-3}$, the concentration C_{AL} is zero.

On the other hand, the mol flux must equal the rate at which the liquid vapourizes:

$$\phi_A'' = \frac{\rho v}{M} \tag{E - 7.46}$$

Combining Equation 7.45 and 7.46 renders for the diffusion coefficient:

$$D = \frac{1}{\ln\dfrac{p}{p-p_v}} \frac{\rho v}{M C} L = 1.0 \times 10^{-5}\ m^2 s^{-1}$$

(E - 7.47)

<table>
<tr><td>

SAQ 7.5

</td><td>

A liquid solvent is placed in a vertical tube of a piece of apparatus similar to that illustrated in Figure 7.4. Using this set up, in the manner described in the example above. Attempt the calculations described below.

Data: if temperature T = 48°C the pressure p = 1 bar. Density of the solvent ρ = 1800 kg m^{-3}. Vapour pressure of the solvent at 48°C is p_v = 282 mm Hg. Molar mass of the solvent M = 180 x 10^3 kg mol^{-1}. Diffusion coefficient of the solvent = 2.0 x 10^{-5} m^{2}s^{-1}. The distance from the liquid level to the horizontal tube is 0.05 m. Universal gas constant R = 8.314 J K^{-1} mol^{-1}. Take 1 bar = 760 mm Hg and 1 bar = 10^5 Nm^{-2}.

Calculate:

- the rate of evaporation (in mol.s^{-1});

- the rate at which the liquid level will fall (in ms^{-1}).

</td></tr>
</table>

In the next section we focus on the transfer of mass across a phase interface.

7.6 Interphase mass transfer

The transfer of material between phases is important in many separation processes in which gases and liquids are involved. When a pure liquid is being evaporated into a gas, only the gas-phase mass transfer needs to be calculated; mass transfer within the pure liquid phase is not involved. Conversly, when a pure gas is being absorbed into a liquid, only the liquid phase mass transfer needs to be considered.

In the Winkelman experiment the mass flux in the vapour phase is pushed by and is equal to the evaporative flux from the liquid phase. It is expressed in terms of a concentration gradient within the vapour phase determined by the concentration just above the phase interface minus the concentration at the rim of the tube and divided by the distance from rim to phase interface. The concentration in the vapour phase just above the interphase, or the partial vapour pressure, is entirely dependant on the temperature of the liquid phase. It may safely be assumed that the molecules are able to instantaneously establish and maintain the equilibrium between liquid and partial vapour pressure at the boundary. Over longer distances within the vapour phase the molecules need more time to effect this equilibrium and that is exactly what is modelled by mass balances and Fick's (or Stefan's) law.

mass balance

Fick's law

7.6.1 Partition coefficients

partition coefficient

In a number of cases, in particular with two liquid phases or with a liquid/vapour system, the ratio of the concentrations in the two phases at either side of the interphase under equilibrium conditions is named the partition coefficient, denoted by m. From this definition it is obvious that the boundary concentrations in the two phases are mutually related by this constant m at all times, while the ratio of the concentrations in the bulk of the two phases approaches m only after a (very) long time. Henry's coefficient (derived from Henry's law) is another form of partition coefficient. Henry's coefficients relate the partial vapour fraction to the mole fraction in the liquid phase. We will not, however expand on Henry's law here.

We can represent the situation of partitioning in the following way.

At the point where we bring the two phases into contact, the concentration of a component in the two phases can be represented by $C_{1,B}$ and $C_{2,B}$ where B represents the initial concentrations and 1 and 2 represent the phases.

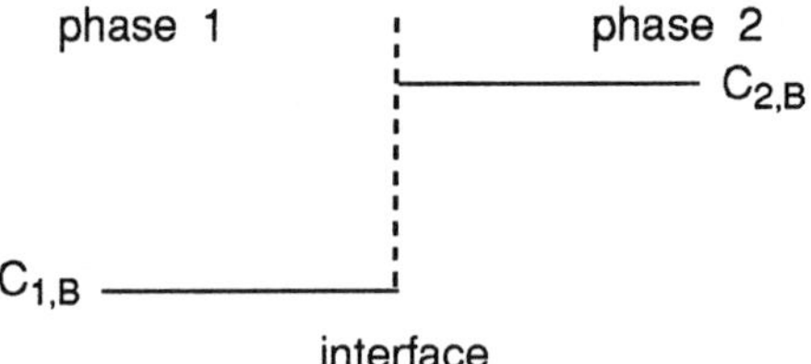

Very quickly, because of diffusion, the concentrations close to the interface will change. After a time, there will be an equilibrium and the ratio of the concentrations close to the interface will become m, the partition coefficient.

Thus:

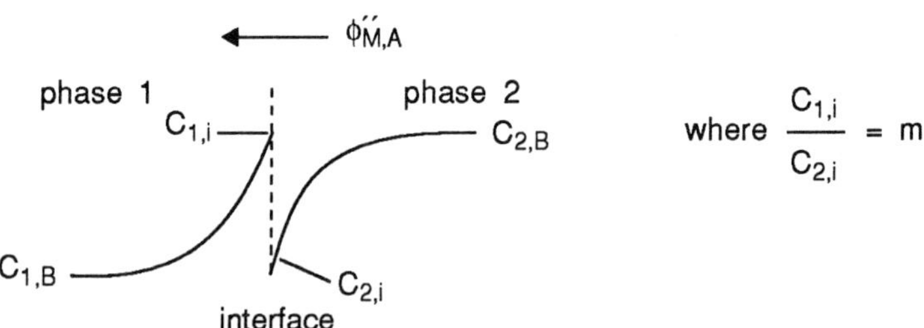

At this early stage the concentrations in the bulk of the phases will remain virtually unaltered. Diffusion, however will continue and gradually the concentrations in the bulk will change. Eventually the ratio of the concentrations in the bulk of the phases ($C_{1,B}$ and $C_{2,B}$) will become $= m$.

Let us work through a numerical example in which we can use the concept of partition coefficients.

Example: Application of partition coefficients to the absorption of gases from bubbles

Small bubbles of a pure gas (initial diameter Do) are dissolved in a liquid. The bubbles rise very slowly. As they rise, the bubbles disappear because gas is absorbed into the liquid. After a time t the bubbles are completely absorbed. We may take Sh = 2 during the whole process (ie the bubbles are spherical).

The partition coefficient is m = 0.5 (defined as concentration in the gas divided by concentration in the liquid). We assume that the pressure and hence the concentration in the bubbles remains constant. Experimentally it is shown that the bubble diameter is a function of the absorption time: $D^2_0 = 25.10^9 \cdot t$. Can we obtain a value for the diffusion coefficient of the gas in the liquid?

Solution

Since the concentration C_g (g represents gas) in the bubble is constant, the same holds for the concentration $C_{L,i}$ in the liquid at the bubble-liquid interface. Since the Sherwood number is given we can write down a mass balance for the bubble as:

$$\frac{d\left(C_g \frac{\pi}{6} D^3\right)}{dt} = - Sh \frac{D}{D} \cdot \pi D^2 \cdot (C_{L,i} - C_{L,B})$$ where $C_{L,B}$ is the concentration in the bulk of the liquid

$$(E - 7.48)$$

Using $C_g = m \cdot C_{L,i}$ and $C_{L,B} = 0$, Equation 7.48 is simplified to:

$$D \frac{dD}{dt} = - \frac{2\, Sh}{m} D$$

$$(E - 7.49)$$

where m is the partition coefficient.

Applying the initial condition $t = 0 \rightarrow D = D_o$ this equation has the solution:

$$D^2 - D_0^2 = - \frac{4\, Sh}{m} D\, t$$

$$(E - 7.50)$$

Thus substituting: at $t = t_e \rightarrow D = 0$, we find indeed:

$$D_0^2 = \frac{4\, Sh}{m} D \cdot t_e$$

$$(E - 7.51)$$

Consequently, the diffusion coefficient equals $1.56\ 10^9\ \text{m}^2\text{s}^{-1}$.

In many biotechnological processes, the availability of oxygen is an important parameter. In many systems including the activated sludge process used in water reclamation, the culture of animal cells to produce monoclonal antibodies as well as in many microbial processes, oxygen is supplied by pumping air into the system. In the following SAQ we ask you to apply what you have learnt about partition coefficients and mass transfer to a problem relating to oxygen transfer between air and water.

SAQ 7.6

1) In order to saturate a small amount of water with oxygen, we contact the water with a large amount of air. After a while the water is saturated and the O_2-concentration is $C_{L,B} = 8$ mgl^{-1}. The air conditions are: temperature $T = 20°C$, pressure $p = 1$ bar, density $\rho = 1.2$ kg m^{-3}, mass fraction of O_2 is 20%. The water temperature is also 20°C. What is the value of the partition coefficient of O_2?

2) The pressure of the air above the water produced in 1) is now reduced to $p = 0.5$ bar

    a) What will be the direction of the O_2 flow?

    b) What will be the quotient (ratio) of the concentrations on both sides of the interface immediately after the pressure is reduced?

    c) What will be the quotient of the concentrations on both sides of the interface a long time after the pressure is reduced?

    d) What will be the equilibrium concentration of O_2 in water at this reduced pressure?

Your answers to SAQ 7.6 should have shown you that (the direction of) mass transfer can not simply be concluded by considering the bulk concentrations. Reducing the oxygen concentration in the gas phase (by lowering the pressure) even resulted in oxygen being transferred from the water into the gas phase, despite the fact that the bulk concentrations in the air were up to 30 times greater than those in water.

In the Winkelman experiment as well as in the previous example the resistance to mass transport is completely in one of the two phases, as the second phase is pure in the component in question. In SAQ 7.6 the resistance to mass transfer is completely in the liquid phase as the mobility of the oxygen molecules is much smaller in the water (where the water molecules are much closer together) than in the low-density gas phase. If the resistance to mass transport is not mainly in one phase, it may be important to consider the resistances to mass transport at both sides of the phase interface. As a matter of fact, however, the fluxes at either side are equal. Along with the concept of the partition coefficient, this is illustrated in the following example.

Example: Application of partition coefficients to liquid: liquid extractions

One carefully puts a layer of benzene (depth 1 cm) on top of a layer of water of 1 cm. Acetic acid is dissolved in the benzene with a concentration of 10^{-2} mol l^{-1}. The partition coefficient m (defined as the quotient of concentration in benzene and in water) is 10. The diffusion coefficients of the acid in water and benzene are:

$$D_w = 1.5 \cdot 10^{-9} \text{ m}^2\text{s}^{-1} \text{ and } D_b = 0.5 \cdot 10^{-9} \text{ m}^2\text{s}^{-1}$$

We want to know what, for short times, the concentration profile at both sides of the interface will be. In order to get some picture of this we shall calculate the interface concentrations, the penetration depths and the concentration gradients at both sides of the interface 10 minutes after the liquid mixtures are contacted.

Solution

The penetration depths are easily calculated:

$$\text{water } x_w = \sqrt{\pi D_w t} = 1.7 \text{ mm} \qquad\qquad (E - 7.52)$$

$$\text{benzene } x_b = \sqrt{\pi D_b t} = 0.97 \text{ mm} \qquad\qquad (E - 7.53)$$

Both are obviously much smaller than the depth of the layers, hence we can use the penetration theory.

To be able to calculate the interface concentrations and the gradients it is handy to introduce a few subscripts: b stand for benzene, w for water, B for a value in the bulk of the liquids and i for a value at the interface. Now the molar fluxes from both sides of the interface at the interface can be written down:

$$\text{water } \phi_w'' = \sqrt{(D_w)/(\pi t)}\ (C_{wi} - C_{wB}) \qquad\qquad (E - 7.54)$$

$$\text{benzene } \phi_b'' = \sqrt{(D_b)/(\pi t)}\ (C_{bB} - C_{bi}) \qquad\qquad (E - 7.55)$$

Clearly both fluxes must be equal (at all times!), since all acid that arrives from the benzene side at the interface must be transported from this interface into the water side. Furthermore the concentrations at the interface are related according to:

$$C_{bi} = m\,C_{wi} \qquad\qquad (E - 7.56)$$

Combining the Equation 7.54, 7.55 and 7.56 together with the bulk values $C_{bB} = 0.01$ mol l^{-1} and $C_{wB} = 0$ yields $C_{bi} = 0.85\ C_{bB}$ and $C_{wi} = 0.085\ C_{bB}$.

Now the concentration gradients at the interface are found from:

$$\text{water}\left[\frac{dC_w}{dx}\right]_i = -\frac{\phi_w''}{D_w} = -\frac{C_{wi} - C_{wB}}{\sqrt{\pi D_w t}} = -1.55 \text{ mol l}^{-1}\text{m}^{-1} \qquad\qquad (E - 7.57)$$

$$\text{benzene}\left[\frac{dC_b}{dx}\right]_i = -\frac{\phi_b''}{D_b} = -\frac{C_{bB} - C_{bi}}{\sqrt{\pi D_b t}} = -0.5 \text{ mol l}^{-1}\text{m}^{-1} \qquad\qquad (E - 7.58)$$

Note, that the equilibrium concentrations C_b and C_w simply follow from an overall mole balance and the value of m:

$$C_b + C_w = C_b(0) \text{ and } C_b = m\,.\,C_w \qquad\qquad (E - 7.59)$$

Thus $C_b = 9.1 * 10^{-3}$ mol l^{-1} and $C_w = 9 *10^{-4}$ mol l^{-1}

From the latter example it is clear now that with interphase mass transfer the concentration profile exhibits a jump across the phase boundary and that the slope of the concentration profiles at either side differs due to at least a difference in diffusion coefficients.

7.6.2 Total mass transfer coefficient

For the steady-state mass transfer of a component A from one phase into another, the concept of a total mass transfer coefficient k is often used. In principle, this mass transfer coefficient is the equivalent of the total heat transfer coefficient U which is made up by the partial heat transfer coefficients and, if present or relevant, the heat resistance of a separation wall. In using the concept of a total mass transfer coefficient k for interphase mass transfer, the role of the partition coefficient m must be taken into account properly, while the phase interface itself does not represent a resistance to mass transport. We will consider the concentration profiles across an interphase (Figure 7.5).

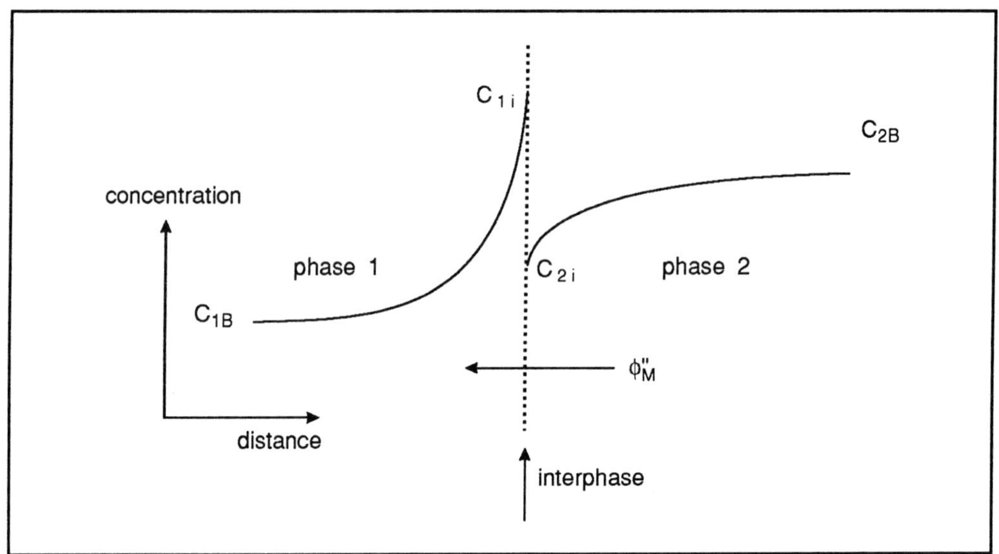

Figure 7.5 Concentration profiles across two phases (see text for a fuller description). C_{1B} = concentration in the bulk of phase 1, C_{1i} = concentration of the interphase film in phase 1, C_{2B} = concentration in the bulk of phase 2, C_{2i} = concentration of the interphase film in phase 2.

In Figure 7.5 the concentration profiles in the two phases look like those in the preceding example frozen at some moment in time. The resistance to interphase mass transfer could be conceived as being concentrated in two films at either side of the phase boundary with a jump in concentration at the boundary. In the steady state of interest the rate at which the component A is withdrawn from (and supplied to) the film at either side of the phase boundary then equals the rate of transfer across the phase boundary. Using the notation as indicated in Figure 7.5 this is expressed is shown in Equation 7.60.

$$\phi_M'' = k_1 \cdot (C_{1i} - C_{1B}) = k_2 \cdot (C_{2B} - C_{2i}) \qquad\qquad (E - 7.60)$$

where:

ϕ_M = mass transfer rate

k_1 = mass transfer coefficient for phase 1

C_{1B} = concentration of phase 1

C_{1i} = bulk concentration of phase 1 at interphase

C_{2i} = concentration of phase 2 at interphase

C_{2B} = bulk concentration of phase 2

k_2 = mass transfer coefficient of phase 2

Eliminating C_{1i} and C_{2i} owing to:

$$C_{1i} = m \cdot C_{2i} \qquad\qquad\qquad\qquad (E - 7.61)$$

transforms Equation 7.60 into:

$$\phi_M'' = \left[\frac{m}{k_2} + \frac{1}{k_1} \right]^{-1} (mC_{2B} - C_{1B}) \qquad\qquad (E - 7.62)$$

$$\phi_M'' = \left[\frac{1}{k_2} + \frac{1}{mk_1} \right]^{-1} (C_{2B} - \frac{C_{1B}}{m}) \qquad\qquad (E - 7.63)$$

The mass transfer flux can be expressed as the product of some total mass transfer coefficient K_1 or K_2 times the difference between the bulk concentrations that drives the mass transfer if the partition coefficient m is included and if K_1 and K_2 are defined by:

$$\frac{1}{K_1} = \frac{m}{k_2} + \frac{1}{k_1} \qquad\qquad\qquad\qquad (E - 7.64)$$

and

$$\frac{1}{K_2} = \frac{1}{k_2} + \frac{1}{mk_1} \qquad\qquad\qquad\qquad (E - 7.65)$$

The total mass transfer coefficients K_1 and K_2 not only comprise the partial mass transfer coefficients k_1 and k_2 but also the coefficient m. When the major contribution to the total interphase mass transfer resistance must be determined, m must be taken into account as well: the ratio of the partial mass transfer resistances also includes m:

$$\frac{\text{resistance in phase 1}}{\text{resistance in phase 2}} = \frac{1/(mk_1)}{1/k_2} \qquad\qquad (E - 7.66)$$

Equations 7.62 and 7.63 further express, in line with the expressions used in the Winkelman experiment, that the concentration difference that drives the transfer process includes the coefficient m as well: the product of m times C_i^*. Hence:

$$\phi_M'' = K (C_B - C_i^*) \qquad\qquad\qquad\qquad (E - 7.67)$$

Finally, the area of the phase interface through which the mass transfer takes place is often expressed in terms of the specific area, ie as the area per unit volume of the two-phase system. It is either denoted by 'S' (in the context of packed or fluidized beds) or by 'a' (in the world of gas/liquid dispersions in bubble columns and stirred vessels).

Let us now apply these principles to an example.

Example: Extraction of H_2S from oil by a gas flow

In an oil-filled reactor one produces H_2S. In order to extract the H_2S (which is dissolved in the oil) from the oil there is a continuous flow (ϕ_v) of H_2 gas through the reactor. The

H$_2$S is transferred from the oil to the gas phase and subsequently leaves the reactor with the gas flow. The reactor is stirred, so the liquid phase and the gas phase may be regarded as ideally mixed. At a certain time (t = 0) the H$_2$S production stops but the H$_2$ flow continues. The concentration H$_2$S in the oil at t = 0 is C_{oil} (0). At equilibrium the H$_2$S concentrations in the oil and H$_2$ gas are related as $C_{oil} = m.C_{gas}$. Our task is to formulate the H$_2$S mass balances for the oil phase and for the gas phase in terms of the concentration of H$_2$S in the oil, in the gas and the total mass transfer coefficient K (based on the oil phase). On the basis of these balances and assuming that the gas phase and the oil phase are practically in equilibrium we would like to calculate the time dependence of the H$_2$S concentration in the oil.

Solution:

There will be an H$_2$S flow out of the oil driven by force (C_{oil} - m C_{gas}). The total surface area over which the transfer takes place is (a.V), where V is the volume of the tank and a the specific area of the oil-gas interphase. Hence the H$_2$S mass balance for the oil reads as:

$$\boxed{\begin{array}{c}\text{rate of change}\\\text{of H}_2\text{S in oil}\\\text{phase}\end{array}} \quad = \quad - \quad \boxed{\begin{array}{c}\text{transfer of}\\\text{H}_2\text{S to gas}\\\text{phase}\end{array}}$$

$$V_{oil} \frac{dC_{oil}}{dt} = - KaV . (C_{oil} - m . C_{gas}) \tag{E - 7.68}$$

For the H$_2$S mass balance for the gas we also have to take the gas flow out of the reactor into account (the incoming flow does not contain H$_2$S), hence:

$$V_{gas} \frac{dC_{gas}}{dt} = - \phi_v . C_{gas} + KaV . (C_{oil} - m . C_{gas}) \tag{E - 7.69}$$

Now in order to calculate C_{oil}(t) we may assume equilibrium, thus $C_{oil} = m.C_{gas}$. Notice that now the driving force has become zero, but this does not mean that the H$_2$S flow from the oil to the gas is zero since we still extract H$_2$S. What happens in reality is that the driving force becomes very small (almost zero) but this can only be the case if the product KaV becomes very large (approaches infinity). Thus we conclude that we can not use Equation 7.68. To overcome this difficulty we add both mass balances to get rid of the internal transport:

$$V_{oil} \frac{dC_{oil}}{dt} + V_{gas} \frac{dC_{gas}}{dt} = - \phi_v C_{gas} \tag{E - 7.70}$$

Now we can use $C_{oil} = m.C_{gas}$ and Equation 7.70 becomes:

$$\frac{dC_{oil}}{dt} = - \frac{\phi_v}{mV_{oil} + V_{gas}} C_{oil} \tag{E - 7.71}$$

Using the initial condition: at t = 0, $C_{oil} = C_{oil}$(0) we finally obtain:

$$C_{oil}(t) = C_{oil}(0) . \exp\left(- \frac{\phi_v}{mV_{oil} + V_{gas}} t\right) \tag{E - 7.72}$$

<table>
<tr><td>

SAQ 7.7

</td><td>

Answer true or false to each of the following:

1) The concentration profile across an interphase is always continuous.

2) If the partition coefficient of a component in two solvents (A and B) is m, and the mass transfer coefficients of the compound in the two solvents are k_A and k_B respectively, the total mass transfer:

$$K_A = \frac{k_B}{m} + k_A.$$

3) If the volume of a solvent in a gas: liquid system is 250 m^3 and the value of a is 0.1 m^2 m^{-3}, then the total interphased area between the two phases is 25 m^2.

</td></tr>
</table>

7.7 Convective mass transfer

Convective mass transfer is usually described in terms of (semi-) empirical relations for the Sherwood number. These relations express in a non-dimensional form how, in a particular situation, the mass transfer coefficient k depends on ρ, <v>, D, η, *D*, etc. In other words:

$$k = k \, (\rho, <v>, D, \eta, D, ...) \qquad\qquad (E - 7.73)$$

is converted into

$$Sh = f \, (Re, Sc, ...) \qquad\qquad (E - 7.74)$$

where:

$$Sh \text{ is Sherwood number} = \frac{kD}{D}$$

$$Re \text{ is Reynolds number} = \frac{\rho <v> D}{\eta}$$

$$Sc \text{ is Schmidt number} = \frac{\eta}{\rho D}$$

by means of dimensional analysis. The appearance of the Reynolds number Re reflects the fact that Re typifies the flow condition. Re represents the ratio of convective transport to molecular transport within the flow domain. This of course is also relevant to mass transfer. The Schmidt number expresses that the boundary layers (or films) for momentum transport and for mass transport are generally not equal. Some of the available correlations for convective mass transfer run as follows:

for mass transfer between the wall and turbulent flow in a pipe:

$$Sh = 0.027 \, Re^{0.8} \, Sc^{0.33} \text{ for } Re > 10^4 \text{ and } Sc > 0.7 \qquad\qquad (E - 7.75)$$

for forced flow around a sphere:

$$Sh = 2.0 + 0.66 \, Re^{0.5} \, Sc^{0.33} \text{ for } 10 < Re < 10^4 \text{ and } Sc > 0.7 \qquad\qquad (E - 7.76)$$

Using empirical relations for Sh to describe convective mass transfer means that such relationships often need to be verified experimentally. In the following example, we illustrate the sort of approach that might be used.

Example: Experiment to verify Sherwood relation for mass transfer from sphere

A student has to verify the following Sherwood relation for mass transfer from a sphere to a flowing medium:

$$Sh = 0.6\ Re^{0.5}\ Sc^{0.33} \tag{E - 7.77}$$

The relation is based on mass transport averaged over the entire sphere surface. In the experiment, a sphere (of pure substance with density $\rho = 2000$ kg.m^{-3}) is suspended in an air flow. The incoming air does not contain component X, has a temperature of 20°C, a velocity v of 5 m s^{-1} and a kinematic viscosity $\upsilon = 1.5\ 10^{-5}$ m^2 s^{-1}. For substance X the following data hold: molar mass M = 120 g mol^{-1}, vapour pressure at the given circumstances $p_v = 0.01$ bar and diffusion coefficient in air $D = 0.5\ 10^{-5}$ m^2 s^{-1}.

The initial diameter of the sphere is $D_0 = 25$ mm. In the experiment the student has to measure how much time it takes before the diameter of the sphere has decreased to 20 mm and compare this result with the time calculated using the relation given above. Let us do that job for him.

Solution:

We first set up a mass balance for the sphere:

$$\rho\ \frac{\pi}{2}\ D^2\ \frac{dD}{dt} = -\pi D^2 . M . k\ (C^*-C_\infty) \tag{E - 7.78}$$

with $C^* = \dfrac{p_v}{RT}$ is the concentration of the vapour of X at the surface and C_∞ is its concentration in the air far from the sphere, thus zero. Now before integration of this equation, we have to take the dependence of k on D into account:

$$k = \frac{ShD}{D} = 0.6\ D\ \left(\frac{v}{\upsilon}\right)^{0.5}\ \left(\frac{\upsilon}{D}\right)^{0.33}\ \frac{1}{D^{1/2}} \tag{E - 7.79}$$

Combining Equation 7.78 and 7.79 yields:

$$D^{1/2}\ \frac{dD}{dt} = -C \quad \text{with}\ C = 1.2\ \frac{M}{\rho}\ \frac{p_v}{RT}\ D^{\,2/3}\ \upsilon^{-1/6}\ v^{1/2} \tag{E - 7.80}$$

Using the initial condition $D(0) = D_0$ renders:

$$\frac{2}{3}\left(D^{3/2} - D_0^{3/2}\right) = -C . t \tag{E - 7.81}$$

Putting in the numerical values we find for the time t at which the diameter is 20 mm: t = 6.0 10^3 s = 1.7 hour. Thus the student could carry out the experiment and if the result was that the sphere decreased its diameter to 20 mm in 1.7 hours, then the relationship between Sh, Re and Sc described at the beginning of the example holds.

This approach to convective mass transfer, along with the above correlations, exhibits a strong resemblance with the approach and the correlations introduced earlier (in section 6.5) for convective heat transfer. Generally:

$$Nu = C \, Re^m \, Pr^n \qquad \text{(E - 7.82)}$$

and

$$Sh = C \, Re^m \, Sc^n \qquad \text{(E - 7.83)}$$

7.7.1 Chilton and Colburn numbers - j_H and j_D

Chilton and Colburn defined two new transfer numbers for heat transfer and mass transfer respectively:

Chilton number
$$j_H = \frac{Nu}{Re \, Pr^{0.33}} \qquad \text{(E - 7.84)}$$

and

Colburn number
$$j_D = \frac{Sh}{Re \, Sc^{0.33}} \qquad \text{(E - 7.85)}$$

For geometrically similar conditions and for not too low values of Re these two transfer numbers j_H and j_D only depend on Re according to:

$$j_H = j_D = C \, Re^{(m-1)} \qquad \text{(E - 7.86)}$$

Combining the Equations 7.84, 7.85 and 7.86 yields:

$$k = \frac{h}{\rho c_p} \cdot \left(\frac{\rho c_p}{\lambda} \cdot D \right)^{2/3} = \frac{h}{\rho c_p} \cdot \left(\frac{D}{\alpha} \right)^{2/3} \qquad \text{(E - 7.87)}$$

The latter equation expresses that the mass transfer coefficient k (which is specific to each new compound and requires specific and tedious concentration measurements) can be calculated from the heat transfer coefficient h (which is known for many practical heat transfer applications due to easy temperature measurements) and a few physical properties of the fluidum in question. This handsome procedure is the direct consequence of the analogy between heat transfer and mass transfer. Hence, neither velocity nor geometrical parameters (like D) occur in Equation 7.87.

Example: Removal of a cake from a vessel

The bottom of a ideal stirred vessel (diameter D = 3 m) is covered with a thick cake. In order to clean the vessel, a pure solvent flows through the vessel at a flow rate $\phi_v = 0.5 \cdot 10^3 \, m^3 \, s^{-1}$. Heat transfer measurements have been performed on the vessel and the following results obtained:

1) the heat transfer coefficient h at the bottom depends on the number of revolutions N of the stirrer: $h \propto N^{2/3}$;

2) the heat transfer coefficient is (for a given N) $h = 3000 \, W \, m^{-2} K^{-1}$;

3)   the power P supplied to the stirrer is also a function of N: $P \propto N^3$.

Numerical data: diffusion coefficient of the 'cake' in solution $D = 10^{-9}\ m^2\ s^{-1}$

density solvent $\rho = 1200\ kg\ m^{-3}$

specific heat solvent $c_p = 3000\ J\ kg^{-1}\ K^{-1}$

coefficient of heat conductivity of solvent $\lambda = 0.5\ W\ m^{-1} K^{-1}$

solubility 'cake' in solvent $C = 50kg\ m^{-3}$

We want to calculate, for a steady state, the value of the mass flow of cake into the solvent for the case when the stirrer rotates at a speed for which the heat transfer coefficient h is $3000\ W\ m^{-2}\ K^{-1}$. Furthermore, we would like to know how this value of the mass flow changes if we doubled the power input of the stirrer.

Solution:

First we have to calculate the mass transfer coefficient k from the data given above. Since the measurements on heat transfer and the cleaning of the vessel obviously are performed for geometrical similar conditions and at the same Re number we can use Equation 7.87 to calculate k. Substitution of the numerical data renders $k = 3.1\ 10^{-5}\ m\ s^{-1}$. In the steady state a mass balance of the cake in the solvent reads as (with C is the concentration of the 'cake' in the bulk of the solvent and C* the concentration at the cake-solvent interface):

$$0 = - \phi_v\ C + k\ \frac{\pi}{4}\ D_2\ (C^* - C) \tag{E - 7.88}$$

Thus the concentration in the solvent is $C = 15.2\ kg\ m^{-3}$. Hence, the mass flow of the cake into the solvent equals:

$$\phi_M = k\frac{\pi}{4} D_2\ (C^* - C) = 7.6\ 10^{-3}\ kg\ s^{-1} \tag{E - 7.89}$$

If we now double the power input: $P_n = 2.\ P$, then along the same lines as given above we find:

$$N_n = 2^{1/3}N, h_n = \left(2^{1/3}\right)^{2/3} h = 2^{2/9}\ h$$

and thus $k_n = 2^{2/9}\ k$. Substituting this new value for k gives $C_n = 16.9\ kg\ m^{-3}$ and $\phi_{M,n} = 8.5\ kg\ s^{-1}$.

SAQ 7.8

Salt water is transported through a pipe. In the course of time a salt layer (thickness $D = 2$ mm) is caked on the wall of the tube. The density of this salt layer (ρ_{salt}) is 2500 kg m^{-3}. In order to remove this salt layer, a flow of distilled water is forced through the tube water under the same flow conditions as for the salt water transport. The salt concentration in the distilled water (C_w) remains negligably small. Heat transfer measurements during the transport of the salt water show that the heat transfer coefficient (h) for transport from the wall to the liquid is 1.0 10^4 W m^{-2}K^{-1}.

Additional data: density of water = 1000 kg m^{-3}; specific heat of water = 4200 J kg^{-1} K^{-1} and $\dfrac{\lambda}{\rho c_p D} = 100$.

The solubility of salt in distilled water is 300 kg m^{-3}.

How long will it take the salt layer to dissolve?

SAQ 7.9

A mass of 200 g salt is spread at the bottom of a vessel. Water is carefully poured onto this layer to a height of 2 cm, using for this purpose 1 l water. The solubility of salt in water is 300 kg m^{-3}, the diffusion coefficient of salt in water is 10^{-9}m^2 s^{-1}.

1) How much time passes before the salt is completely dissolved? (To solve this problem you will need to establish a concentration profile. You will find that this can be treated in an analogous way to establishing the temperature profile and heat transfer described by Figures 6.10 and 6.11)

2) If the same experiment is done, but the water is carefully stirred so that the mass transfer coefficient is 10^{-3} m s how long does it then take for the salt to be completely dissolved?

SAQ 7.10

A vessel with a cross-sectional area A and height H is half filled with water. The space above the water is filled with pure CO_2. This is absorbed by the water. The pressure of the CO_2 above the water-CO_2 interface is kept constant via supply from outside. The water is carefully stirred, so that there is no waving at the liquid interface, but the mixing in the water is ideal. The water is continuously refreshed with a flow ϕ_v. The system is in a steady state.

Deduce an expression from which the partition coefficient in the liquid-phase can be calculated if:

the CO_2-concentration in the gas-phase is C_g, in the water-phase at the exit is C_w and the partition coefficient m is defined as:

$$m = \left[\frac{CO_2 - \text{concentration in gas–phase}}{CO_2 - \text{concentration in water–phase}} \right]_{\text{equilibrium}}$$

SAQ 7.11

Ca^{++} ions have to be removed from a water flow of 100 1 h^{-1}. The initial concentration is 500 mg 1^{-1}, the desired final concentration 10 mg 1^{-1}. The process takes place in a column filled with ion exchanger. This exchanger consists of sphere-shaped grains. The active surface for the ion-exchange is, per m^3 of the column, 1.5 10^{2}m^2.

The Ca^{++}-ion concentration, which is in equilibrium with the exchanger, can be neglected, The step that determines the rate of the ion-exchange is the exterior transport, from water to grain surface. For the total concentration range the following relation holds:

$$\frac{kd_0}{D} = 1.4 \left(\frac{v_0\, d_0}{v}\right)^{\frac{1}{2}} \left(\frac{v}{D}\right)^{\frac{1}{3}} \text{ for } Re > 10$$

where: k = mass transport coefficient

D = diffusion constant of Ca^{++}-ions in water ($=10^{-9}$m^2 s^{-1})

v = kinematic viscosity of water ($=10^{-6}$ m^2 s^{-1})

d_o = grain diameter (= 2 mm)

v_o = velocity of the water flow, refering to an empty column

The question is to design the column. The characteristic properties, that need to be determined, are: the length of the column (accuracy 1 cm), the diameter of the column (accuracy 2 mm). The design is satisfactory if the exit concentration is in the range from 8 to 10 mg 1^{-1}.

1) Draw a sketch of the situation.

2) Set up a mass balance for the Ca^{++}-ions over a slab of the column between z and z + dz (z coordinate parallel to the column axis). Show that this balance can be written in the form:

$$\frac{dC}{dz} = - A\, C \text{ (C is the } Ca^{++}\text{-ion concentration); determine } A.$$

3) Use the result from 2 to determine the ratio between entrance and exit concentration as a function of column length and diameter.

4) Determine values for the column length and diameter that satisfy the demands.

SAQ 7.12

A dry air stream has to be saturated with water vapour up to 99%. This can be achieved by passing the air flow through a wet-wall-column. This is a tube with a water film at the inside flowing downwards; the thickness of the liquid film can be neglected compared to the column diameter. The temperature in the column is constant. The entire column is operated in a steady state.

Data:

$\rho_{air} = 1 \text{ kg m}^{-3}$

$\phi_{v,air} = 6.9 \ 10^{-3} \text{ m}^3 \text{ s}^{-1}$

$D_{inner} - 5 \ 10^{-2} \text{ m}$

$\eta_{air} = 17 \ . \ 10^{-6} \text{ Ns m}^{-2}$

$D \text{ H}_2\text{O}_{\text{ in air}} = 2.5 \ 10^{-5} \text{ m}^2 \text{ s}^{-1}$

For the transport coefficient in the column the following holds:

$Sh = 0.023 \ Re^{0.83} \ Sc^{0.44}$ for $2.10^3 < Re$, $3.5 \ 10^4$ and $0.60 < Sc < 2.5$.

1) Set up a mass balance for the water vapour in the air over a slab between z and z + dz of the column. (z - coordinate parallel to the column).

2) Give explicit the boundary conditions and solve the equation from 1) and determine the necessary length of the column.

SAQ 7.13

A physicist suspects that the column length from SAQ 7.11 could be much smaller if the gas flow recirculates. Show that this presumption is wrong for the case in which the same amount of gas recirculates as the amount that leaves the device (see Figure below). To do so, answer the same two questions as in SAQ 7.11 and thus finally determine the new length.

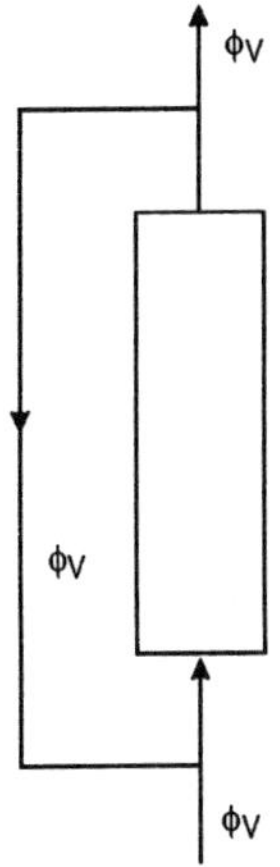

Summary and objectives

In this chapter, we have examined mass transport. We began by showing that mass and heat transport can be treated similarly. We then examined mass transport by diffusion processes in a variety of geometric systems and introduced the concept of Sherwood numbers and Stefan's law. This led into a discussion of mass transfer across interphases, partial and total transfer coefficients. In the final part of the chapter we considered convective mass transfer. The SAQ's associated with this chapter were used to demonstrate that the principles described can be applied to a large variety of processes involving mass transfer.

Now that you have completed this chapter you should be able to:

- use supplied data to calculate mass flow rates in a wide variety of situations;

- use supplied data to calculate mass transfer coefficients;

- use supplied data to calculate penetration distances of diffusable components;

- use supplied data to calculate the transfer of components from one phase to another;

- use supplied data to determine evaporation and dissolution rates;

- describe differences in heat and concentration profiles at interphases;

- use supplied data to calculate the dimensions of the system needed to achieve desired mass transfers.

Introduction to mathematical modelling of biological processes

Introduction to mathematical modelling of biological processes

8.1 Introduction

The purpose of this and the remaining chapters of this text is to enable you to design simple models of a microbiological process based on available chemical, biochemical and physiological knowledge and to use these models to adequately describe growth and product formation in the fermentation process.

Although entitled mathematical modelling, the level of mathematical knowledge and skills needed are quite modest. To help you, we have divided the text into small sections. In this chapter, we simply introduce the principles which underpin the modelling process. For convenience, we have provided a list of the symbols, subscripts and superscripts used in the remaining chapters of this text as an appendix at the end of this text. For clarity, we have separated these from the symbols used in Chapters 1 to 7.

8.1.1 The modelling cycle

Microbial systems are very complex in their appearance and their time dependent (dynamic) behaviour. In order to design, by calculation, production processes in which such complex systems are involved, one needs to use the invaluable tool of mathematical modelling. The procedure to develop such a mathematical model can best be illustrated by the so-called model cycle (Figure 8.1).

The fact that the procedure is represented as a cycle indicates that a mathematical model is not created, but developed over time with adjustment according to deviations from reality.

steps in model cycle

The steps in the 'model cycle' speak for themselves. It is however important to realise that the most significant element is the translation of available verbal knowledge into a mathematical model. This chapter will concentrate on this aspect.

Some comments about the other elements in the model cycle are relevant.

- One should solve the model equations as soon as possible to get an early, global impression about the performance of the model. If, on solving the equations, the results bear no resemblence to reality, then the model is either wrong (and should be discarded) or it requires some modification to bring it closer to reality.

- The parameters used in the model equations influence the model result. However the effect of parameter variation can range from very minor to very large.

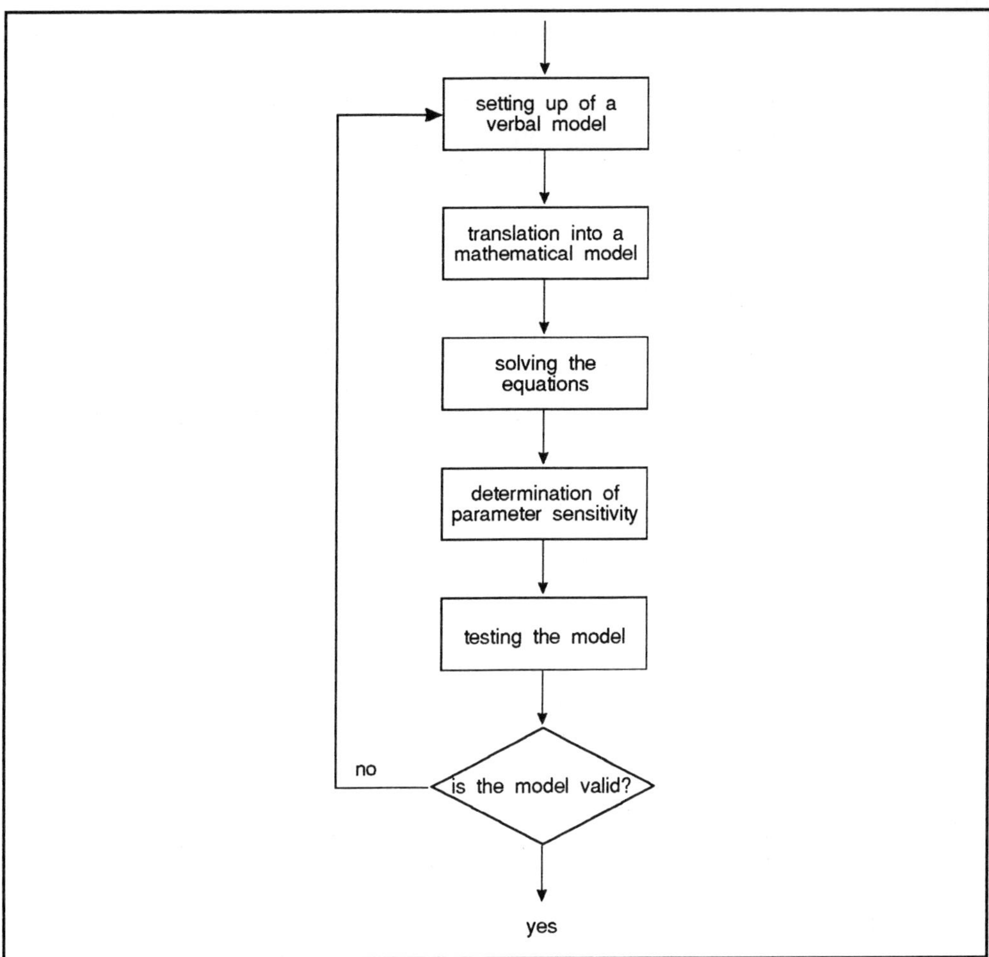

Figure 8.1 The mathematical model cycle.

An early impression of the sensitivity to particular parameters provides clues to the necessary experimental set-up and for the required accuracy of estimation of the parameter value. For example if the growth rate of an organism is very sensitive to temperature changes, then we need to be able to control and measure the temperature very accurately. If, on the other hand, the growth rate of the organism is not greatly influenced by salt concentration, we will not need to measure or estimate salt concentration very accurately.

As an illustration of the above concepts we will consider the phenomenon of microbial growth. We begin by using some 'verbal' information about microbial growth and then translate this into a mathematical form.

Verbal information states:

- microbial growth occurs auto-catalytically (2,4,8,16,32 etc) by cell division;

- in the absence of substrate, growth does not occur;

- with an increase in substrate concentration, the growth rate increases;

- at very high concentrations of substrate, the growth rate reaches a maximum.

The mathematical representation of these facts leads to:

- the concept of a biomass specific growth rate μ, where:

$\mu = \dfrac{r_{Ax}}{C_x}$ in which C_x = concentration of biomass, r_{Ax} = rate of accumulation of biomass;

- a hyperbolic function between μ and C_s (C_s = substrate concentration).

Monod equation We need to make some additional points here concerning these relationships. Firstly, let us briefly enlarge on the concept of specific growth rate. From the equation give above, we can define the specific growth rate (μ) as the rate of increase in biomass per unit of biomass. The specific growth rate has the units of (time)$^{-1}$ (eg h^{-1} or day^{-1}). The second point is that Monod showed that the specific growth rate (μ) of a micro-organism is related to the substrate concentration in the following way:

$$\mu = \frac{\mu^{max} C_s}{K_s + C_s}$$

where μ^{max} is the maximum specific growth rate of the organism on the substrate.

K_s is the substrate saturation constant.

These facts are represented graphically in Figure 8.2. A comment needs to be made about the nomenclature used in these equations. A common convention in biochemistry and microbiology is to represent substrate concentration by the symbol [S] (the square bracket meaning concentration - thus [H$^+$] means the concentration of hydrogen ions). From a technology/engineering standpoint, however, the use of square brackets in this way can lead to some confusion because of a more extensive need to use brackets to signify mathematical relationships. Here we will use the technology orientated notation of representing concentration by C and using a subscript to define which concentration we are considering. Note also that the convention in biochemistry/microbiology is to describe the maximum specific growth rate as μ_{max} whilst in technology the convention is to often use μ^{max}. No wonder there has often been confusion between biologists and engineers!

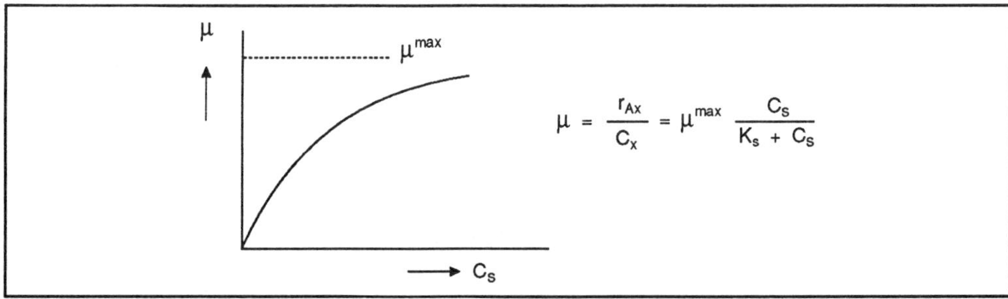

Figure 8.2 Mathematical growth rate as a function of C_s as described by Monod.

Let us return to our central theme. What we have done so far is to convert verbal information about microbial growth into a mathematical representation.

What should be our next step? (Look back at Figure 8.1 to give yourself a clue).

Our next step is to determine parameter sensitivity. In this case, how sensitive is μ to C_s?

parameter sensitivity

The parameter sensitivity shows that:

- at high C_s ($C_s \gg K_s$) μ is not influenced by K_s (since $\dfrac{\mu^{max} C_s}{K_s + C_s}$ approaches μ^{max});

- at low C_s ($C_s \ll K_s$) one needs only one parameter ($= \dfrac{\mu^{max}}{K_s}$) to express the relation $\mu \leftarrow C_s$. (Since $\dfrac{\mu^{max} C_s}{K_s + C_s}$ approaches $\dfrac{\mu^{max} \cdot C_s}{K_s}$).

Hence in systems with high C_s only experiments for the determination of μ^{max} are relevant; K_s has no influence. However, in systems with low C_s one needs the 'lumped parameter' $\mu^{max/K}{}_s$. For intermediate values of C_s one needs both μ^{max} and μ^{max}/K_s.

If you re-examine Figure 8.2 you will see that at low C_s values, μ changes rapidly with increases in C_s. If C_s is increased still further then μ changes more slowly with increases in C_s.

$\prod$ Using the above information, show that the specific growth rate: $\mu = \dfrac{\mu^{max}}{2}$ when

$$C_s = K_s$$

You should have used the following type of approach

Since $\mu = \dfrac{\mu^{max} C_s}{K_s + C_s}$ then under the conditions stated in the problem

$$\dfrac{\mu^{max}}{2} = \dfrac{\mu^{max} C_s}{K_s + C_s}. \text{ Thus } \dfrac{1}{2} = \dfrac{C_s}{K_s + C_s}$$

Then $2C_s = K_s + C_s$ and $K_s = C_s$

In other words, a specific growth rate of $\dfrac{\mu^{max}}{2}$ is achieved when $C_s = K_s$ (in fact K_s is defined as C_s when $\mu = \dfrac{\mu^{max}}{2}$).

The relationship can be used to gain a rough estimate of K_s from experiments. In practice, however, it is not straight forward. In order to determine the substrate concentration which supports a specific growth rate of $\dfrac{\mu^{max}}{2}$, it is necessary to know the value of μ^{max}. For a variety of reasons, it is not always easy to determine μ^{max}. In practice it is much more usual to use a reciprocal plot of $\dfrac{1}{\mu}$ against $\dfrac{1}{C_s}$. Using the Monod relationship of $\mu = \dfrac{\mu^{max} C_s}{K_s + C_s}$ then a reciprocal plot of $\dfrac{1}{\mu}$ against $\dfrac{1}{C_s}$ should give a straight line in which the slope of the line $= \dfrac{K_s}{\mu^{max}}$ and which cuts the $\dfrac{1}{\mu}$ axis at $\dfrac{1}{\mu^{max}}$ and the $\dfrac{1}{C_s}$ axis at $-\dfrac{1}{K_s}$ (see Figure 8.3).

Note that since $\mu = \dfrac{\mu^{max} C_s}{K_s + C_s}$ then $\dfrac{1}{\mu} = \dfrac{K_s}{\mu^{max}}\left(\dfrac{1}{C_s}\right) + \dfrac{1}{\mu^{max}}.$

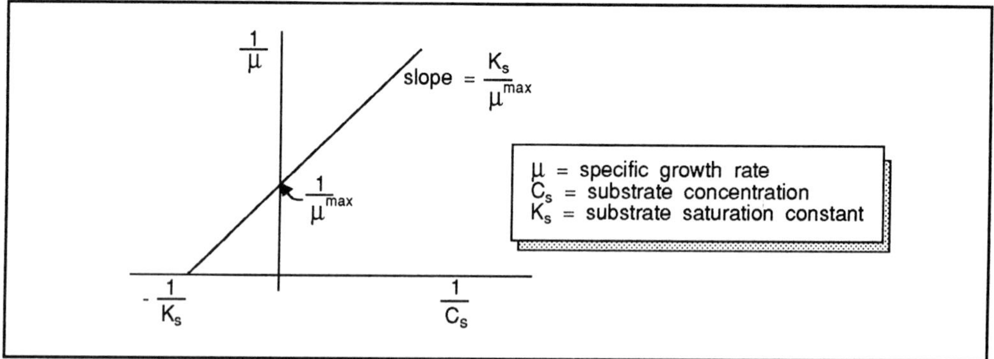

Figure 8.3 A double reciprocal plot of growth rate against substrate concentration.

Michaelis-Menten equation

Lineweaver Burke plot

For those familiar with enzyme reaction kinetics, this type of relationship between substrate concentration and velocity of reaction (ie specific growth rate) is entirely analogous with Michaelis-Menten kinetics and the double reciprocal plot parallels the Lineweaver-Burk plot used in enzyme kinetics.

The advantages of using the double reciprocal plot is that μ^{max} need not be known in order to determine K_s. By determining μ at several C_s values, K_s and μ^{max} may be determined graphically from the intercepts on the $\dfrac{1}{\mu}$ and $\dfrac{1}{C_s}$ axis or from the slope.

8.1.2 System definition and types of systems

In scientific investigations one always studies only a part of the complex reality.

Let us introduce some terms:

- the system; is the part of reality which is studied;

- the system boundary; separates the system from it surroundings;

- through this system boundary, exchange (Figure 8.4a) with the environment occurs.

Inside the system is the complex reality which is to be studied.

Now it is useful to consider 3 types of systems:

- isolated system where there is no energy or mass exchange with the surroundings;

- closed system where there is only energy exchange with the surroundings;

- open system, which allows mass and energy exchange with the surroundings.

These are represented in Figure 8.4b.

It should be realised that these descriptions are idealisations. In practice an isolated system does not exist; there is always some exchange with the surroundings. However the rate of exchange is so slow that it can be neglected.

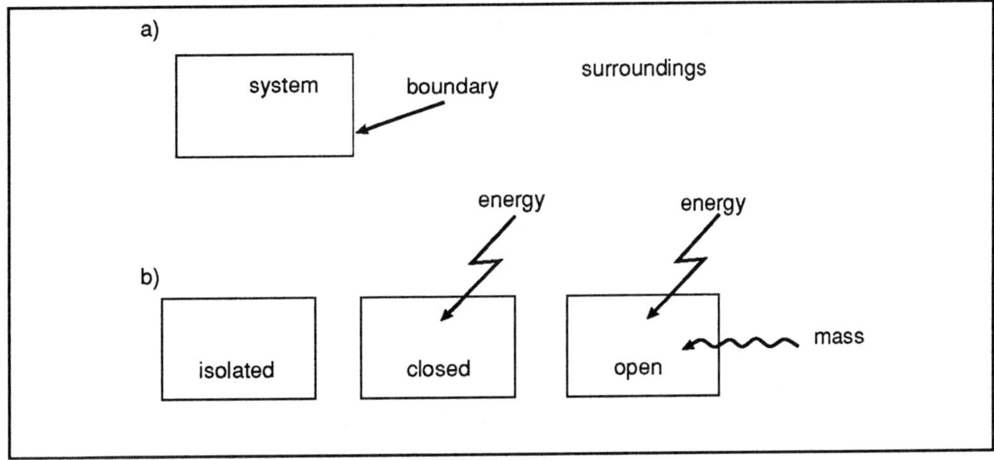

Figure 8.4 a) System definitions. Note that the 'system' is separated from its 'surroundings' by a system boundary. b)Three types of systems, isolated, closed and open. Note that in isolated systems nothing crosses the system boundary. In a closed system only energy crosses the boundary whereas in an open system both mass and energy cross the boundary.(See text).

The most important aspect of microbial systems is that they are open systems because a micro-organism exchanges mass and energy with its environment.

Π Which specific criterion regarding chemical transformations should be fulfilled in an isolated or closed system?

This was not an easy question to answer so do not worry if you could not come up with one. The answer we were looking for was that isolated and closed systems do not exchange matter with their surroundings hence there must be a system for recycling matter within the system. For example, on Earth, photosynthesis produces carbohydrates (such as starch). Without recycling all carbon would eventually be locked up in that form. For stability, recycling is essential in a closed/isolated system.

<table>
<tr><td>SAQ 8.1</td><td>

Assign the following to its appropriate system type (isolated, closed, open):

1) the planet Earth;

2) our galaxy;

3) a closed anaerobic fermentor in a constant temperature room;

4) an anaerobic fermentor whose temperature is controlled by a thermostat and in which CO_2 is vented.

</td></tr>
</table>

Having defined a specific system and its boundary there remains the task of formulating a mathematical model for the complex reality inside the system.

8.1.3 The macroscopic approach

As stated before, the most important step in setting up a mathematical model is the translation of the verbal information into a mathematical model. However, in microbiological systems we are faced with an enormous complexity which could lead to countless model parameters.

For example there occur hundreds of biochemical reactions, catalysed by hundreds of enzymes between hundreds of chemical compounds inside billions of microbial cells. To illustrate this statement one should consider that typically a bioreactor contains 10^{10} cells ml^{-1}, where each cell contains 1000-2000 different enzymes and about 1000 different chemicals.

allowable reductions, approximations, macroscopic approach

Therefore the major task is to obtain an allowable reduction of the complex real world into a mathematical model with minimal complexity which is still an adequate description of the microbial system. In other words there is a need for approximations of some kind. The main approximation which is generally applied in engineering sciences is the macroscopic approximation.

It is assumed that, although our reality is in principle made up of randomly moving particles in the form of atoms and molecules, it is acceptable to neglect these stochastic and particle aspects (stochastic = random).

The practical conditions which allow such a simplification are:

- the system which we consider is large and contains many so called elementary volumes;

elementary volume

- such an elementary volume is small compared to the system, is large compared to its particles, is ideally mixed and contains so many particles (like molecules or gas bubbles or microbial cells etc) that the particle properties can be described by their average value only. This average is a system property. Its stochastic variation is so small that one may neglect this.

Such system properties are called macroscopic properties.

Usually this macroscopic approach is justified and gives rise to workable solutions.

For example in a fermentor of 100m^3 which contains 10^{11} micro-organisms ml^{-1} of 1 μm diameter, an elementary volume of 1 cm^3 is a suitable elementary volume with which to define macroscopic averages. However there are cases where the macroscopic approach is not allowed (try SAQ 8.2).

<table>
<tr><td>SAQ 8.2</td><td>In a sterilisation procedure cells need to be totally killed, hence their number must be brought down from a contaminant level of, for example, 10^6 ml^{-1} to 0 ml^{-1}. What are the risks of the application of a macroscopic description? At what level would the macroscopic approach not be justified?</td></tr>
</table>

Basically we can define in each elementary volume a set of macroscopic properties which fall into 2 distinct classes (Table 8.1).

intensive and extensive properties

These are intensive properties, all of which are somehow related to a system quality and extensive properties which relate to system quantities.

Table 8.1 gives a list of the features and some examples of intensive and extensive properties.

Intensive properties	Extensive properties
- relates to system quality	- relates to system quantities
- cannot be added up	- can be added up
- no balance possible	- balance allowed
- Examples:	- Examples:
temperature, pressure, concentration, specific heat	volume, mass energy, entropy, electric charge

Table 8.1 Macroscopic description and macroscopic properties

Let us see if we can explain why temperature is an intensive and not an extensive property.

If we have 1 kg of water at 100°C and we add to it another 1 kg of water at 100°C, we do not get water at 200°C. Temperature is not additive and is therefore an intensive property. On the other hand, if we add 1 kg of water to 1 kg of water, we get 2 kg of water. Mass is additive and is, therefore, an extensive property.

<table>
<tr><td>SAQ 8.3</td><td>Use the properties listed in Table 8.1 to help you identify intensive and extensive properties from the following list: colour, shape, particle diameter, particle number, system length, system area, amount of biomass and glucose, concentration of biomass, concentration of ammonia.</td></tr>
</table>

The main attractive feature of this approach is that one can make so called macroscopic balances for the extensive properties.

One cannot make a balance over an intensive property!!

Having defined and illustrated the characteristics of intensive and extensive properties, there remains the important question of identifying which set of macroscopic properties fully describes any system. The answer comes from experience.

For chemical systems a complete macroscopic specification is assured if the following set is provided:

- volume;

- temperature, pressure, number and kind of phases (solid, liquid, gas);

- concentration of all chemical compounds in each phase.

Although this set of variables seems simple and short it will be clear that for microbiological systems one is faced with a nearly impossible task because there are thousands of chemical compounds. We will come back to this problem in the next chapter.

8.1.4 The importance of macroscopic balances

In the preceeding sections a complex reality has been approximated by:

- definition of a system;

- the macroscopic approach;

- the specification of a complete set of intensive and extensive properties.

An apparent characteristic of any real world system is that there occur changes in time and space of the intensive and extensive properties.

For example in a fermentor, the substrate and biomass concentrations change, the pH, temperature and pressure vary and the broth volume changes.

The question then arises, how can we model such changes. The answer is by using the macroscopic balance, which is, as we saw in earlier chapters, the fundamental relation in engineering.

A macroscopic balance can be formulated for each extensive property.

If we use the symbol E to represent an extensive property, then the macroscopic balance states that the accumulation inside the system, represented by the time derivative $\overset{o}{E}$ (the ° means the time derivative, thus: $\overset{o}{E} = \dfrac{dE}{dt}$), is due to the overall result of:

- conversions, P_E;

- transfer over the boundary, ϕ_E.

This is represented diagrammatically in Figure 8.5.

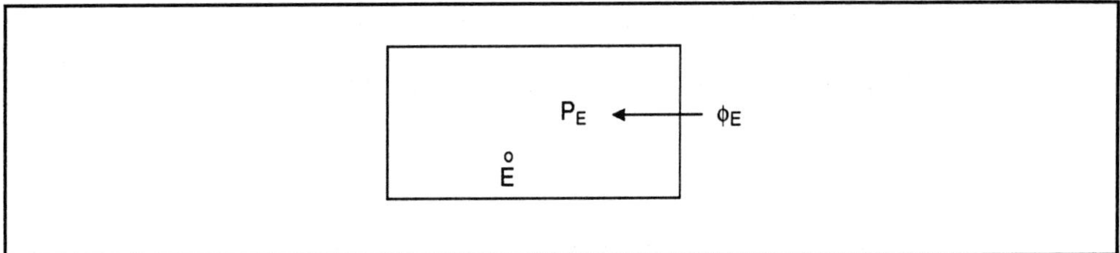

Figure 8.5 The macroscopic balance for an extensive property

In words we can describe this as:

Accummulation = net transfer plus net conversion

Mathematically, this becomes:

$$\overset{o}{E} = \phi_E + P_E \quad \text{or} \quad \frac{dE}{dt} = \phi_E + P_E$$

$$(E - 8.1)$$

If the system boundary is chosen such that the content is ideally mixed then we may formulate Equation 8.1.

macroscopic balance is principal tool of engineering

As we have learnt in earlier chapters, the importance of the macroscopic balance cannot be valued too highly. It is the principal tool of engineering.

8.1.5 Observations

In order to use Equation 8.1 one needs for each extensive property (eg moles of substrate in the system):

- a kinetic relation for net-conversion of substrate;

- a kinetic relation for the net-transfer of substrate.

The kinetic relations for conversion, P_E, are specific for the micro-organism or enzyme under consideration and are determined mainly by the time evolution of the system. For example as biomass increases with time, the rate of production or removal of an extensive property increases.

The kinetic relations for transfer, ϕ_E, are specific for the reactor and are determined mainly by the spatial evolution of the system.

Now we have available all the basic tools to start the mathematical modelling of complex biological processes. Attempt the two SAQs before moving on to Chapter 9.

<table>
<tr><td>

SAQ 8.4

</td><td>

Let us examine your knowledge of the terms and concepts raised in this brief chapter.

1) A system in which only energy is exchanged with the surroundings is called a [   ] system.

2) Are intensive properties additive?

3) We have a small scale biochemical experiment. Its total volume is 1 ml. The reactants are used in concentrations of 0.1 mol l^{-1}. The progress of the reaction is monitored taking 5 µl portions. Is a macroscopic approach justified in analysing this reaction?

4) For which of the following can we formulate macroscopic balances: temperature, volume, energy, mass?

5) Use a macrobalance approach to formulate a relationship between accummulation, net conversion and net transfer of a chemical in a closed system.

</td></tr>
<tr><td>

SAQ 8.5

</td><td>

Formulate the macroscopic balance for ethanol in a vessel where ethanol is chemically produced at a rate of 1 kg h^{-1} and where there occurs stripping of ethanol into sparged air at a rate of 0.2 kg h^{-1}. Finally calculate the increase in ethanol inside the vessel after 10 hours.

</td></tr>
</table>

Summary and objectives

In this chapter, we have introduced the rationale behind the modelling of biological processes, especially relating to the growth of micro-organisms. We have also defined various types of systems, and introduced the terms extensive and intensive properties and outlined the macroscopic approach to modelling.

Now that you have completed this chapter you should be able to:

- identify open, closed and isolated systems;

- identify circumstances in which a macroscopic approach is or is not allowed;

- identify intensive and extensive properties;

- set up simple macroscopic balances and calculate values from supplied data.

Modelling of microbiological processes and kinetic conversion rates

Modelling of microbiological processes and kinetic conversion rates

9.1 The macroscopic description of microbiological processes

9.1.1 System definition

As we described in the previous chapter, microbiological processes occur generally in open systems ie open to exchanges of energy and mass across boundaries. However, the temperature and pressure in bioreactors are often fairly homogeneous and constant.

This means that a full system definition requires the specification of:

*   the volume of the system;

*   all chemical compounds which are somehow reacting, (For the arguments discussed below we will assume that about 800 compounds are involved - this could well be a gross underestimation).

Furthermore for each chemical compound one needs relations for the:

*   kinetics for chemical conversion;

*   kinetics for transfer over the system boundary.

9.1.2 The problem of large numbers of model parameters

If one considers the above requirements of a system definition in the light of the large number (eg 800) of chemical compounds which can be found in microbial systems then there arises a problem. A mathematical model which contains 2 x 800 kinetic relations for conversion and transport is obviously of no use. The number of model parameters is so large (about 1600) that it is impossible to obtain numerical values by parameter fitting of the model upon the actual data or to perform enough experiments to determine these parameters.

∏   How can we reduce such a complex system to one which is manageable?

The answer is that one needs a second round of approximations.

These approximations apply to:

*   reduction of the required number of model parameters in the conversion kinetics (Section 9.1.3);

*   reduction in the number of required model parameters in the transfer kinetics (Section 9.1.4).

∏   Can you remember what was the first round of approximations?

The first approximation is that we can use a macroscopic approach to represent the system (Chapter 8), that is, we use average values.

9.1.3 A general approach to reduce the number of model parameters in conversion kinetics

As stated above, there is a need to minimise the number of model parameters. Several possibilities exist to achieve this:

- pseudo-steady state assumptions;

- definition of only the relevant compounds;

- conservation principles;

- definition of reactions which occur.

In the next sections we will examine these possibilities, but first we will need to define the system description.

9.2 A system description of microbial growth

Let us consider microbial cells (eg *Escherichia coli*) which grow aerobically in a reactor.

$\prod$ Write down a list of items that you think need to be specified.

intracellular and intracellular parameters

According to Section 9.1.1, we need to specify the reactor volume and all of the chemical compounds involved. You may have concluded that the micro-organism uses a substrate (glucose), oxygen (O_2) and ammonia (NH_3), (these we can call inputs). The organism uses these inputs to produce biomass, CO_2, H_2O and secondary products. so we need to include these.

Intracellularly we arrive at a number of about 800 chemical compounds which play some role in the metabolism (eg glucose 6-P, pyruvate, glyceraldehyde-P, acetyl CoA, 21 amino acids, 4 nucleotides, numerous sugars and lipids etc). Inputs are converted to these products. Thus in principle there are a very large number of items that need to be specified.

The processes which occur inside the cells can be represented in a box which represents the system boundaries (Figure 9.1).

Glucose, O_2 and NH_3 enter the micro-organism and, through a large number ($I_1 \rightarrow I_{800}$) of intermediate compounds, are converted into H_2O, CO_2 biomass and products. In summary this system definition contains 807 relevant compounds (ie 3 inputs, 4 products and 800 intermediates).

Even this large number is an underestimation because we have not considered the input of a large number of mineral salts (Mg^{2+}, Ca^{2+}, K^+, PO_4^{3-}, SO_4^{2-} etc). We will, however continue by not including these for the moment.

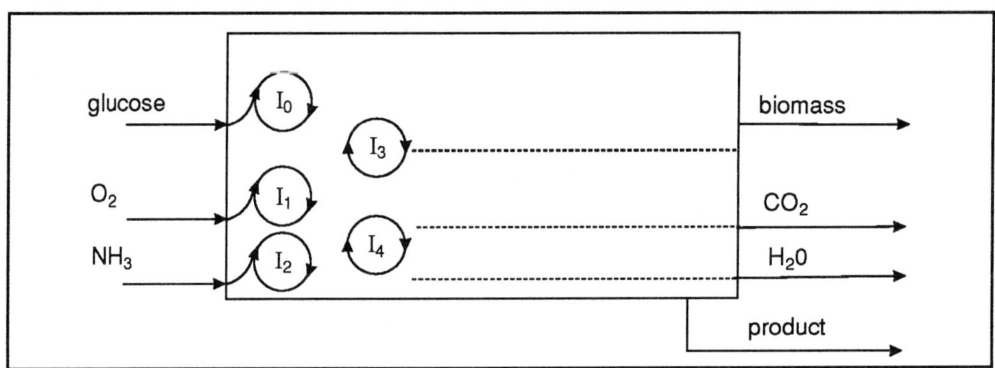

Figure 9.1 System definition of aerobic microbial growth with product formation. I_0, I_1, I_2 are intermediates.

9.3 The macroscopic balances for the defined chemical compounds

In order to describe this defined system one needs a macroscopic balance for each defined chemical compound. Hence, because there are $800 + 7 = 807$ compounds, we arrive at 807 macroscopic balances.

In order to describe this system exactly, one would need at least 807 kinetic relations for conversion and 807 kinetic relations for transport for 800 intermediates and 7 extracellular compounds (Biomass, H_2O, substrate, N-source, CO_2, O_2, product).

Now let us write a macroscopic balance for such a compound i where $i = 1 \rightarrow 807$

Change in the amount of the compound in the system is given by:

$$\frac{dC_iV}{dt} = \frac{C_idV}{dt} + \frac{VdC_i}{dt}$$

But this must be the result of the transport of the compound into the system and its synthesis within the system.

Thus $C_i\dfrac{dV}{dt} + \dfrac{VdC_i}{dt} = \phi_{A,i} V + r_{Ai}V \ldots\ldots$

$$(E - 9.1)$$

where:

V = volume of the system (m^3).

C_i = concentration (k mol m^{-3}).

r_{Ai} = the volume specific net conversion rate of compound i. The term net conversion rate stresses the fact that in general a compound may be involved in multiple reactions where production and/or consumption of the compound occurs. Thus r_{Ai} is the net result of the conversion in all of these processes per unit volume of the system.

ϕ_{Ai} = the net transfer of the compound per unit volume of the system.

We would like you to note one further convention. We could write Equation 9.1 as:

$$C_i \overset{o}{V} + V\overset{o}{C_i} = r_{Ai} V + \phi_{Ai} V$$

$\overset{o}{C_i}$ and $\overset{o}{V}$ are the time derivatives of C_i and V. Thus $\overset{o}{C_i}$ is another form of writing $\dfrac{dC_i}{dt}$ and $\overset{o}{V} = \dfrac{dV}{dt}$

Note we can also write $\dfrac{dC_i V}{dt}$ for $\dfrac{C_i dV}{dt} + \dfrac{V dC_i}{dt}$

The convention is that substrates are assigned a negative value and products a positive value of the net-conversion rate r_{Ai} described in Equation 9.1.

∏ Write down an explanation as to why products and substrates are given opposite signs.

The sign reflects the fact that the production of a metabolite represent the formation of the product and is positive. By definition therefore, the utilisation of substrates, which leads to their disappearance, is negative.

Equation 9.2 below can be written, by dividing Equation 9.1 by V, a condition which would be valid in a constant volume system.

Thus:

$$\overset{o}{C_i} + C_i \frac{\overset{o}{V}}{V} = r_{Ai} + \phi_{Ai} \left(\text{or } \frac{dC_i}{dt} + \frac{C_i}{V}\frac{dV}{dt} = r_{Ai} + \phi_{Ai} \right) \qquad (E - 9.2)$$

| SAQ 9.1 | A microbial culture uses $4gl^{-1}$ glucose in 4 hours and increases its biomass concentration by $2gl^{-1}$. What are the net-conversion rates of substrate and biomass (per hour). |

It is now useful to draw up a separate set of macroscopic balances for the intracellular compounds, which are complementary to the macroscopic balances of the extracellular compounds.

9.4 The macroscopic balance for the intracellular compounds

It is useful to derive the macroscopic balances for chemical intermediates inside the biomass (of which there are 800 out of 807).

Before we derive these balances, we should first introduce some further conventions. So far we have given the term concentration the symbol C. It is conventional to assign biomass the symbol X.

Using these conventions then we could represent the concentration of biomass as C_x.

It is often convenient to think of compounds in terms of concentration relation to biomass concentration. Thus for compound i we could write $\frac{C_i}{C_x}$ to represent the concentration of i in relation to the concentration of biomass.

biomass specific concentration

The term $\frac{C_i}{C_x}$ is often written as X_i. Thus X_i simply implies the concentration of i in relation to biomass concentration. It is usual to define biomass in terms of moles of carbon. Thus X_i has the units mol/C-mol biomass. We can regard X_i as the biomass specific concentration of i.

biomass specific set conversion

Also we have previously used the symbol $r_{A,i}$ to represent the volume specific net conversion rates. In discussing biomass we more frequently think in terms of specific net conversion rates per unit of biomass. We can define the term q_{Ai} as the biomass specific unit conversion rate $= \frac{r_{Ai}}{C_x}$. It is more frequent in biological literature to refer to this as the metabolic quotient (hence q) for compound i. q_{Ai} therefore represents the rate of conversion per unit biomass concentration (units are mol/Cmol biomass h).

biomass specific net transfer rate

Finally we have defined the net transfer rate ϕ_{Ai}; by analogy we can define the biomass specific net transfer rate over the cell membrane as $\frac{\phi_{Ai}}{C_x}$. (This we can give the symbol ψ_{Ai} and it usually has the units of mol/C-mol biomass h).

We can write the specific growth rate (μ) as the rate of increase in biomass per unit of biomass. If we define biomass in terms of the volume of the biomass V then as

$$\mu = \frac{\frac{dV}{dt}}{V} = \frac{\overset{o}{V}}{V}$$

We can now write a macroscopic balance for an intermediate i as:

$$\overset{o}{X_i} + X_i\,\mu = q_{Ai} + \psi_{Ai} \tag{E - 9.3}$$

You might like to derive this yourself. Use Equation 9.2 and divide through by C_x. Then substitute in X_i, q_{Ai} and ψ_{Ai}. Equation 9.2 also remains valid for extracellular compounds.

9.5 The pseudo-steady state approximation for the intracellular compounds

It should be realised that q_{Ai} is the net result of production processes and consumption processes of intermediates inside the cell. Examples of important intracellular intermediates are:

- ATP;

- $NADH_2$;

- precursors for biomass synthesis (eg amino acids).

In the case of ATP and NADH$_2$, these are produced linked to the catabolic breakdown of organic substrates. They are consumed during the synthesis of new cell constituents (anabolism). Thus anabolic metabolism consumes these intermediates, catabolic metabolism produces them.

With respect to ATP one can distinguish for example the anabolic consumption of ATP, q_{ATP}^{ana}, which leads to biomass and the catabolic production of ATP, q_{ATP}^{cat}, which originates from substrate combustion. For the net-conversion of ATP we can write:

$$q_{A,ATP} = q_{ATP}^{cat} + q_{ATP}^{ana}$$

∏ Write down answers to the following:

1) is ATP passed into and out of the cell;

2) is the concentration of ATP inside of a cell high;

3) does the concentration of ATP inside of a cell fluctuate greatly with time?

The answer to each of these questions is no. The same is true for many other biochemical intermediates.

Thus from biochemical and microbial knowledge the following facts are apparent for biochemical intermediates:

- there is no mass transfer over the system (=biomass) boundary (the system is closed for intermediates). Hence: $\psi_{Ai} = 0$

- the concentration of intermediates is generally very low 10^{-3} - 10^{-5} mol l^{-1}, hence $X_i \approx 0$ (a dilute system);

- the concentration of intermediates is mostly fairly constant due to the fact that micro-organisms tend to maintain a constant internal environment (homeostasis, ie there is no net accumulation).

This leads to the pseudo-steady state assumption (Equation 9.4).

Pseudo-steady state assumption

Since the composition of the cells remain constant, then $\dfrac{C_i}{C_x}$ = constant. But $\dfrac{C_i}{C_x} = X_i$ and

$$\overset{o}{X_i} = \frac{dX_i}{dt}. \text{ Thus } \frac{dX_i}{dt} = 0$$

pseudosteady
state
assumption

Thus we can represent the pseudosteady state assumption by:

$$\overset{o}{X_i} = 0 \qquad\qquad\qquad\qquad\qquad (E - 9.4)$$

Remember $\overset{\circ}{X}_i$ is a time dependent parameter. The combination of these assumptions, together with Equation 9.3 leads then to the important result that Equation 9.5 holds for intracellular intermediates only, ie the sum of catabolic and anabolic reactions is zero.

$$q_{Ai} = \frac{r_{Ai}}{C_x} = 0$$

$$(E - 9.5)$$

Summarising, the pseudo-steady state assumption of microbial intracellular intermediates leads to the conclusion that in biochemical terms:

- $\overset{\circ}{X}_i = 0$, which means that the chemical composition of the biomass is time independent, and thus is constant.

- we can neglect all intracellular intermediates because it appears that:

$$r_{Ai} = C_x * q_{Ai} = 0 \text{ and } \phi_{Ai} = C_x * \psi_{Ai} = 0$$

black box approach

This approximation of pseudo-steady state is called the black box approach for the obvious reason that all intermediates have been neglected, ie we have treated what is going on inside of the cells as unseen and unknown. The only chemical compounds left in Figure 9.1 of importance are the compounds which enter and leave the biomass box (ie inputs and products).

It should be clear, now, that the pseudo-steady state assumption is a most powerful method to reduce the number of parameters.

In fact one has eliminated 2 x 800 parameters.

The question may arise why one uses the term pseudo-steady state and not steady state. The reason is that in many microbial systems concentration changes do occur, but the time interval over which the change occurs is small and is described in Figure 9.2.

We will illustrate this with a specific example.

A cell may have a single copy of its genetic material (DNA). Thus a cell contains a particular amount of DNA. Before the cell can divide, however, an extra copy of the DNA needs to be made (ie the cell now contains two copies of the DNA). At cell division, each daughter cell receives a single copy of the DNA. Average amounts of DNA per cell for the culture appears to be more-or-less constant (ie in a steady state) when in reality, the content of DNA in each cell oscillates between 1 and 2 copies. Such a system is of course not in a steady state but in a pseudo-steady state.

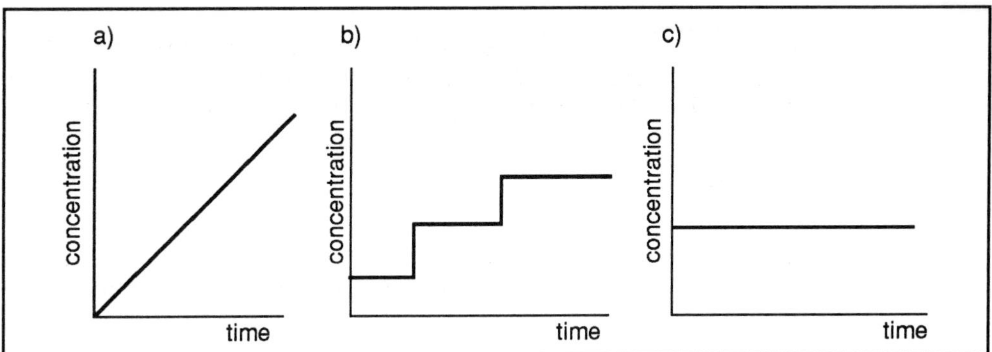

Figure 9.2 Dynamic (a), pseudo-steady state (b), steady-state situations (c). See text for details.

∏ For an aerobic micro-organism the specific growth-rate μ changes linearly over time from $0.5 \rightarrow 0.1\ h^{-1}$ in 4 hours. This leads to a decrease of the intracellular ATP-concentration from 5mmol/C-mol-biomass to 1 mmol/C-mol-biomass.

The ATP needed for biomass synthesis from glucose (anabolism) is 2.7 mol ATP/C-mol-biomass. This ATP is generated from glucose metabolism (catabolism).

Calculate the required anabolic ATP generation per C-mol biomass per h at a growth rate of $0.3\ h^{-1}$ (which occurs in the middle of the above time interval). Use this to calculate the required catabolic ATP generation for the 2 cases: when the pseudo-steady state is used and when it is not used.

The answers we should come to are using none pseudo-steady state assumptions, the required ATP production is $8099 * 10^{-4}$ mol ATP/C-mol biomass h. Using pseudo-steady state we calculate the rate to be $8100 * 10^{-4}$ mol ATP/C-mol biomass.h. Do not worry if at this stage you could not do this, we will show you how to do it. It is the final conclusion we reach which is important.

The way we calculated this is as follows:

We used Equation 9.3.

The change of ATP with time is:

X_{ATP} (at time 0) = 5 mmol/C-mol biomass

X_{ATP} (4 hours later) = 1 mmol/C-mol biomass

Therefore:

$$\overset{o}{X}_{ATP} = \frac{-4 * 10^{-3}}{4} = -10^{-3}\ \text{mol ATP/C-mol biomass h}$$

and at $\mu = 0.3\ h^{-1}$: $X_{ATP} = 3 * 10^{-3}$ mol ATP/C-mol biomass

Furthermore $\psi_{ATP} = 0$ (intracellular)

From Equation 9.3 we thus obtain the net-conversion rate of ATP in the following way.

Since $\overset{o}{X} + X_i * \mu = q_{Ai} + \psi_{Ai}$ and in this case $\psi_{Ai} = 0$

We can write:

$$\overset{o}{X}_{ATP} + X_{ATP} * \mu = q_{A,ATP}$$

Thus:

$$q_{A, ATP} = -10^3 + 3 * 10^3 * 0.3 = -1.0 \times 10^4 \text{ mol ATP/C-mole biomass.h}$$

$q_{A, ATP}$ is the net-conversion rate of ATP, which is obviously the net result of anabolic ATP-consumption for biosynthesis and ATP-production from glucose catabolism.

Hence: $q_{A,ATP} = q_{ATP}^{cat} + q_{ATP}^{ana}$

Remember that q_{ATP}^{ana} is the rate of conversion of ATP in relation to the biomass concentration $(= \frac{r_{A,ATP}}{C_x})$.

At $\mu = 0.3$ h^{-1} the ATP-consumption for anabolism is equal to $q_{ATP}^{ana} = -0.3 * 2.7$ mol ATP/C-mol h $= -8100 * 10^4$ mol ATP/C-mol.h (because it was specified that there is an ATP-need of 2.7 mol ATP/C-mol for biomass synthesis).

Since $q_{A, ATP} = -1.0 * 10^4$ mol/C-mol biomass this leads to:

$$q_{ATP}^{cat} = -1.0 * 10^4 + 8100 * 10^4 = 8099 * 10^4 \text{ mol ATP/C-mol biomass.h}$$

Thus we have calculated the rate of ATP generation in this system $= 8099 * 10^4$ mol ATP/C-mol biomass h.

But what would we have calculated it at, if we had assumed a pseudo-steady state?

For a pseudo-steady state we would have written $q_{A, ATP} = 0$. Thus:

if the pseudo-steady state assumption is followed, $(q_{A,ATP} = 0)$, then $q_{ATP}^{cat} = 8100 * 10^4$ mol ATP/C-mol biomass h.

What you have shown by this calculation is that the pseudo-steady state assumption leads to a negligible error ($< 1\%$) and is therefore allowed.

9.6 Definition of the relevant extracellular chemical compounds in a black box description of microbial growth

In the preceding section it has been shown that in general it is justified to simplify microbial systems enormously by the pseudo-steady state assumption for the biochemical intermediates (ie a black box description).

identification of
relevant
compounds

This means that one can limit the specification of the chemical compounds to those which enter and leave the micro-organism. The question then arises what are the relevant compounds.

Some simple considerations will lead to a short list. First it is obvious that biomass is made up for 90% of C, H, O and N with only minor amounts of S, P and other elements (trace elements). This means that one has to at least specify:

- a N-source eg NH_3, HNO_3, N_2 or organic N-compound;

- a C-source eg CO_2 or an organic compound;

- a H- and O-source eg H_2O;

- biomass.

This leads to a set of 4 chemical compounds purely from an elemental point of view (C, H, O and N). Secondly, it is well known that micro-organisms grow on simple compounds, which can cross the cell membrane. These compounds are in general present at low concentrations, (eg 10^{-3} mol carbon 1^{-1}), and are then polymerised into a concentrated product which appears in biomass.

For example wet biomass contains in general about 90% water and 10% dry matter of which about 50% is carbon. This is equivalent to a concentration of 4.16 mol C l^{-1}.

Now let us perform some thermodynamic calculations.

The process of biomass synthesis requires polymerisation and concentration and can be represented by the following simple equation, where A can be thought of as biomass precursor and A_2 as biomass:

$$2 * A \rightarrow A_2 + H_2O \qquad\qquad \Delta G_R^o = +10 \text{ kJ}$$

Typical concentration values for A and A_2 are:

$$A = 10^3 \text{ mol } l^{-1} \qquad\qquad A_2 = 4.16 \text{ mol } l^{-1}$$

The polymerisation with H_2O release in an aqueous environment is in itself already thermodynamically unfavourable, with a typical positive standard Gibbs free energy of reaction, $\Delta G_R^o = + 10$ kJ (a negative ΔG_R^o will lead to a large positive value for the equilibrium constant for the reaction). The actual free-enthalpy difference is even greater due to concentration differences. Using the well known relation:

$$\Delta G_i = \Delta G_i^o + RT\ln C_i \qquad\qquad (E\ 9.6)$$

we find for the actual value of the free Gibbs energy of reaction ΔG_R:

$$\Delta G_R = \Delta G_R^o + RT\ln \frac{[A_2]}{[A]^2} \qquad\qquad (E\ 9.7)$$

Substitution of the concentrations given above at $T = 25°C$ and $\Delta G_R^o = 10kJ$ leads to:

$$\Delta G_R = +10 + 8.314 \times 10^{-3} \times 298 \times \ln\left[\frac{4.16}{(10^{-3})^2}\right] = +10 + 37.8 = +47.8 \text{ kJ}$$

It would appear, therefore, thermodynamically the process of biomass synthesis from precursor is completely to the left and cannot proceed to any extent as such!

Thermodynamically this process of polymerisation and concentration increase, therefore requires a net input of Gibbs free energy in order to push the equilibrium to the right (biomass synthesis). Thus an energy input is required. This means that in each microbial system one needs a source of Gibbs free energy.

electron donors and acceptors

The most general source of this free energy is the transfer of electrons from a donor couple to an acceptor couple. This means that, besides the 4 compounds already mentioned (an N-source; a C-source; H and O-source and biomass), one needs to specify 4 redox compounds. (You might ask, why 4? We will explain this in the following way). Consider a compound A that can be oxidised. As it is oxidised, something else (B) is reduced. Thus A_{red} = electron donor, B_{ox} = electron acceptor.

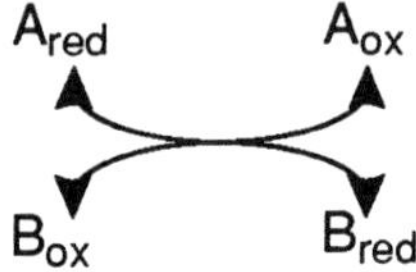

For the transfer of electrons one therefore may need to specify up to 4 chemical compounds because an electron donor is converted into a conjugated electron acceptor. The same holds for an electron acceptor which is converted into a conjugated electron donor.

So in general the following relevant compounds are required for microbial growth systems:

- C-source;

- N-source;

- H_2O;

- biomass;

- electron donor/conjugated acceptor couple;

- electron acceptor/conjugated donor couple.

We have represented these in Figure 9.3.

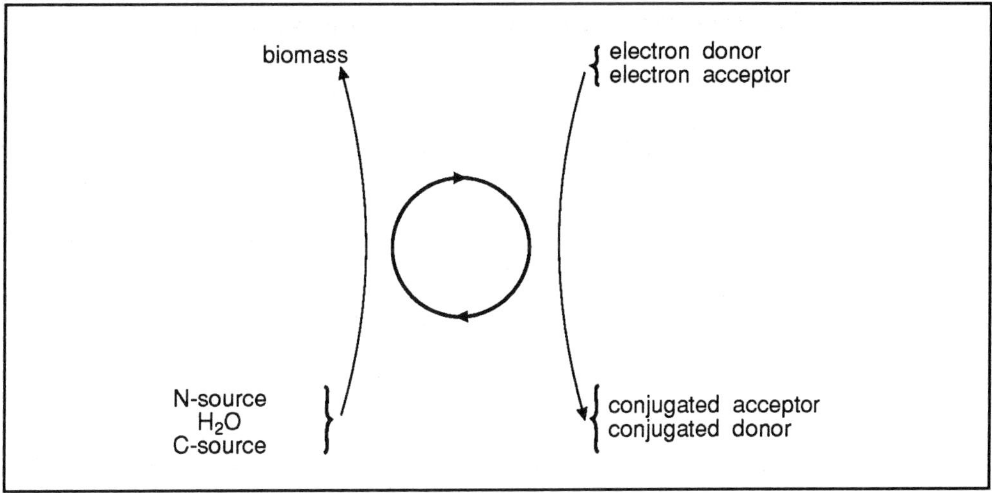

Figure 9.3 Relevant compounds needed for microbial growth by conversion of simple inputs to biomass.

In using this list, it should be realised that some compounds can have multiple functions. For example:

- HNO_3 can be not only a N-source but also an electron acceptor which is reduced to N_2;

- H_2O is also the conjugated donor for O_2 as electron acceptor;

- Organic compounds can function as C-source and electron donor.

In practise it will be observed that a list of 6-8 compounds is sufficient to describe the black box growth system. In Table 9.1 we have defined these and the symbols used to represent them. We have also included the symbols used to represent their net-conversion rates, net transfer and their concentrations.

Relevant compound	Symbol	Net-conversion rate	Net-transfer rate	Concentration
biomass	x	r_{Ax}	ϕ_{Ax}	C_x
substrate	s	r_{As}	ϕ_{As}	C_s
product	p	r_{Ap}	ϕ_{Ap}	C_p
O_2	o	r_{Ao}	ϕ_{Ao}	C_o
H_2O	w	r_{Aw}	ϕ_{Aw}	C_w
CO_2	c	r_{Ac}	ϕ_{Ac}	C_c
N-source	n	r_{An}	ϕ_{An}	C_n

Table 9.1 System definition, aerobic microbial growth and product formation. Note that in this case substrate acts as both carbon source and electron donor.

Using the information provided in Table 9.1 see if you can write down the macroscopic balances for each of these compounds.

Remember that the total amount of a compound in a system at any time = concentration of the compound . the volume of the system, ie = C_iV.

Remember also that the macroscopic balance includes both the net conversion rate ($r_{Ai}V$) and the input ($\phi_{Ai}V$) where ϕ_{Ai} is the net transfer ratio.

Thus $\overset{o}{C_i}V = r_{Ai}V + V\phi_{Ai}$ (see Equation 9.1).

Thus you should have written the macrobalances for each of the components above as:

$$(\overset{o}{C_x}V) = V\,r_{Ax} + V\,\phi_{Ax}$$
$$(\overset{o}{C_s}V) = V\,r_{As} + V\,\phi_{As}$$
$$(\overset{o}{C_p}V) = V\,r_{Ap} + V\,\phi_{Ap}$$
$$(\overset{o}{C_o}V) = V\,r_{Ao} + V\,\phi_{Ao}$$
$$(\overset{o}{C_w}V) = V\,r_{Aw} + V\,\phi_{Aw}$$
$$(\overset{o}{C_c}V) = V\,r_{Ac} + V\,\phi_{Ac}$$
$$(\overset{o}{C_n}V) = V\,r_{An} + V\,\phi_{An}$$

In setting up such a system definition of microbial growth it is of the utmost importance that one checks extensively whether all relevant compounds are indeed specified. This of course depends on the nature of the organism's metabolism.

application of elemental balances

This can be done simply by measuring all compounds and then draw up the elemental balances (we will deal with this in the next section in a little more detail). If the elemental balances fit, then one can be confident that all relevant compounds are indeed identified.

It should be realised that this section about relevant compounds is only concerned with the macro-chemical compounds. It is assumed that the micro-organism has sufficient access to micro-compounds such as vitamins, hormones, trace elements. Micro-compounds are those which are required in small amounts.

Now, based on the above considerations a system description for growth and product formation can be produced. We shall examine more fully how to do this in the next section. Attempt SAQ 9.2 before reading the next section.

SAQ 9.2

Draw up a table of relevant compounds for the following microbial growth systems, assign the proper function (like C- or N- source, electron donor, conjugated acceptor, electron acceptor, conjugated donor) and provide the total number of relevant compounds. (Use Table 9.1 to help you).

1) aerobic growth on glucose and NH_3;

2) aerobic growth on sulfide and HNO_3;

3) denitrification on glucose and NH_3;

If you have difficulty with this question because you do not have a lot of microbiological experience, you should find the response helpful.

9.7 Application of conservation principles

In the preceding sections it has been shown that in a microbial system one only needs to consider the inputs and products. Nevertheless this still amounts to a large number (6-8) of relevant compounds.

conservation principles

It is possible to reduce the number of parameters even further by the application of conservation principles. These are:

* conservation of mass;

* conservation of elements;

* conservation of electric charge;

* conservation of energy (1st law of Thermodynamics).

For illustrative purposes we will elucidate here the use of elemental conservation in relation to Figure 9.4. If we neglect the formation of product for the sake of simplicity, it appears that this microbial growth system contains 6 compounds (Figure 9.3, Table 9.2).

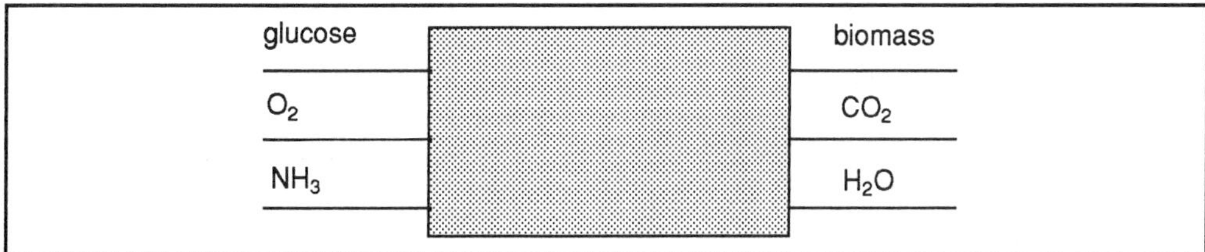

Figure 9.4 Black box description of aerobic microbial growth (system definition is given in Table 9.2).

As displayed these compounds contain only 4 elements C, H, O, N.

This means that we may formulate 4 elemental conservation relations. Table 9.2 contains the relevant chemical compounds and their composition.

Relevant compounds	Symbol	Net-conversion rate	Net-transfer rate	Composition
Biomass	x	r_{Ax}	ϕ_{Ax}	$CH_{1.8}O_{0.5}N_{0.2}$
Glucose	s	r_{As}	ϕ_{As}	CH_2O
O_2	o	r_{Ao}	ϕ_{Ao}	O_2
CO_2	c	r_{Ac}	ϕ_{Ac}	CO_2
H_2O	w	r_{Aw}	ϕ_{Aw}	H_2O
NH_3	n	r_{An}	ϕ_{An}	NH_3

Table 9.2 Relevant compounds for aerobic growth on glucose with NH_3 as N-source and with no product formation as described by Figure 9.4.

Some remarks are applicable to Table 9.2:

- carbonaceous compounds are defined in the 1-C-mole convention; hence glucose $C_6H_{12}O_6 \rightarrow CH_2O$;

- the biomass composition is assumed to be constant. This appears in general a reasonable approximation and in accordance with the black box approach we are using here.

conservation relations
Using Table 9.2 the following elemental conservation relations can be formulated for the net-conversion rates only.

$$\text{C-conservation: } 1\,r_{Ax} + 1\,r_{As} + 1\,r_{Ac} = 0 \qquad \text{(E - 9.8a)}$$

$$\text{H-conservation: } 1.8\,r_{Ax} + 2\,r_{As} + 2\,r_{Aw} + 3\,r_{An} = 0 \qquad \text{(E - 9.8b)}$$

$$\text{O-conservation: } 0.5\,r_{Ax} + r_{As} + 2\,r_{Ao} + 2\,r_{Ac} + r_{Aw} = 0 \qquad \text{(E - 9.8c)}$$

$$\text{N-conservation: } 0.2\,r_{Ax} + r_{An} = 0 \qquad \text{(E - 9.8d)}$$

These relations are obvious because they simply use the fact that there is no net conversion of elements. It should also be clear that the numbers which occur in the above relations are determined by the elemental compositions shown in Table 9.2. Hence the biomass conversion r_{Ax} has a coefficient of:

1 in the C-equation;

1.8 in the H-equation;

0.5 in the O-equation;

0.2 in the N-equation.

For C, H, O, N in the substrate conversion (glucose, $C_6H_{12}O_6 \cong CH_2O$), the values are 1,2,1 and 0 respectively and these values are used as the multipliers of r_{As} in each conservation Equation (9.8 a-d).

Now the elemental conservation relations show 4 relations between 6 r_{Ai}'s. Mathematically this means that only 2 net-conversion rates are independent while the remaining 4 can be related to these 2.

In general terms, we can say that for a microbial system with a total of n relevant compounds, involving k elements, the degree of freedom d_c follows Equation 9.9.

$$d_c = n - k \qquad \text{(E - 9.9)}$$

One can choose the independent rates at will, but the choice is generally based on convenience.

For example water (r_{Aw}) is very difficult to measure. Biomass is an important product and the substrate is often an expensive compound. Hence in general one chooses r_{Ax} and r_{As} as the independent rates.

After some manipulations one obtains the following relation from 9.8 a-d:

$$r_{Ac} = - r_{Ax} - r_{As} \qquad\qquad\qquad\qquad \text{(E - 9.10a)}$$

$$r_{An} = - 0.2\, r_{Ax} \qquad\qquad\qquad\qquad \text{(E - 9.10b)}$$

$$r_{Ao} = + 1.05\, r_{Ax} + r_{As} \qquad\qquad\qquad\qquad \text{(E - 9.10c)}$$

$$r_{Aw} = - 0.6\, r_{Ax} - r_{As} \qquad\qquad\qquad\qquad \text{(E - 9.10d)}$$

Thus the net-conversion rates for 4 components are expressed in terms of the rates for 2 other, selected variables.

∏ It would provide useful experience if you derived Equations 9.10a-d for yourself using Equation 9.8 a-d.

(It might be fun to see if you can derive all of these yourself).

In principle, therefore, if we measure r_{As} (rate of substrate use) and r_{Ax} (rate of biomass production) and we know the elemental composition of the biomass and substrates, we can calculate the other values.

One should remember that compounds which disappear (like substrate, O_2, N-source) have negative rates, while the compounds which appear (r_{Ax}, r_{Ac}, r_{Aw}) have positive rates.

In conclusion we have found that the application of elemental conservation leads to reduction in the number of required model parameters.

Below we have devised some in-text activities and SAQ's to provide you with some experience in handling these equations and to demonstrate some of the applications of the elemental relationships described above. These have been designed to:

*   illustrate the effect of different chemicals in the system definition of Table 9.2 (SAQ 9.3) on the numerical coeffcents in Equations 9.10a-9.10d;

*   illustrate the effect of addition of other relevant elements like S and P (SAQ 9.4);

*   illustrate the effect of product formation (SAQ 9.5);

*   illustrate the use of elemental conservation as a check on the proposed microbial system definition (SAQ 9.6).

To help you, we have worked through an example for you.

∏ An aerobic microbial system shows a measured biomass production of 1 g l^{-1}h^{-1}. The N-source is NH_3 and the biomass composition is according to Table 9.2.

The substrate is glucose, which disappears with a rate of 1.83 g l^{-1}h^{-1}.

Calculate the oxygen consumption, the carbon dioxide production, the ammonia consumption and the water production for this culture.

Equations 9.10a-9.10d can be applied because the system definition was taken from Table 9.2. In order to use these relations one need to convert the provided rates to a C-molar basis. For biomass production the mass of 1-C-mol can be calculated from its composition.

Since its elemental formula is $CH_{1.8}O_{0.5}N_{0.2}$ then:

$1 \times 12 + 1.8 \times 1 + 0.5 \times 16 + 0.2 \times 14$g of cells contain 12 g (ie 1 mole) of C. Thus 24.6g of biomass contain 1 mole of C. Hence a biomass production of 1 $gl^{-1}h^{-1}$ leads to:

$r_{Ax} = 40.6 \times 10^{-3}$ C-mol $l^{-1}h^{-1}$

For r_{As} we find:

$$r_{As} = \frac{-1.83}{30} = -61 \times 10^{-3} \text{ C-mol } l^{-1} h^{-1} \text{ (NB 30g of glucose contains 1 mole of C)}.$$

Note that r_{As} is negative because of the convention to assign compounds which are consumed a negative value. Furthermore r_{As} and r_{Ax} are in C-mol units.

Application of Equation 9.10a-9.10d leads to:

$r_{Ac} = +20.4 \times 10^{-3}$ C-mol $l^{-1}h^{-1}$

$r_{An} = -8.12 \times 10^{-3}$ mol $l^{-1}h^{-1}$

$r_{Ao} = -18.37 \times 10^{-3}$ mol $l^{-1}h^{-1}$

$r_{Aw} = +36.64 \times 10^{-3}$ mol $l^{-1}h^{-1}$

SAQ 9.3	How do you expect that the elemental relations for the microbial system in Table 9.2 change and calculate this change when: 1) glucose is replaced by acetic acid; 2) glucose is replaced by ethanol; 3) NH_3 is replaced by HNO_3; 4) the biomass composition changes to $CH_{1.7}O_{0.8}N_{0.1}$.

SAQ 9.4	The biomass composition of Table 9.2 is only a reasonable approximation. It is well known that biomass contains also significant amounts of sulphur and phosphorous-compounds. A more precise composition is $CH_{1.8}O_{0.5}N_{0.2}S_{0.01}P_{0.02}$. In general the S and P comes from sulphate and phosphate in the medium. How does one modify Table 9.2 with respect to chemical compounds and how do the resulting elemental relations change (r_{Ao}, r_{Ac}, r_{Aw}, r_{An} as functions of r_{Ax} and r_{As}).

<table>
<tr><td>

SAQ 9.5

</td><td>

Suppose that the microbial system definition according to Table 9.2 is extended to include a product, which is ethanol.

The symbol for product is p, and its net-conversion rate is r_{Ap}.

1) How many independent rates can now be chosen?

2) Calculate the set of relations which is analogous to Equation 9.10a-d.

</td></tr>
<tr><td>

SAQ 9.6

</td><td>

Consider a microbial system where growth occurs on glucose as C and energy source and NH_3 as N-source. The biomass composition is according to Table 9.2.

One obtains the following measurements:

Biomass production = 1 kg organic dry matter h^{-1}

Substrate consumption = 3kg h^{-1}

O_2-consumption = 1 kg h^{-1}

CO_2-production = 1.85 kg h^{-1}

Check the elemental conservation relations and provide an indication if products other than biomass and CO_2 are made.

</td></tr>
</table>

From the experiences you have gained from the SAQs on the use of the conservation of elements you should have learnt that:

- a number of k extra relations are obtained (k = number of elements) between the net conversion rates r_{Ai}, which decreases the number of model parameters required considerably;

- the chemical composition of the defined relevant compounds (such as those in Table 9.2) in the microbial system definition determines the numerical value of the coefficients which appear in the elemental equations (in Equations 9.10a-d);

- the required conservation of elements can be used to check whether a proposed system definition is missing an unknown compound.

Nevertheless it may also have become clear that the necessary calculations are cumbersome. A generalised treatment will be offered in a later section which also will contain the conservation of energy and the balance of Gibbs free energy.

9.8 Application of metabolic information

In the preceding section the 'black box' approach has been demonstrated. It has been shown that due to the application of the pseudo-steady state approximation all intracellular intermediates can be neglected.

Nevertheless it often happens that there exists considerable knowledge about metabolic pathways. This knowledge can sometimes lead to extra relations between net-conversion rates and therefore to a decrease in the number of kinetic model parameters required.

A simple example is the anaerobic denitrification process.

According to SAQ 9.2 (3) this denitrification system contains 7 compounds (glucose, NH_3, biomass, H_2O, CO_2, N_2, HNO_3) and 4 elements (C, H, O, N). This means that there are 3 degrees of freedom. From the point of view of metabolic pathways it is however known that all HNO_3 is converted into N_2 - gas. This leads then to an extra relation which decreases the degrees of freedom from 3 to 2.

identification of reactions

The question is then how do we represent metabolic knowledge. The answer is that we specify a number of metabolic chemical reactions. If we return to the microbial growth system of Table 9.2 we can specify a set of reactions which are based on biochemical knowledge. After some deliberation with biochemists one arrives at the following 4 main reaction groups.

Reaction 1	Production of biomass precursors, which have the same composition as biomass, from the C- and N-source. This is called anabolism.
Reaction 2	Production of biomass from the precursors by polymerisation.
Reaction 3	Production of reducing equivalents by oxidation of the electron donor.
Reaction 4	Production of ATP by electron transport phosphorylation where the reducing equivalents react with the electron acceptor.

It is pointed out that these reactions:

• have to be extended to 5 reactions if a product is made in addition to biomass;

• are specific for each microbial system with respect to stoichiometric coefficients, which follow from specific biochemical knowledge.

Note also that the overall format of precursor synthesis; polymerisation; electron donor catabolism to $NADH_2$; electron acceptor coupled to ATP generation; and product reaction can be universally used.

For the microbial growth system of Table 9.2 we can now specify the reactions, described above. We have added a commentary to help you understand these.

Reaction 1 - Anabolism

First we will write out an equation and then explain how it has been produced and what it means.

$$- 1.095 \ CH_2O - 0.2 \ NH_3 - 0.051 \ ATP + \text{'}CH_{1.8} \ O_{0.5} \ N_{0.2}\text{'}$$

$$+ 0.095 \ CO_2 + 0.405 \ H_2O + 0.090 \ NADH_2 = 0$$

Note: '$CH_{1.8} O_{0.5} N_{0.2}$' represents biomass precursors.

- As a convention consumed reactants have negative stoichiometric coefficients, reactants produced have positive stoichiometric coefficients.

- In each reaction there can be assigned a stoichiometric coefficient of +1 to an arbitrarily chosen reactant. By convention we choose a stoichiometric coefficient of +1 for the main product.

- In each reaction the elemental balances must fit. Check eg that C, H, N, O are conserved.

- This reaction contains 4 elements (C, H, O, N). This means that at the most 4 + 1 stoichiometric coefficients are fixed. However in this reaction there are 7 coefficients. This means that 2 coefficients have had to be specified on the basis of biochemical knowledge.

Indeed, the CO_2-production (0.095) in this reaction is based on the decarboxylation reactions which occur in the pathways of synthesis of for example amino acids and other cellular building blocks.

Furthermore the ATP coefficient (-0.051) is calculated from the ATP-involvement in known pathways.

- This reaction contains 3 intracellular relevant intermediates (ATP, precursor, $NADH^2$) and 4 extracellular relevant compounds (glucose, NH_3, CO_2, H_2O).

Reaction 2 - Polymerisation

-1 '$CH_{1.8}O_{0.5}N_{0.2}$' - 1.7 ATP + 1 $CH_{1.8}O_{0.5}N_{0.2}$ = 0;

Note: '$CH_{1.8}O_{0.5}N_{0.2}$' is biomass precursors $CH_{1.8}O_{0.5}N_{0.2}$ is biomass

- biomass is the product, hence a coefficient of +1 is assigned to it;

- here the ATP-coefficient of - 1.7 is based on the fact that polymerisation from precursors at low concentration (typically values of 10^{3} mol l^{-1}) to a concentrated polymer (typically = 3 mol l^{-1}) needs Gibbs free energy input which is provided by ATP-hydrolysis.

The number 1.7 mol ATP/C-mol biomass is derived from the fact that generally about 15 g biomass can be synthesised from precursors per mol of ATP (consider that 1-C-mol biomass = 25 g dry biomass).

Reaction 3 - Oxidation of electron donor (glucose; CH₂O)

-0.5 CH_2O - 0.5 H_2O + 0.5 CO_2 + 0.33 ATP + 1 $NADH_2$ = 0.

- This reaction leads to a production of 'Gibbs free energy rich' electrons which are 'carried' by the biochemical electron carrier $NADH_2$. In stoichiometric terms $NADH_2$ is equivalent to H_2, but $NADH_2$ is the main product of this reaction and its stoichiometric coefficient is +1.

- From C, H and O-conservation we obtain the coefficient for substrate CH_2O, H_2O and CO_2. (We will do this as an in-text activity in a moment).

- The stoichiometric coefficient for ATP is due to the substrate phosphorylation from glucose in glycolysis (being 2 ATP/mol glucose = $\frac{2}{6}$ mol ATP/C-mol glucose).

Let us try using elemental conservation to calculate some stoichiometric coefficients.

Π Calculate the stoichiometric coefficients a_1, a_2, a_3 in the following reaction:
$a_1\, CH_2O + a_2 H_2O + a_3 CO_2 + H_2 = 0$

In the reaction there are 3 elements involved. Hence one may write out 3 elemental conservation relations, which allow the calculation of 3 stoichiometric coefficients:

C-conservation: $1\, a_1 + 1\, a_3 = 0$

H-conservation: $2\, a_1 + 2\, a_2 + 2 = 0$

O-conservation: $a_1 + a_2 + 2\, a_3 = 0$

Solving these relations leads to:

$a_1 = -0.5;\ a_2 = -0.5;\ a_3 = +0.5$

Reaction 4 - ATP production by transfer of electrons to electron acceptor (= O_2)

$-0.5\, NADH_2 - 0.25\, O_2 + 0.5\, H_2O + 1\, ATP = 0;$

- in this aerobic case O_2 is the electron acceptor and H_2O is the conjugated electron donor;

- the electrons from the carrier are transferred to O_2 and the released Gibbs free energy is conserved in ATP;

- the reaction contains 2 elements (H and O) and therefore only 2 coefficients can be calculated;

- the main product is ATP, hence this is given a coefficient of +1.

The amount of ATP is based on biochemical evidence, where a P/O ratio (=mol ATP/atom O) of 2 is assumed. This leads to a coefficient of -0.25 for O_2. The coefficients of H_2O and $NADH_2$ follow from the H and O conservation.

From the above considerations it follows that:

- the stoichiometric coefficients of CO_2 and ATP are based on biochemical arguments and are therefore specific for the micro-organism. In general their values depends on type of C-source; type of electron donor; type of electron acceptor and type of N-source.

The actual values of these stoichiometric coefficients should therefore be provided by biochemists/microbiologists who have knowledge about the biochemistry of the micro-organism under consideration.

You should note that:

- the remaining stoichiometric coefficients are calculated from elemental conservation which should fit for each reaction;

- the intracellular intermediate compounds should occur in at least 2 reactions. This is to ensure that their net-conversion rate is zero. (Or, more simply, a metabolic description with reactions where for example ATP is generated but where ATP is nowhere consumed cannot be correct);

- the extracellular compounds should occur at least in 1 reaction.

Table 9.3 summarises the proposed metabolic model description of the microbial system originally described in Table 9.2.

In this table these reactions have been assigned a rate of reaction r_i (eg mol $l^{-1}h^{-1}$) which can be given a set of dimensions.

This rate of reaction (in mol $l^{-1}h^{-1}$) is to be distinguished strictly from the net-conversion rate of a chemical compound r_{Ai}.

Reaction 1 -rate r_1

$- 1.095\ CH_2O - 0.2\ NH_3 - 0.051\ ATP + \text{'}CH_{1.8}\,O_{0.5}\,N_{0.2}\text{'}$

$+\ 0.095\ CO_2 + 0.405\ H_2O + 0.090\ NADH_2 = 0$

Reaction 2 - rate r_2

$- 1\ \text{'}CH_{1.8}\,O_{0.5}\,N_{0.2}\text{'} - 1.7\ ATP + 1\ CH_{1.8}\,O_{0.5}\,N_{0.2} = 0$

Reaction 3 - rate r_3

$- 0.5\ CH_2O - 0.5\ H_2O + 0.5\ CO_2 + 0.33\ ATP + 1\ NADH_2 = 0$

Reaction 4 - rate r_4

$- 0.5\ NADH_2 - 0.25\ O_2 + 0.5\ H_2O + 1\ ATP = 0$

Table 9.3 Metabolic description of the aerobic microbial growth system described in Table 9.2.

relation between r_{Ai} and r_i Of course there is the basic relation between r_{Ai} and r_i which states that for each compound involved in the reactions one can write that the net-conversion rate r_{Ai} is the sum of the conversions involved in all the reactions, ie $r_{Ai} = \Sigma\ r_{(1-4)i}$ in this case.

Table 9.3 contains 9 chemical components (CH_2O, NH_3, O_2, H_2O, CO_2, biomass and ATP, $NADH_2$, precursor). Therefore we can write down 9 relations. Furthermore the first 6 of the mentioned compounds are extracellular, while 3 are intracellular for which the pseudo-steady state condition holds such that $r_{Ai} = 0$ (Equation 9.5).

The 9 relations do follow straightforwardly from Table 9.3 by summing the rate for each component synthesised by the appropriate stoichiometric coefficient.

For example for substrate (CH_2O) which occurs in reaction equations r_1 and r_3.

r_{As} = (stoichiometric coefficient (CH_2O) in r_1) $* r_1$ + (stoichiometric coefficient (CH_2O) in r_3) $* r_3$

$= -1.095 r_1 - 0.5 r_3$

Note that the signs associated with the stoichiometric coefficients in the reactions are kept in the summations.

Thus using this approach we obtain:

substrate:	$r_{As} = -1.095\ r_1 - 0.5\ r_3$	(E - 9.11a)
biomass:	$r_{Ax} = +1\ r_2$	(E - 9.11b)
ammonia:	$r_{An} = -0.2\ r_1$	(E - 9.11c)
oxygen:	$r_{Ao} = -0.25\ r_4$	(E - 9.11d)
CO_2:	$r_{Ac} = +0.095\ r_1 + 0.5\ r_3$	(E - 9.11e)
water:	$r_{Aw} = +0.405\ r_1 - 0.5\ r_3 + 0.5\ r_4$	(E - 9.11f)
precursor:	$0 = r_1 - r_2$	(E - 9.11g)
$NADH_2$:	$0 = 0.09\ r_1 + r_3 - 0.5\ r_4$	(E - 9.11h)
ATP:	$0 = -0.051\ r_1 - 1.7\ r_2 + 0.33\ r_3 + 1\ r_4$	(E - 9.11 i)

This set of 9 relations contains 10 unknown rates (6 times r_{Ai} plus 4 times r_i). Hence it is now possible to relate all rates to 1 independent rate, for which we choose r_{As} (as conversion of substrate may be easy to measure in practice).

After some algebraic manipulations we are able to obtain Equations 9.12 a-e. These give the net conversion rate for each component relative to that of the substrate.

You might like to check these - although we will warn you that the algebra is a little cumbersome.

$$r_{Ax} = -0.7\ r_{As} \qquad (E - 9.12a)$$

$$r_{An} = +0.14\ r_{As} \qquad (E - 9.12b)$$

$$r_{Ao} = +0.26\ r_{As} \qquad (E - 9.12c)$$

$$r_{Ac} = -0.30\ r_{As} \qquad (E - 9.12d)$$

$$r_{Aw} = -0.58\ r_{As} \qquad (E - 9.12e)$$

$$r_1 = -0.70 \; r_{As} \tag{E - 9.12f}$$

$$r_2 = -0.70 \; r_{As} \tag{E - 9.12g}$$

$$r_3 = -1.06 \; r_{AS} \tag{E - 9.12h}$$

$$r_4 = -0.47 \; r_{As} \tag{E - 9.12 i}$$

If we compare this set of relations to those already available from our original black box conservation relations (Equations 9.10a-d) the following points emerge:

- compared to the black box description with 2 degrees of freedom, the metabolic description has 1 degree of freedom. All rates are related to r_{As}. The reason for this is that an extra restriction has been specified, the ATP-balance (Equation 9.11i).

- a complete kinetic description is now available if 1 kinetic relation is introduced (for example r_{As}).

- the metabolic description leads to one extra relation compared to the black box description (this is Equation 9.12a).

If Equation 9.12a is combined with the black box relations (Equation 9.10a - Equation 9.10d) one obtains (Equation 9.12b - Equation 9.12e). This is logical because elemental conservation is 'built in' in the specified stoichiometric coefficients in the reactions. Hence we may call Equation 9.12a a metabolic relation because it is the extra result of the metabolic description above the black box description.

It is clear that Equation 9.12a gives the numerical value of the biomass yield, which is then 0.70 C-mol biomass/C-mol glucose.

The calculated rates of reaction r_i (Equation 9.12f - Equation 9.12i) are all positive due to the sign convention of the reaction specification. (Note that r_{As} will be negative in all cases because substrate is used up, not produced).

The important message to take from this is that the formulation of a metabolic description is a very useful tool to generate from biochemical knowledge. It provides additional relationships between the net-conversion rates compared to that obtained by just using straight forward conservation relationships.

⫪ It would be a useful form of revision to draw up a list of the features of the use of elemental conservation relationships to describe the kinetics of a process and to compare this with the use of a metabolic description.

Let us now test our ability to apply what we have learnt about using metabolic descriptions.

<table>
<tr><td>

SAQ 9.7

</td><td>

Glucose can be fermented to ethanol ($CH_3O_{0.5}$) under anaerobic conditions. Glycerol ($CH_{2.66}O$) is a by-product.

Based on biochemical evidence the following metabolic reactions are postulated. The biomass composition is $CH_{1.8}O_{0.6}N_{0.20}$ and NH_3 is N-source.

Reaction 1 - anabolism of precursors

$- 1.095\ CH_2O - 0.2\ NH_3 - 0.051\ ATP$

$+ 0.19\ NADH_2 + 'CH_{1.8}O_{0.6}N_{0.2}' + 0.095\ CO_2 + 0.305\ H_2O = 0$

Reaction 2 - polymerisation of biomass

$- 'CH_{1.8}O_{0.6}N_{0.2}' - 1.7\ ATP + CH_{1.8}O_{0.6}N_{0.2} = 0$

Reaction 3 - ethanol production

$- 1.5\ CH_2O + CH_3\ O_{0.5} + 0.5\ CO_2 + 0.5\ ATP = 0$

Reaction 4 - glycerol production

$- 1\ CH_2O + CH_{2.66}\ O - 0.33\ NADH_2 - 0.33\ ATP = 0$

This metabolic model contains 7 extracellular compounds (glucose, biomass, NH_3, CO_2, H_2O, ethanol, glycerol) and 3 intracellular compounds ($NADH_2$, precursors, ATP).

1) Calculate the degrees of freedom.

2) Choose r_{As} as the independent rate and derive the following relations.

   r_{Ax} versus r_{As} (biomass x);

   r_{Ae} versus r_{As} (ethanol e);

   r_{Ag} versus r_{As} (glycerol g).

3) Calculate the yield of:

   biomass on glucose (in terms of C-mol biomass / C-mol glucose);

   ethanol on glucose (C-mol ethanol / C-mol glucose);

   glycerol on glucose (C-mol glycerol / C-mol glucose).

</td></tr>
</table>

<table>
<tr><td>

SAQ 9.8

</td><td>

Summarise the various possibilities to reduce the number of required model parameters for the conversion relations.

</td></tr>
</table>

Conclusion

The complex process of microbial growth, which involves a multitude of intra- and extracellular chemical compounds, has been reduced to the requirement of 1 single kinetic relation with which the total microbial process of growth is specified.

In the previous sections this has been obtained by the application of:

- the definition of the microbial system by the relevant compounds;

- pseudo-steady state assumption (= black box model);

- the conservation principles of elements;

- the metabolic description of the biochemical knowledge.

We now leave the conversion kinetics for a moment and move on to a more general approach to transfer kinetics.

9.9 Modelling of transfer kinetics

9.9.1 A general approach to reduce the number of model parameters in transfer kinetics

The macroscopic balance (Figure 9.4) for chemical compounds in a bioreactor contains:

- a conversion part r_{Ai};

- a transfer part ϕ_{Ai}.

In the previous sections we dealt with the conversion part. In this section, we will examine the transfer part.

Here however only those expressions of relevance to biotechnology will be examined. The broad subject of calculation and modelling of the transfer term ϕ_{Ai} has been treated in earlier chapters. This problem is the central domain of physical technology.

bioreactor system specifies transfer kinetics biocatalyst specifies conversion kinetics

It has been already been indicated (Figure 9.4) that the transfer kinetics are specific properties of the bioreactor system, while the conversion kinetics are strongly determined by the biocatalyst. To remind you, we have re-drawn Figure 9.4 in a different way (Figure 9.5). It will be apparent that, because we can formulate for each chemical compound a macroscopic balance, we need for each chemical compound an expression for the transfer kinetic equation. Therefore, also here numerous model parameters occur, and simplifications are called for.

9.9.2 Simplifications

From the pseudo-steady state approximation it may be concluded that we may limit ourselves to the extracellular chemical compounds listed in Tables 9.1 and 9.2.

$$(C_i^o V) = r_{Ai}V + \phi_{Ai}V$$

↓   ↓

biocatalyst   reactor
specific   specific

Figure 9.5 Macroscopic balance (the rate of chemical change reflects the amount of cells/enzymes present; the transfer part depends on reactor design).

For all the intracellular compounds $\phi_{Ai} = 0$, as they never leave the cells of the biomass. In general these are the only simplifications which are possible with respect to transfer kinetics.

transfer kinetics not subject to conservation principles It should also be realised that the net-transfer kinetics ϕ_{Ai} as such are not subject to the conservation principles. These hold only for the conversion terms r_{Ai}. Thus we cannot use this approach to simplify the model parameters when we are considering transfer kinetics.

There are, however, two particular situations which do allow a simplification of the transfer kinetics:

- the reactor system is in steady state;

- the chemical compound is not transferred at all (this is generally called the batch situation).

In the steady state situation, one can apply conservation principles to ϕ_{Ai}. The reason is simple.

In a steady state the whole system maintains a constant temperature, pressure, volume, concentration, etc.

Hence: $\overset{o}{V} = 0; \quad \overset{o}{C_i} = 0$

This means that we may write for Equation 9.1.

Steady state: $r_{Ai}^+ + \phi_{Ai}^+ = 0$

Note: superscript + indicates a steady state, superscript * indicates an equilibrium state. This means that, only in a steady state, ϕ_{Ai}^+ is equal but opposite to r_{Ai}^+. Hence we may also apply the conservation principles to ϕ_{Ai}^+.

It is stressed that only in a steady state are we allowed to apply the conservation principles to transfer terms. A well known example is the Continuous Flow Stirred Tank Reactor (CFSTR).

In the batch-situation one can conclude that the chemical compound is not transported outside the system. So for a batch system:

$$\phi_{Ai} = 0 \qquad\qquad\qquad\qquad (E - 9.13)$$

This situation often occurs.

Examples are:

- an aerobic batch fermentation, which already contains from time zero all C-, N-, P- and S-sources, while biomass remains in the reactor. Hence the transfer kinetics of these compounds are zero.

- a fed-batch fermentation, where eg C-source and N-source is continuously added, but where there is no removal of product or biomass, Hence the transfer terms ϕ_{Ax} and ϕ_{Ap} are zero.

Now for practical reasons it is useful to present some kinetic relations for a number of common transport processes.

9.10 Expressions for transfer kinetics

We remind you of earlier chapters of this text in which we explained that the net-transfer of material, heat etc, into or out of defined system can occur as follows:

- within one phase (liquid or gas);

    - (convective transport);

    - (diffusive transport);

    - (a combination of both).

- between phases (between gas $\leftarrow \rightarrow$ liquid or liquid $\leftarrow \rightarrow$ solid etc); the transfer occurs over the interphase area.

Let us examine some examples:

Convective transport

One of the most important types of bioreactor with convective transport is the continuous culture (Figure 9.6).

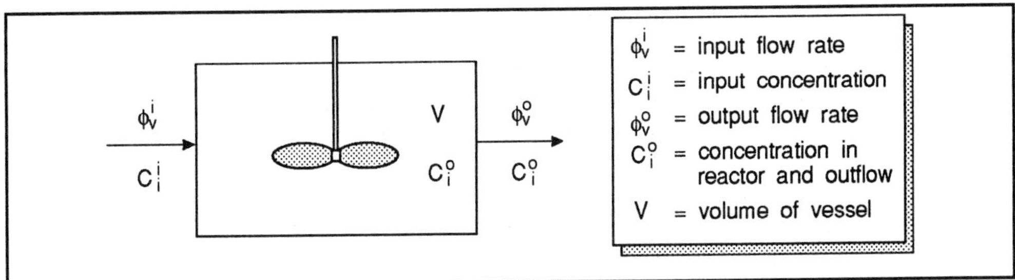

Figure 9.6 Transfer kinetics by convection in a continuous culture (Note ϕ_v = volumetric flow rate).

dilution rate   By definition dilution rate $D = \dfrac{\phi_v^o}{V}$, ie dilution rate = $\dfrac{\text{flow rate out}}{\text{volume}}$

Thus the transport is given by:

$$\phi_{Ai} = \frac{(\phi_v^i C^i - \phi_v^o C_i^o)}{V} = \left\{ \frac{\phi_v^i}{\phi_v^o} C_i^i - C_i^o \right\} D$$

Thus the net convective transport of the compound through the system is the difference between the amount of input ($\phi_v^i C^i$) and the output ($\phi_v^o C_i^o$). Since we usually like to express the transfer component (ϕ_{Ai}) in terms of unit volume, we have divided by V.

If, and only if, the volumetric inflow ϕ_v^i and outflow ϕ_v^o are equal then:

$$\phi_{Ai} = (C_i^i - C_i^o) D \qquad\qquad (E - 9.14)$$

This is an important relationship although it is usually written in this form:

$$\phi_{Ai} = D (C_i^i - C_i^o) \qquad\qquad (E - 9.15)$$

This only holds if the volumetric in- and outflow are equal. This is often the situation, but not always so.

Π Why can the volumetric inflow and outflow be different? (See if you can write down two or more examples).

There are many examples you could have used. For examples a flow of CO_2-gas into a reactor can totally 'disappear' due to absorption into the through-flowing liquid. Another example is that the substrate might be added as a concentrate (ie in a small volume) and is diluted by another stream in the bioreactor.

Thus in this case ϕ_v^i will be small, ϕ_v^o will be large. We can represent this by:

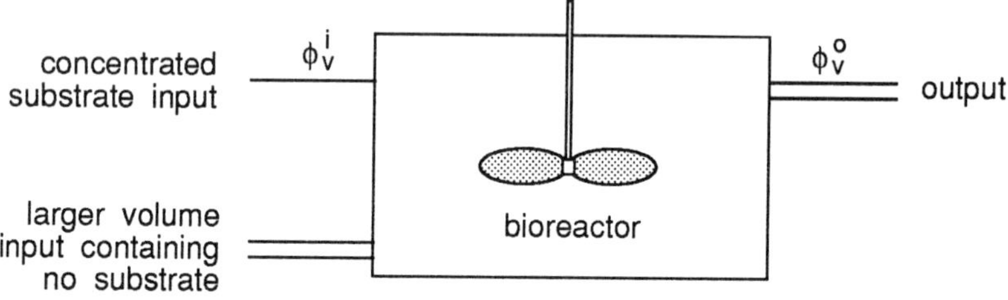

Π Although relevant volumetric inflow and outflow rates can be different, can inflowing and outflowing masses be different?

A moments thought should have enabled you to remember that masses cannot be destroyed, it can only be converted from one form to another. Thus the inflowing and outflowing masses are related through the principle of mass conservation.

Diffusive transport

Diffusion occurs if there are concentration gradients in a reactor. We dealt with diffusive transport in earlier chapters. The general expression is from Fick's law (see Equation E - 3.1 Chapter 3) is:

$$\phi_{Ai} = -D \, \frac{d\,C_i}{dx}$$

(E - 9.16)

This relationship is relevant in the transfer inside solid catalyst which contain enzymes, micro-organisms or inorganic catalysts. This is a feature of so called immobilised systems in which the biocatalyst is held on or by an immobilised matrix. This type of system is dealt with in the Biotol texts 'Operational Modes of Bioreactors' and 'Bioreactor Design and Product Yield'.

Interphase transfer

transfer relation   In many cases there is an exchange of material or energy into or out of a system though an interface with surface A (m^2). Such an interface separates distinct phases like gas/liquid, liquid/liquid, liquid/solid, solid/gas etc. Important examples are the transfer of O_2 from the gas to a liquid in an aerated reactor, the heat transfer from the fermentor broth to a cooling coil etc.

Generally the net-transfer relation is then provided by:

$$\phi_{Ai} = \frac{A}{V} \, k_L \, (C_i^* - C_i)$$

Where:

k_L is a transfer coefficient, which depends mainly on hydrodynamic factors (ms^{-1});

A is the surface area over which the transport occurs (like bubble area, cooling coil area) in m^2;

V is the system volume (m^3);

C_i^* is the equilibrium concentration in for example the gas phase;

C_i is the concentration in for example the liquid phase;

$k_L \, \dfrac{A}{V}$ is commonly simplified to $k_L a$ where a is the area per system volume (a = A/V).

$\prod$ Write out the expression for ϕ_{Ai} in a liquid system where convective transport and transfer over an interface from the gas phase occurs. Assume that convective in- and outflow volumes are equal and that the liquid is ideally mixed.

We can represent this system in the following way:

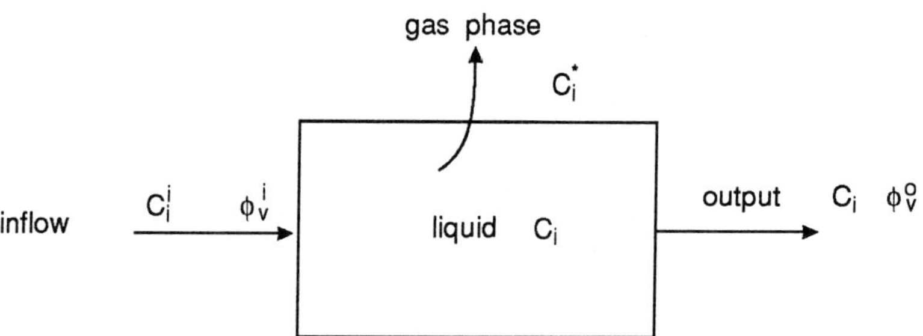

Because ϕ_{Ai} is the net-transfer rate, we should summarise all transfer processes. Hence:

$$\phi_{A,i} = \underbrace{\frac{A}{V} \, k_L \, (C_i^* - C_i)}_{\text{interphase transfer}} + \underbrace{D(C_i^i - C_i^o)}_{\text{convective transport}}$$

<table>
<tr><td>

SAQ 9.9

</td><td>

For what chemical compounds do we need transfer kinetics in the following systems?

1) anaerobic closed bioreactor

2) aerobic sparged batch bioreactors

3) anaerobic closed bioreactor with a pH-control

4) a fed batch aerobic bioreactor with pH-control and in which NH_3 and glucose are being fed in.

</td></tr>
</table>

Summary and objectives

In this chapter we have shown that microbial growth can be described using a macroscopic approach. Central to this approach is the macroscopic balance for each chemical compound. The number of required chemical compounds to describe microbial growth can be reduced to a small number of extracellular compounds (6-10) by introduction of the pseudo-steady state assumption for the intracellular compounds. This results in a so-called black box description. This simplified description is only satisfactory if all relevant extracellular compounds have been identified. This can be assured by requiring that the elemental conservation is indeed experimentally satisfied.

Producing such a system description enables a significant additional reduction of the required number of kinetic expressions for conversion processes using:

- elemental conservation;

- specification of metabolic reactions which are specific for a micro-organism.

The complex process of microbial growth can then be described by a single kinetic conversion rate relation, despite the hundreds of reactions which are occuring.

A decrease in the number of required kinetic relations for the transfer processes is in general possible in steady state and batch situations. It is stressed that the conservation principles do not apply to transfer rates.

Now that you have completed this chapter you should be able to:

- calculate net-conversion rates from supplied data;

- assign functions to the relevant compounds needed to follow microbial growth;

- use the principles of conservation to determine elemental relations in describing microbial growth;

- use the principles of elemental conservation to check the definition of a system;

- use metabolic information to calculate yields from supplied data;

- identify the chemical compounds for which transfer kinetics are needed in a variety of systems.

A generalised treatment of aerobic microbial growth and product formation

A generalised treatment of aerobic microbial growth and product formation

In the preceding chapter, the basic aspects of the macroscopic approach to model the microbial growth system have been provided using a specific aerobic growth example (Table 9.2).

In practice many industrial microbial systems are aerobic, but substrates other than glucose are used, and a variety of products may be produced. Therefore it is useful to generalise the results of Chapter 9 for an arbitrary aerobic system definition. This will enable also the application of both the 1st law and the 2nd law of thermodynamics.

Because the necessary calculations are completely analogous to those described in Chapter 9, only the results will be given without derivation.

In many ways this chapter has a structure similar to that of Chapter 9. We will begin by developing a generalised approach to system definition and then we will apply the principles of elemental conservation to model this system. Subsequently we will examine how energy conservation and Gibbs free energy balances may be applied to these systems. Finally we will examine the application of metabolic modelling to these systems.

10.1 A generalised system definition for aerobic growth and product formation on organic substrates

standard
enthalpy

standard free
enthalpy

By analogy with the discussions described in Chapter 9 (Section 9.6) the relevant compounds are defined in Table 10.1. We have also included in this Table some energetic quantities. Thus for each compound, we have defined its standard enthalpy and its standard Gibbs energy. Note also that we have defined chemical composition in general terms. Make certain you have looked at the coefficients used and understand the nomenclature used. Thus a subscript cx signifies coefficient of carbon (c) in biomass (x); subscript op signifies coefficient of oxygen (o) in the product (p).

∏ Given the substrate CH_2O what is the value of the coefficient hs?

You should conclude hs = 2 as hs signifies hydrogen in the substrate.

Furthermore, as in every chemical process, one can define the heat production or consumption ϕ_H and the Gibbs energy dissipation Ds. If there is heat production, a microbial system is said to be exothermic, if there is heat consumption it is called endothermic. In principle both situations can occur, hence ϕ_H can be positive or negative respectively. If heat is produced ϕ_H is positive, its units are $kJ\ m^{-3}\ s^{-1}$.

The Gibbs energy dissipation rate D_s is always positive, in accordance with the 2nd law of thermodynamics.

relevant compound or quantity	net conversion rate	chemical composition	standard enthalpy	standard Gibbs free energy
biomass	r_{Ax}	$C_{cx}H_{hx}O_{ox}N_{nx}$	h_x^o	μ_x^o
substrate	r_{As}	$C_{cs}H_{hs}O_{os}N_{ns}$	h_s^o	μ_s^o
product	r_{Ap}	$C_{cp}H_{hp}O_{op}N_{np}$	h_p^o	μ_p^o
CO_2	r_{Ac}	CO_2	h_c^o	μ_c^o
H_2O	r_{Aw}	H_2O	h_w^o	μ_w^o
N-source	r_{An}	$C_{cn}H_{hn}O_{on}N_{nn}$	h_n^o	μ_n^o
O_2	r_{Ao}	O_2	h_o^o	μ_o^o
heat	ϕ_H	-	-	-
Gibbs energy dissipation	D_s^o	-	-	-

Table 10.1 Defined relevant compounds/quantities for aerobic microbial systems with growth and product formation.

In discussing the application of the principles of elemental conservation and the first and second laws of thermodynamics you should realise that this system description contains:

- 7 chemical conversion rates;
- 1 heat production rate ϕ_H;
- 1 dissipation rate D_s.

for which there are available the following relations:

- 4 elemental conservation relations (C,H,O,N);
- 1 energy conservation relation (1st law);
- 1 Gibbs energy balance (2nd law).

Hence there are 9 unknown rates and 6 relations. This means that there are 3 degrees of freedom. In the subsequent paragraphs we will use:

r_{As}, r_{Ax}, r_{Ap} as independent rates for practical reasons.

This means that, utilising the 6 relations, we can relate:

r_{Ac}, r_{An}, r_{Aw}, r_{Ao}, ϕ_H, D_s^o r_{As}, r_{Ax}, r_{Ap}.

∏ Why are r_{As}, r_{Ax} and r_{Ap} good net conversion rates to use?

You should have realised that in many systems, the rate of production of biomass and product or consumption of substrate can be measured. Therefore, these are good rates to choose.

10.2 Application of the principle of element conservation

As stated above one can relate r_{Ac}, r_{An}, r_{Ao}, r_{Aw} to r_{As}, r_{Ap} by using the 4 elemental conservation relations. The calculations are exactly analogous to the case treated in Section 9.7, except now we have extended them to include product formation and assumed arbitary chemical composition of biomass, product, carbon source and nitrogen source.

Below we have produced a series of relationships derived from applying elemental conservation similar to those reported in Equations 9.8 a-d. Note, however, that in this case we have included a term for a product and have used α, β γ and δ to represent the coefficients. Obviously these coefficients depend upon the composition of the biomass, product, substrate and other components involved in setting up the elemental conservation relationships. We will examine these coefficients more closely a little later.

The results of applying elemental conservation are:

$$\alpha_n\, r_{An} + \alpha_x\, r_{Ax} + \alpha_s\, r_{As} + \alpha_p\, r_{Ap} = 0 \qquad\qquad \text{(E - 10.1a)}$$

This is in effect a nitrogen balance and gives the rate of nitrogen source consumption in terms of the rate of biomass production, product synthesis and substrate use.

$$\beta_w\, r_{Aw} + \beta_x\, r_{Ax} + \beta_s\, r_{As} + \beta_p\, r_{Ap} = 0 \qquad\qquad \text{(E - 10.1b)}$$

This in effect gives us the water production rate in relation to the rate of biomass production, product synthesis and substrate used.

$$\gamma_o\, r_{Ao} + \gamma_x\, r_{Ax} + \gamma_s\, r_{As} + \gamma_p\, r_{Ap} = 0 \qquad\qquad \text{(E - 10.1c)}$$

This is an important relationship as it gives us the oxygen consumption rate (r_{Ao}) as a function of the substrate used, biomass produced and the product made. It is important because oxygen transfer is often a limiting factor in large scale bioreactors.

$$\delta_c\, r_{Ac} + \delta_x\, r_{Ax} + \delta_s\, r_{As} + \delta_p\, r_{Ap} = 0 \qquad\qquad \text{(E - 10.1d)}$$

This equation gives us the carbon balance.

Now let us deal with the coefficients.

First let us deal with the α terms. α coefficients relate to the nitrogen terms. α_n is the coefficient of nitrogen in the nitrogen source. From Table 10.1 we can write $\alpha_n = nn$. Similarly we can write $\alpha_x = nx$, $\alpha_s = ns$ and $\alpha_p = np$.

In a more general form we could write:

$$\alpha_i = ni \qquad\qquad \text{(E - 10.2a)}$$

We will not go through the derivation of all of the other coefficients in detail, but they too can be related to the coefficients in Table 10.1. These relationships are:

$$\beta_i = hi - \frac{ni}{nn} * hn \tag{E - 10.2b}$$

$$\gamma_i = 4 \times ci + hi - 2 * oi - \frac{ni}{nn}(4 * cn + hn - 2 * on) \tag{E - 10.2c}$$

$$\delta_i = ci - \frac{cn}{nn} * ni \tag{E - 10.2d}$$

We do not expect you to derive these for yourself although you might have fun verifying them - it would prove an interesting test of your algebra!

It is important that you can apply these equations.

By substitution of the appropriate composition-coefficients it is straight forward to find the α, β, γ and δ coefficients in Equation 10.1a - Equation 10.1d from Equation 10.2a - Equation 10.2d for each component. Let us work through an example together.

$\prod$ We are asked to calculate the coefficients of Equation 10.1 a-d for the microbial system defined in Table 9.2 using Equation 10.2 a-d and to compare the results to demonstrate that the general relationships described by Equation 10.1 a-d and Equation 10.2 a-d are true! Before doing this examine the next table.

For convenience we have repeated the data of Table 9.2 here.

Relevant compound	Composition
biomass	$CH_{1.8} O_{0.5} N_{0.2}$
glucose	CH_2O
O_2	O_2
CO_2	CO_2
H_2O	H_2O
NH_3	NH_3

From the table the following composition indices are found (note product is absent):

Biomass: $cx = 1$; $hx = 1.8$; $ox = 0.5$; $nx = 0.2$;

Substrate: $cs = 1$; $hs = 2.0$; $os = 1.0$; $ns = 0$;

N-source: $cn = 0$; $hn = 3$; $on = 0$; $nn = 1.0$.

Furthermore we can write for:

O_2: $co = 0$; $ho = 0$; $oo = 2$; $no = 0$;

CO_2: $cc = 1$; $hc = 0$; $oc = 2$; $nc = 0$;

H_2O: $cw = 0$; $hw = 2$; $ow = 1$; $nw = 0$.

Substitution of these values in Equation 10.2 a-d to calculate α_i, β_i, γ_i, δ_i. Here we have written out these calculations. You should follow each.

substrate:

$$\alpha_s = ns = \underline{0}$$

$$\beta_s = hs - \frac{ns}{nn} * hn = 2.0 - \frac{0}{1.0} * 3 = \underline{2.0}$$

$$\gamma_s = 4 * cs + hs - 2 * os - \frac{ns}{nn} (4 * cn + hn - 2 * on)$$

$$= 4 * 1 + 2.0 - 2 * 1 - \frac{0}{1.0} (4 * 0 + 3 - 2 * 0) = \underline{4}$$

$$\delta_s = cs - \frac{cn}{nn} * ns = 1.0 - \frac{0}{1.0} 0 = \underline{1.0}$$

biomass:

$$\alpha_x = nx = \underline{0.2}$$

$$\beta_x = hx - \frac{nx}{nn} * hn = 1.8 - \frac{0.2}{1.0} * 3 = \underline{1.2}$$

$$\gamma_x = 4 * cx + hx - 2 * ox - \frac{nx}{nn} (4 * cn + hn - 2 * on)$$

$$= 4 + 1.8 - 1 - 0.6 = \underline{4.2}$$

$$\delta_x = cx - \frac{cn}{nn} * nx = 1 - \frac{0}{1.0} * 0.2 = \underline{1}$$

O_2:

$$\alpha_o = no = \underline{0}$$

$$\beta_o = ho - \frac{no}{nn} * hn = 0 - \frac{0}{1} * 3 = \underline{0}$$

$$\gamma_o = 4 * C_o + ho - 2 * 00 - \frac{no}{nn} (4 * cn + hn - 2 * on)$$

$$= 4 * 0 + 0 - 2 * 2 - \frac{0}{1} (0 + 3 - 2 * 0) = \underline{-4}$$

$$\delta_o = co - \frac{cn}{nn} * no = 0 - \frac{0}{1} * 0 = \underline{0}$$

CO_2:

$$\alpha_c = nc = \underline{0}$$

$$\beta_c = hc - \frac{nc}{nn} * hn = 0 - \frac{0}{1} * 3 = \underline{0}$$

$$\gamma_c = 4 * cc + hc - 2 * oc - \frac{nc}{nn}(4 * cn + hn - 2 * on)$$

$$= 4 * 1 + 0 - 2 * 2 - \frac{0}{1}(4 * 0 + 3 - 2 * 0) = 4 - 4 = 0$$

$$\delta_c = cc - \frac{cn}{nn} * nc = 1 - \frac{0}{1} * 0 = \underline{1}$$

H_2O:

$$\alpha_w = nw = \underline{0}$$

$$\beta_w = hw - \frac{nw}{nn} * hn = 2 - \frac{0}{1} * 3 = \underline{2}$$

$$\gamma_w = 4 * cw + hw - 2 * ow - \frac{nw}{nn}(4 * cn + hn - 2 * on)$$

$$= 4 * 0 + 2 - 2 * 1 - \frac{0}{1}(4 * 0 + 3 - 2 * 0) = 2 - 2 = 0$$

$$\delta_w = cw - \frac{cn}{nn} * nw = 0 - \frac{0}{1} * 0 = \underline{0}$$

N-source:

$$\alpha_n = nn = \underline{1}$$

$$\beta_n = hn - \frac{nn}{nn} * hn = 3 - \frac{1}{1} * 3 = \underline{0}$$

$$\gamma_n = 4 * cn + hn - 2 * on - \frac{nn}{nn}(4 * cn + hn - 2 * on)$$

$$= 4 * 0 + 3 - 2 * 0 - \frac{1}{1}(4 * 0 + 3 - 2 * 0)3 - 3 = \underline{0}$$

$$\delta_n = cn - \frac{cn}{nn} * nn = 0 - \frac{0}{1} * 1 = \underline{0}$$

Substitution of the various α, β, γ and δ-values into Equation 10.1a - Equation 10.1d leads to:

$$1 * r_{An} + 0.2\, r_{Ax} = 0$$

$$2 * r_{Aw} + 1.2\, r_{Ax} + 2\, r_{As} = 0$$

$$-4 * r_{Ao} + 4.2\, r_{Ax} + 4\, r_{Ao} = 0$$

$$r_{Ac} + r_{Ax} + r_{As} = 0$$

These 4 equations are indeed the same as Equation 9.10 a-d. Thus we can conclude that the generalised Equations are applicable.

We remind you that Equation 10.1a is in fact an N balance. Equation 10.1b gives the water production rate (since the system is usually in an aqueous environment r_{Aw} is not usually experimentally determined).

Equation 10.1a is an important relation because it gives us the oxygen consumption of the microbial system as a function of substrate used, biomass produced and product made; because oxygen transfer is often a limiting factor in large scale fermentations, this equation is very relevant.

degree of reduction

To stress this importance the coefficients γ_i in this relation have been given a name of their own; the degree of reduction. From Equation 10.2c it clearly appears that γ_i depends on: the composition of compound i and the composition of the N-source of the microbial system.

However, it is also clear from Equation 10.2c that for organic compounds which do not contain N as an element ($n_i = 0$), the value of γ_i is independent of the N-source. For oxygen the value $\gamma_o = -4$; for CO_2, H_2O and N-source $\gamma_i = 0$.

The term degree of reduction becomes clear if the value for γ_i is calculated for a large number of common organic substrates (Table 10.2).

It appears that γ_i for organic compounds can be identified as the number of electrons per C-mol which are released upon combustion to CO_2, N-source and H_2O.

To illustrate the effect of N-source on γ_i it is useful to calculate γ_x (biomass) for various N-sources.

Π Calculate γ_x for biomass (composition $CH_{1.7}O_{0.4}N_{0.25}$) in microbial systems which contain NH_3.

From the information provided since:

$$\gamma_x = 4 * cx + hx - 2 * ox - \frac{nx}{nn} (4 * cn + hn - 2 * on), \text{ we can write:}$$

biomass:

$cx = 1; hx = 1.7; ox = 0.4; nx = 0.25$

NH_3 as N-source:

$cn = 0 ; hn = 3; on = 0; nn = 1$

From Equation 10.2c we can calculate:

$$\gamma_x = 4 * 1 + 1.7 - 2 * 0.4 - \frac{0.25}{1} (0 + 3 - 0) = \underline{4.15}$$

SAQ 10.1

Calculate γ_x for the microbial system (biomass composition $CH_{1.7}O_{0.4}N_{0.25}$) using HNO_3, N_2 or aspartic acid ($C_4H_7NO_4 = CH_{1.75}N_{0.25}O$) as N-source.

What these calculations should have shown you is that γ_x is dependant upon the nature of the nitrogen source used by the micro-organism.

compound	composition	1-C-mole formula	γ_i	β_i	h^o_{ci} (kJ / C-mol)	μ^o_{ci} (kJ / C-mol)
methane	CH_4	CH_4	+8	+4	892	818
n-alkane	$C_{15}H_{32}$	$CH_{2.13}$	+6.13	+2.13	700	670
methanol	CH_4O	CH_4O	+6	+4	728	693
ethanol	C_2H_6O	$CH_3O_{0.5}$	+6	+3	684	659
glycerol	$C_3H_8O_3$	$CH_{2.66}O$	+4.67	+2.66	554	548
mannitol	$C_6H_{14}O_6$	$CH_{2.33}O$	+4.33	+2.33	508	514
acetic acid	$C_2H_4O_2$	CH_2O	+4	+2	438	447
lactic acid	$C_3H_6O_2$	CH_2O	+4	+2	456	459
glucose	$C_6H_{12}O_6$	CH_2O	+4	+2	468	479
formal-dehyde	CH_2O	CH_2O	+4	+2	572	501
gluconic acid	$C_6H_{12}O_7$	$CH_2O_{1.17}$	+3.67	+2		443
succinic acid	$C_4H_6O_4$	$CH_{1.5}O$	+3.5	+1.5	373	400
citric acid	$C_6H_8O_7$	$CH_{1.33}O_{1.17}$	+3	+1.33	327	358
malic acid	$C_4H_6O_5$	$CH_{1.5}O_{1.25}$	+3	+1.5	332	361
formic acid	CH_2O_2	CH_2O_2	+2	+2	255	281
oxalic acid	$C_2H_2O_4$	CHO_2	+1	+1	123	163
pyruvic acid	$C_3H_4O_3$	$CH_{1.33}O_{1.00}$	+3.33	1.33		380
biomass	$CH_{1.8}O_{0.5}N_{0.2}$		+4.2	1.2	490.2	474
oxygen	O_2		-4	0	0	0
carbon dioxide	CO_2		0	0	0	0
water	H_2O		0	2	0	0
ammonia	$NH3$		0	0	0	0
glyoxylate	$C_2H_2O_3$	$CHO_{1.5}$	+2	+1		+256
tartrate	$C_4H_6O_6$	$CH_{1.5}O_{1.5}$	+2.5	+1.5		
malonate	$C_3H_4O_4$	$CH_{1.33}O_{1.33}$	+2.7	+1.33		
acetoin	$C_4H_9O_2$	$CH_{2.25}O_{0.5}$	+5.25	+2.25		+587
butanediol	$C_4H_{10}O_2$	$CH_{2.5}O_{0.5}$	+5.5	+2.50		+608

Table 10.2 Organic growth substrates and their γ_i and β_i, h^o_{ci}, μ^o_{ci} values for NH_3 as N-source.

Of the group of equations (Equation 10.1a - Equation 10.1d) we have so far examined Equations 10.1a - Equation 10.1c. The remaining equation (Equation 10.1d) is in fact a carbon balance which allows estimation of the CO_2 production. The equation is also important because CO_2 removal can sometimes be critical for product formation (ventilation problems).

For illustrative purposes the values of β_i and γ_i have been assembled in Table 10.2 for a number of organic substrates assuming NH_3 as N-source.

γ_i = number of moles of electrons per C-mol combusted

We remind you that γ_i, can be identified as the number of moles of electrons, which are released per C-mol oxidised to CO_2 and H_2O and N-source. If the substrate does not contain N, we can re-state this as γ_i is equal to the number of moles of electrons which are released per C-mol oxidised to CO_2 and H_2O. You will see from Table 10.2 that the more oxidised the substrate, the fewer electrons released during the combustion of the substrate to CO_2 and H_2O, the lower the γ_i value. Note that the values of h_{ci}^{o} and μ_{ci}^{o} are lower for more oxidised components. In other words, less energy is released during their complete oxidation to carbon dioxide and water.

From the calculations we have carried out it should be clear that the generalised equations (Equation 10.1a - Equation 10.1d and Equation 10.2a - Equation 10.2d) are applicable to any microbial system whether or not organic products other than biomass are produced.

Many experiments with a wide variety of organic substrates have provided evidence for the wide applicability of these relations.

growth yield

Y_{sx}

Y_{ox}

Finally in this section let us turn to another aspect of the application of the principle of element conservation. Growth yield (Y) can be defined in a number of ways. For example, biomass yield on a substrate (Y_{sx}) can be defined as the amount of biomass produced as a result of consumption of a known amount of substrate. Similar biomass yield on oxygen (Y_{ox}) can be defined as the amount of biomass produced as the result of the consumption of a known amount of oxygen. We can also define Y_{sx} and Y_{ox} in terms of their respective rates of consumption as described in Equations 10.3 and 10.4. (Remember the sign convention).

$$Y_{sx} = \frac{r_{Ax}}{-r_{As}} \text{ (ie rate of biomass production/rate of substrate utilisation)} \tag{E-10.3}$$

$$Y_{ox} = \frac{r_{Ax}}{-r_{Ao}} \text{ (ie rate of biomass production/rate of oxygen utilisation)} \tag{E 10.4}$$

If we combine Equations 10.3, 10.4 and 10.1c (without product formation) we obtain a relation between Y_{sx} and Y_{ox} (Equation 10.5).

$$Y_{ox} = \frac{-\gamma_o}{\gamma_x} * \frac{Y_{sx} * \frac{\gamma_x}{\gamma_s}}{1 - Y_{sx} * \frac{\gamma_x}{\gamma_s}} \tag{E-10.5}$$

You should be able to carry out this derivation for yourself.

Since $\gamma_o\, r_{Ao} + \gamma_x\, r_{Ax} + \gamma_s\, r_{As} + \gamma_p\, r_{Ap} = 0$ (Equation 10.1c)

and since $\gamma_p\, r_{Ap} = 0$ as no product is formed, we can write:

$\gamma_o\, r_{Ao} + \gamma_x\, r_{Ax} + \gamma_s\, r_{As} = 0$

Then we can convert r_{Ao}, r_{Ax} and r_{As} into terms of Y_{sx} and Y_{ox}. For example:

$$\gamma_o\, \frac{r_{Ax}}{(-Y_{ox})} + \gamma_x\, r_{Ax} + \gamma_s\, \frac{r_{Ax}}{(-Y_{sx})} = 0$$

$$\frac{\gamma_o}{(-Y_{ox})} + \gamma_x + \frac{\gamma_s}{(-Y_{sx})} = 0$$

Rearranging, gives Equation 10.5.

From this relation it is clear that, because:

- for a constant biomass composition and a defined N-source (eg NH_3) $\gamma_x \approx 4.2$;

- $\gamma_o = -4$.

There exists a specified relation between Y_{ox} and $Y_{sx} * \frac{\gamma_x}{\gamma_s}$ (ie substrate and oxygen utilisation rates are related and both effect the amount of biomass produced).

This relation is shown in Figure 10.1 together with experimentally measured data, and there appears fair agreement.

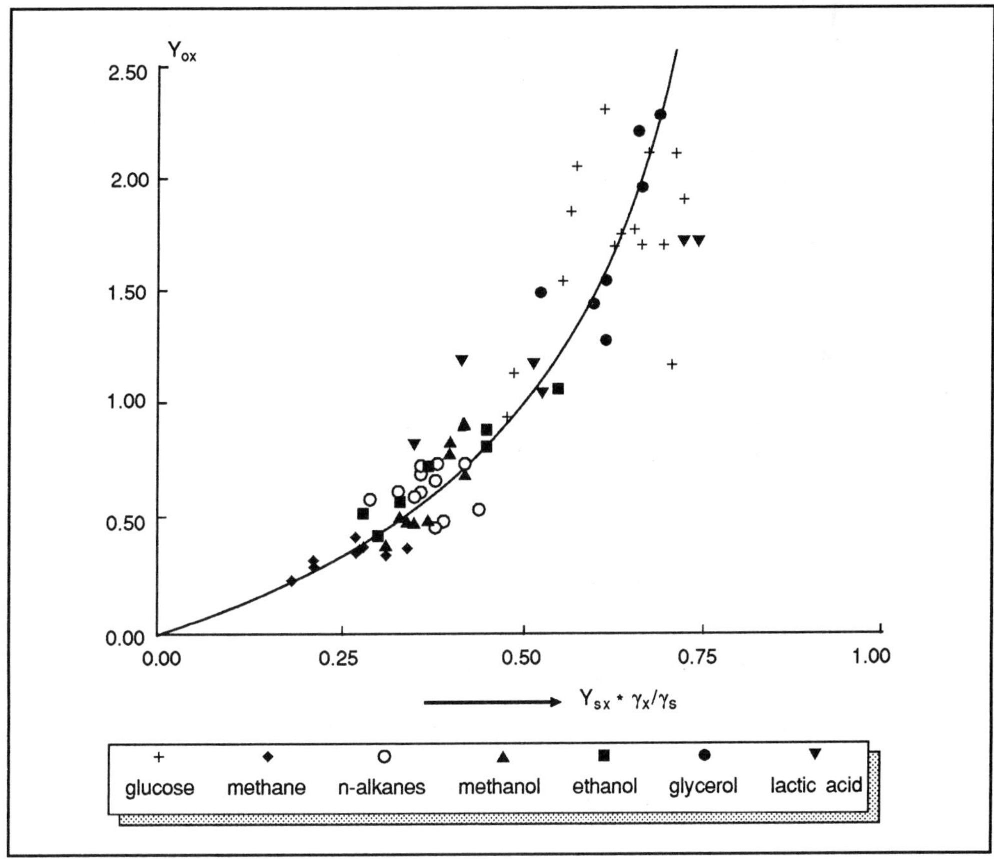

Figure 10.1 Calculated and actual relation between Y_{ox} and $Y_{sx} * \frac{\gamma_x}{\gamma_s}$ (experimental data is plotted as symbols, the continuous line plots the calculated value of $Y_{sx} * \frac{\gamma_x}{\gamma_s}$.

It should also be clear that, for the situation where no product formation occurs, the biomass yield Y_{sx} has well established relations to other yields of biomass which one may define. Equation 10.5 is just one example.

<table>
<tr><td>

SAQ 10.2

</td><td>

The yields of biomass on CO_2, N-source and water are defined as:

$$Y_{cx} = \frac{r_{Ax}}{r_{Ac}} \; ; \; Y_{nx} = \frac{r_{Ax}}{-r_{An}} \; ; \; Y_{wx} = \frac{r_{Ax}}{r_{Aw}}$$

Find the relations with Y_{sx} in analogy to Equation 10.5.

For example, for CO_2 you will need to combine the defined biomass yields on CO_2 and the definition of Y_{sx} (Equation 10.3 and Equation 10.1d) to find the relationship of Y_{cx} to Y_{sx}

Likewise the relationship between Y_{nx} and Y_{sx} can be established by combining the defined biomass yield on N-source (Equations 10.1a and 10.3).

</td></tr>
</table>

10.3 Application of the principle of energy conservation

10.3.1 Heat production

In an analogous way as for elements one can formulate the conservation principle of energy (1st law of thermodynamics).

For the defined system (Table 10.1) the conservation principle simply states that the biological heat production ϕ_H is due to the enthalpy of conversion of the chemical compounds.

In mathematical terms, using the standard enthalpy of the chemical compounds as defined in for example Table 10.1, we can produce Equation 10.6.

$$\phi_H + r_{Ax} * h_x^o + r_{As} * h_s^o + r_{Ap} * h_p^o + r_{Ac} * h_c^o$$
$$+ r_{Aw} * h_w^o + r_{An} * h_n^o + r_{Ao} * h_o^o = 0 \qquad \text{(E-10.6)}$$

This relation can be combined with the relations Equation 10.1a - Equation 10.1d to eliminate r_{An}, r_{Aw}, r_{Ao}, r_{Ac} in order to obtain the relationship between ϕ_H and the chosen independent rates r_{As}, r_{Ax}, r_{Ap}.

After some manipulations one obtains Equation 10.7 (it might be quite a challenge for you to do this, but it would be useful practice).

$$\phi_H + r_{Ax} * h_{cx}^o + r_{As} * h_{cs}^o + r_{Ap} * h_{cp}^o = 0 \qquad \text{(E-10.7)}$$

We can calculate h_{ci}^o ($i = x, s, p$) using Equation 10.8.

$$h_{ci}^o = h_i^o - \frac{\gamma_i}{\gamma_o} h_o^o - \frac{\alpha_i}{\alpha_n} h_n^o - \frac{\beta_i}{\beta_w} h_w^o - \frac{\delta_i}{\delta_c} h_c^o \qquad \text{(E-10.8)}$$

The value of h^o_{ci} can be calculated for each compound of the microbial growth system definition in Table 10.1 based on the available chemical composition and enthalpy values of the various compounds and the N-source of the system. Let us work through a calculation together.

$\prod$ Let us assume we wish to calculate the enthalpy coefficient of biomass h^o_{cx}, using the biomass composition $CH_{1.8}O_{0.5}N_{0.2}$, for NH_3 as N-source.

From thermodynamic tables the following molar enthalpies are available (note that for C-containing compounds h^o_i is the enthalpy per C-mol). Thus:

biomass, h^o_x = -89.7 kJ/C-mol; CO_2, h^o_c = -393 kJ/C-mol; water, h^o_w = -285 kJ/mol; NH_3 (aq.), h^o_n = -79.4 kJ/mol; O_2, h^o_n = 0 kJ/mol.

The way to calculate this is to use Equation 10.8. We obtain for h^o_{cx} (where i = x).

$$h^o_{cx} = h^o_x - \frac{\gamma_x}{\gamma_o} h^o_o - \frac{\alpha_x}{\alpha_n} h^o_n - \frac{\beta_x}{\beta_w} h^o_w - \frac{\delta_x}{\delta_c} h^o_c$$

From our earlier ITA or from the general formula Equation 10.2 a - 10.2 d one can obtain the following values for α_x, β_x, γ_x, δ_x, α_n, β_w, γ_o and δ_c :

$$\alpha_x = 0.2 \,;\; \beta_x = 1.2 \,;\; \gamma_x = 4.2 \,;\; \delta_x = 1 \,;\; \alpha_n = 1 \,;\; \beta_w = 2 \,;\; \gamma_o = -4 \,;\; \delta_c = 1$$

Substitution of these values, with the molar enthalpies gives us:

$$h^o_{cx} = -89.7 + \frac{4.2}{4} * 0 - \frac{0.2}{1} * (-79.4) - \frac{1.2}{2} (-285) - \frac{1}{1} (-393)$$
$$= +490.2 \text{ kJ/C-mole}$$

If we apply Equation 10.8 it will be found that for water, CO_2 and N-source the value of $h^o_{ci} = 0$.

Furthermore, it can be found that for N-containing compounds the value of h^o_{ci} depends on the N-source of the system.

Table 10.2 contains, for a number of organic compounds, the calculated values of h^o_{ci}. On a closer inspection it can be seen that h^o_{ci} is in fact the enthalpy released upon oxidation of the organic compound to H_2O, CO_2 and the N-source of the system.

A further important observation from Table 10.2 is that h^o_{ci} for organic compounds in systems with NH_3 or N_2 as N-source shows a good correlation with the calculated γ_i-data. It can be shown that the following correlation holds to a good approximation.

$$h^o_{ci} = 115 \, \gamma_i \qquad\qquad (E\text{-}10.9)$$

This relation is well worth remembering. It is a mathematical expression of the fact that highly reduced organic compounds, eg CH_4, have a high degree of reduction ($\gamma = 8$) and a high combustion enthalpy.

It is stressed that Equation 10.9 is a correlation. For exact calculation of the heat production Equation 10.6 or Equation 10.7 using the appropriate values of h_i^o or h_{ci}^o is to be preferred.

heat production is proportional to oxygen consumed

An important relation can now be obtained, by combining Equation 10.9, Equation 10.7 and Equation 10.1c. It then follows that:

$$\phi_H + 460\, r_{Ao} = 0 \qquad \text{(E - 10.10)}$$

(ie the total rate of heat production is directly related to the rate of oxygen consumption).

Equation 10.10 states that for aerobic fermentations (utilising NH_3 or N_2 as N-source) the heat production follows approximately from the oxygen consumption where for each mole O_2 consumed there is a heat production of 460 kJ. Equations 10.7 and 10.10 are extremely important because they give direct access to the heat production during fermentor operations.

Because such reactors are often run at a moderate temperature ($20\text{-}40^\circ$), and oxygen consumption rates are quite high, cooling is often a limiting factor. This means that it is important to minimise heat production per unit of biomass produced or product made, hence it is relevant to study this heat production per C-mole biomass.

SAQ 10.3

1) If the rate of heat production in a bioreactor in which an aerobic organism was being cultivated, was determined to be 4410 kJh^{-1}, at what rate is oxygen being consumed? (Assume NH_3 is being used as N-source).

2) If the rate of oxygen consumption was determined to be 2 mole O_2 h^{-1}m^{-3} of reactor, how much heat would be generated every hour by a reactor with a working volume of 50m^3.

10.3.2 Heat production and growth yields

Now, in complete analogy to Equation 10.5 and the equations found in SAQ 10.2, it is possible to derive the relation between heat production per unit of biomass produced and the biomass yield Y_{sx} (for the case of no product formation). If we define the yield of biomass in terms of heat produced (Y_{Hx}) as Equation 10.11.

$$Y_{Hx} = \frac{r_{Ax}}{\phi_H} \qquad \text{(E - 10.11)}$$

We can combine Equation 10.11 with Equation 10.7 and the definition of Y_{sx} (Equation 10.3), to generate Equation 10.12.

$$Y_{Hx} = \frac{Y_{sx}}{h_{cs}^o - h_{cx}^o * Y_{sx}} \qquad \text{(E - 10.12)}$$

Furthermore Equation 10.12 can be recalculated using the correlation Equation 10.9 to give Equation 10.13.

$$Y_{Hx} = \frac{Y_{sx}}{115\,(\gamma_s - \gamma_x Y_{sx})} \qquad \text{(E-10.13)}$$

From this equation it appears that in aerobic systems Y_{Hx} is a function of Y_{sx} and γ_s only, because γ_x is more or less constant (≈ 4.2).

It is possible to calculate the heat production per unit biomass produced ($= Y_{Hx}^{-1}$) for the various organic substrates for which there are available yield data (represented in Figure 10.2). These yield data (Y_{sx}) are provided in Table 10.3, together with the calculated heat production from Equation 10.12.

substrate	γ_s	Y_{sx} (C-mole/C-mol)	ϕ_H/r_{Ax} (kJ/C-mol)	D^o_s/r_{Ax} (kJ/C-mol)
methane	8	0.49	1330	1195
n-alkanes	6.13	0.53	830	790
methanol	6	0.54	776	809
ethanol	6	0.51	851	818
glycerol	4.67	0.67	337	344
mannitol	4.33	0.56	417	444
acetic acid	4.0	0.41	578	616
lactic acid	4.0	0.51	404	426
glucose	4.0	0.61	277	311
formaldehyde	4.0	0.47	727	592
gluconic acid	3.67	0.51	-	394
succinic acid	3.50	0.40	443	526
citric acid	3.0	0.365	406	507
malic acid	3.0	0.375	395	489
formic acid	2.0	0.250	530	650
oxalic acid	1.0	0.070	1267	1855

Table 10.3 Characteristic values of Y_{sx}, $\dfrac{\phi_H}{r_{Ax}}$ and $\dfrac{D^o_s}{r_{Ax}}$ for aerobic growth on organic substrates for NH_3 as N-source.

In this table we have also reported a value D^o_s/r_{Ax} where D_s is the Gibbs free energy dissipation rate. We will discuss the free energy dissipation rate in the next section.

This heat production /C-mole biomass is represented as a function of the degree of reduction of the substrate γ_s in Figure 10.2.

From the data presented in Table 10.3 and Figure 10.2 it appears that the heat production is at a minimum for substrates with a degree of reduction γ around 4.2.

∏ Examine the data in Table 10.3 again and see if you can come to any conclusions about the heat produced with substrates with equal γ_s values (compare for example formaldehyde, acetic acid, lactic acid and glucose).

Formaldehyde contains one carbon atom, (HCHO); acetic acid contains two (CH_3COOH); lactic acid three ($CH_3CHOHCOOH$) and glucose six ($C_6H_{12}O_6$). They all conform to the general formula $(CH_2O)_x$. Examination of the data presented in Table

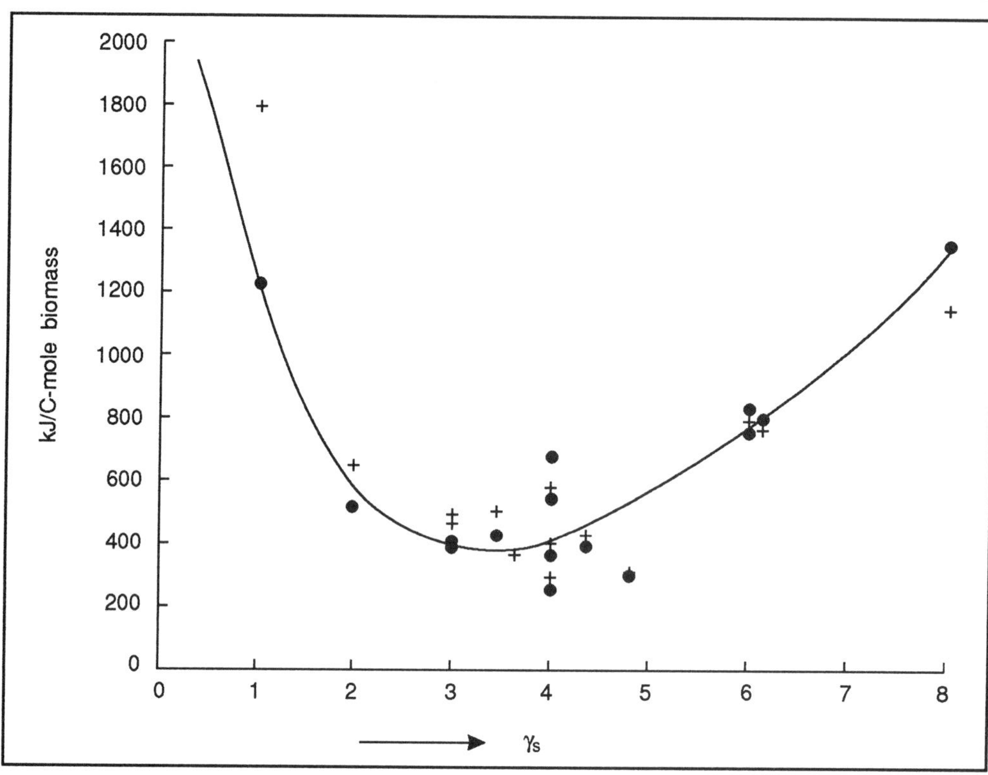

Figure 10.2 Aerobic fermentation (NH$_3$ as N-source). Heat production and Gibbs free energy dissipation per C-mole produced biomass as a function of the degree of reduction (γ_s) of the organic substrate.

● are determined from $\gamma_s = \dfrac{\phi H}{r_{Ax}}$ and + are determined from $\gamma_s = \dfrac{D_s^o}{r_{Ax}}$ (see text for details).

10.3, shows that the shorter the chain length, the greater the heat production (did you recognise this?).

We might ask ourselves, why this should be so? In broad terms, the reasons behind this are as follows.

Biomass is made up of precursor molecules (amino acids monosaccharides, nucleotide bases etc) with an average of about 5C-atoms. So we can recognise two types of energy changes when substrates are converted into biomass.

First the C-sources must be brought to the same degree of reduction as the biomass. Thus, if the C-sources are more reduced than biomass, the C-sources have to be oxidised. If the C-sources are more oxidised than biomass then they will have to be reduced.

In our example above formaldehyde, acetic acid, lactic acid and glucose are all at the same oxidation state (CH$_2$O), so we cannot explain the difference in heat production on the basis of the oxidation state of the substrate.

Second, the C-sources must be brought to the same degree of polymerisation. If the C-sources have a shorter chain length than 5C-atoms, then metabolism must result in C-atoms being joined together. Thus when formaldehyde is used as a C-source, then metabolism has to effectively make 4C-C bonds to make biomass precursor molecules with a chain length of about 5C-atoms. We can stylised this as:

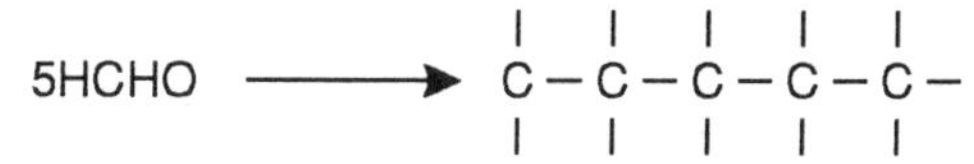

$$5HCHO \longrightarrow \overset{|\ \ |\ \ |\ \ |\ \ |}{\underset{|\ \ |\ \ |\ \ |\ \ |}{C-C-C-C-C-}}$$

formaldehyde   biomass precursor

For acetic acid, on average 2.5 C-C bonds need to be produced:

$$5H-\overset{\overset{H}{|}}{\underset{\underset{H}{|}}{C}}-\overset{\overset{O}{\|}}{C}-OH \longrightarrow 2-\overset{|\ \ |\ \ |\ \ |\ \ |}{\underset{|\ \ |\ \ |\ \ |\ \ |}{C-C-C-C-C-}}$$

<table>
<tr><td>

SAQ 10.4

</td><td>

We have continually referred to the fact that γ_x is more-or-less constant in aerobic systems at about 4.2. Use the information in Table 10.3 and equation 10.13 to show that γ_x is more-or-less constant. To do this, use the data for methane, methanol and lactic acid.

</td></tr>
</table>

<table>
<tr><td>

SAQ 10.5

</td><td>

An industrial fermentor has a cooling capacity of 10^6 kJ h^{-1}. The fixed costs of fermentor operation (overhead, depreciation etc) is 100 \$ h^{-1}. The fermentor produces biomass. Substrates of choice are methanol and glucose. The cost of methanol is \$0.20 kg^{-1} and of glucose \$0.60 kg^{-1}. Which substrate gives the lowest cost price of biomass. (Use the data of Table 10.3).

</td></tr>
</table>

10.3.3 Concluding remarks on the application of the first law of thermodynamics

The most important conclusions are that the 1st law of thermodynamics:

* provides us with a relation for the heat production in a microbial system (Equation 10.6);

* that the heat coefficient is not a compound property, but a property of the system (Equation 10.8);

* that the heat coefficient is correlated to the degree of reduction (γ), of the substrate (Equation 10.9);

* that the heat production is proportional to O_2-consumption in aerobic systems (Equation 10.10);

* that the heat production/unit biomass produced is directly related to the biomass yield on the substrate (Equation 10.12 or Equation 10.13);

- that the heat production per unit biomass is very dependent on the type of C-source; a longer chain substrate (C_6) with a degree of reduction equal to biomass gives the lowest heat production.

$\prod$ It might be a useful form of revision to write out the list of conclusions given above together with the relevant equations to produce a kind of summary sheet.

10.4 Application of the Gibbs energy balance

10.4.1 Dissipation of Gibbs energy

In a completely analogous way to that for the heat production we can apply the balance of Gibbs energy, which gives us the 'dissipation of Gibbs free energy' from the chemical processes. Based on the system definition (Table 10.1) we can then write for standard conditions (superscript°):

$$D_s^o + r_{Ax} * \mu_x^o + r_{As} * \mu_s^o + r_{Ap} * \mu_p^o + r_{Ac} * \mu_c^o$$
$$+ r_{Aw} * \mu_w^0 + r_{An} * \mu_n^0 + r_{Ao} * \mu_o^0 = 0 \qquad \text{(E 10-14)}$$

Again this relation can be combined with Equation 10.1 a - Equation 10.1d to eliminate r_{An}, r_{Aw}, r_{Ao}, r_{Ac} in order to obtain the relation between D and the chosen independent rates r_{As}, r_{Ax}, r_{Ap}.

After some manipulations one obtains Equation 10.15.

$$D_s^o + r_{Ax} * \mu_{cx}^o + r_{As} * \mu_{cs}^o + r_{Ap} * \mu_{cp}^0 = o \qquad \text{(E-10.15)}$$

We may describe μ_{ci}^o as the Gibbs free energy coefficient of compound i under standard conditions. Its value follows (in complete analogy to Equation 10.8) that:

$$\mu_{ci}^o = \mu_i^o - \frac{\gamma_i}{\gamma_o} \mu_o^o - \frac{\alpha_i}{\alpha_n} \mu_n^o - \frac{\beta_i}{\beta_w} \mu_w^o - \frac{\delta_i}{\delta_c} \mu_c^o \qquad \text{(E-10.16)}$$

The value of μ_{ci}^o can be calculated straightforwardly for each compound of the microbial growth system defined in Table 10.1, based on the chemical composition and Gibbs free energy values μ_{ci}^o of the chemical compounds. As for h_{ci}^o, the value of μ_{ci}^o is zero for water, CO_2 and N-source, while for N-containing compounds the value of μ_{ci}^o depends on the specified N-source in the system (as do h_{ci}^o and γ_i).

Table 10.2 contains the calculated μ_{ci}^o - values for a number of organic compounds.

On a close inspection, it can be seen that μ_{ci}^o is the standard Gibbs energy which is released upon oxidisation with O_2, of 1-C-mole of organic compound to CO_2, H_2O and the N-source of the microbial system. It appears that μ_{ci} is lower for the more oxidised compounds. Thus from Table 10.2 it appears that, for organic compounds, μ_{ci}^o is correlated to the degree of reduction of the substrate γ_s.

The relationship is:

$$\mu_{ci}^0 = 94.4\,\gamma_i + 86.6\ kJ/C\text{-}mol \qquad\qquad (E\text{-}10.17)$$

It is now possible to derive a relation between the biomass yield Y_{sx} and the Gibbs energy dissipation per C-mole biomass (D_s^0/r_{Ax}), (when there is no product formation ie $r_{Ap} = 0$)

Combination of Equation 10.15 and 10.3 gives:

$$\frac{D_s^{\,o}}{r_{Ax}} = \frac{\mu_{cs}^o}{Y_{sx}} - \mu_{cx}^o \qquad\qquad (E\text{-}10.18)$$

Using this equation one can calculate $\dfrac{D_s^o}{r_{Ax}}$ for the various aerobic growth systems with NH_3 as N-source for organic substrates.

The results are presented in Table 10.3 and Figure 10.2.

There is a strong correlation between $\dfrac{D_s^o}{r_{Ax}}$ and $\dfrac{\phi_H}{r_{Ax}}$, which shows that in aerobic systems the dissipation of Gibbs free energy is mainly by heat production.

It appears that the lowest Gibbs energy dissipation occurs for long chain C-sources with a degree of reduction of about 4. The reason is simply that these C-sources require the least modification for biomass synthesis since each modification requires a certain effort (energy), which is quantitatively expressed as dissipation of Gibbs free energy. It is clear that for substrates that require extensive modification requirement (like CH_4 or oxalic acid) the dissipation is high.

Π Using the values of $\dfrac{D_s^o}{r_{Ax}}$ and Y_{sx} from Table 10.3 and values of μ_{ci}^o for each substrate from Table 10.2 to calculate μ_{cx}^o using methane and lactic acid as substrates and ammonia as N-source.

Your calculations should have shown that $\mu_{cx}^0 = +474\ kJ/C\text{-}mole$ biomass in both cases. They should of course, be the same since biomass is the same in each case. We did these calculations in the following way:

$$\text{Since: } \frac{D_s^o}{r_{Ax}} = \frac{\mu_{cs}^o}{Y_{sx}} - \mu_{cx}^o \qquad\qquad (E\text{-}10.18)$$

$$\text{then for methane: } 1195 = \frac{818}{0.49} - \mu_{cx}^o$$

therefore: $\mu_{cx}^o = +\,474\ kJ/C\text{-}mole$

$$\text{for lactic acid: } 426 = \frac{459}{0.51} - \mu_{cx}^o, \text{ leading to the same value for } \mu_{cx}^o$$

10.4.2 Maximum biomass yields

It is well established that conservation principles *do not* place restriction on the process direction. For example from a conservation point of view it would be perfectly allowable to produce O_2 and biomass from heat, H_2O, CO_2, NH_3. However the 2nd law of thermodynamics provides the direction of the process by stating that:

$D_s = 0$ at equilibrium and that
$D_s > 0$ for each process

$$\text{(E-10.19a)}$$

If we assume that we do not introduce a serious error (which can indeed be shown to be correct) by using standard conditions (1 mol l^{-1}, 1 atmosphere pressure and 298K) we can combine Equations 10.15 and 10.19a to a limit for Y_{sx}, which is set by the 2nd law.

$$r_{Ax}\,\mu^{o}_{cx} + r_{As}\,\mu^{o}_{cs} \leq 0 \qquad\qquad \text{(E-10.19b)}$$

This means that for $Y_{sx}\,(= -\frac{r_{Ax}}{r_{As}})$ one can write under standard conditions.

$$Y_{sx} \leq \frac{\mu^{o}_{cs}}{\mu^{o}_{cx}} \qquad\qquad \text{(E-10.20)}$$

Equation 10.20 clearly shows us that the maximum biomass yield Y^{max}_{sx} is dependent on the value of μ^{o}_{cs}, assuming that μ^{o}_{cx} is constant (Table 10.2).

We have shown (Equation 10.17) that μ^{o}_{cs} is correlated to the degree of reduction of the substrate γ_s.

Combination of Equation 10.20 and 10.17 for an aerobic growth system with NH_3 as substrate (for which $\mu^{o}_{cx} = 474$ kJ/C-mole - we calculated this earlier), enables us to generate Equation 10.21 to calculate the thermodynamic maximum growth yield.

$$Y^{max}_{sx} \leq \frac{94.4\ \gamma_s + 86.6}{474} \qquad\qquad \text{(E-10.21)}$$

∏ Calculate some values of the maximum Y_{sx} values for some substrates using the data presented in Table 10.3. For example: glucose $\gamma_s = 4$. Thus maximum Y_{sx} values is just under 1. Carry out this calculation for a few other substrates and compare these maxima with real Y_{sx} values (Table 10.3).

If we compare these maxima with the actual values of Y_{sx} it appears that generally aerobic micro-organisms have a yield which is approximately 60% of the theoretical maxima.

Finally, it is relevant to stress the fact that Equation 10.14 is the basic relation to calculate D_s, which can be derived directly from a proposed system definition.

Equations 10.15 - 10.21 are useful for the examining aerobic microbial growth.

10.4.3 Application to anaerobic system

As an illustration of Equation 10.14 let us try to calculate the Gibbs free energy dissipation and heat production for an anaerobic microbial system which produces ethanol from glucose. Experimentally a biomass yield $Y_{sx} = 0.133$ C-mol/C-mol glucose is found. The system definition is as follows:

- Relevant compounds - glucose, biomass, CO_2, H_2O, NH_3, ethanol;

- 6 compounds, 4 elements.

This means that we can choose 2 independent rates to express the other 4 rates. Let us choose substrate and biomass (r_{As} and r_{Ax}) and let r_{Ap} be the net-conversion rate of ethanol. We will use a biomass composition of ($CH_{1.7}O_{0.5}N_{0.2}$).

After some manipulations with the elemental conservation relations one obtains (using the usual biomass composition):

$$+ r_{Ac} = -0.30\, r_{Ax} - 0.333\, r_{As}$$

$$+ r_{Ap} = -0.70\, r_{Ax} - 0.666\, r_{As}$$

$$- r_{An} = 0.2\, r_{Ax}$$

$$+ r_{Aw} = +0.45\, r_{Ax}$$

The values for molar enthalpies and Gibbs free energies, in kJ C-mol^{-1}, are:

H_2O	$h_w^o =$	-285	$\mu_w^o =$	-237
CO_2	$h_c^o =$	-393	$\mu_i^o =$	-394
NH_3	$h_n^o =$	-79.4	$\mu_n^o =$	-26.7
biomass	$h_x^o =$	-90.7	$\mu_x^o =$	-67.1
glucose	$h_s^o =$	-210.5	$\mu_s^o =$	-152.4
ethanol	$h_p^o =$	-138.5	$\mu_p^o =$	-87.1

The relations for ϕ_H and D_s^o run in analogy to Equation 10.6 and 10.14 as:

$$\phi_H + r_{Ax} * h_x^o + r_{As} * h_s^o + r_{An} * h_n^o + r_{Ac} * h_c^o + r_{Aw}\, h_w^o + r_{Ap} * h_p^o = 0$$

$$D_s^o + r_{Ax} * \mu_x^o + r_{As} * \mu_s^o + r_{An} * \mu_n^o + r_{Ac} * \mu_c^o + r_{Aw} * \mu_c^o + r_{Ap} * \mu_p^o = 0$$

Substitution of the h_i^o and μ_i^o values, and elimination of $r_{Ac}, r_{Ap}, r_{An}, r_{Aw}$ with the above elemental relationships leads to the relationship of ϕ_H and D_s^o as function of r_{As} and r_{Ax}

$$\phi_H + 11.8\, r_{Ax} + 12.6\, r_{As} = 0$$

$$D_s^o + 10.8\, r_{Ax} + 36.8\, r_{As} = 0$$

Because the biomass yield on glucose is 0.133 C-mol/C-mol $\left(= \dfrac{r_{Ax}}{-r_{As}} \right)$, we can eliminate r_{As} and obtain $\dfrac{\phi_H}{r_{Ax}}$ and $\dfrac{D_s^o}{r_{Ax}}$

For the heat production /C-mole produced biomass we can calculate that:

$$\phi_H / r_{Ax} = 83 \text{ kJ/C-mol}$$

Note that in this anaerobic case there is still considerable heat-production despite the absence of O_2-consumption. It also shows that Equation 10.10 is only applicable to aerobic systems.

For the Gibbs energy dissipation/C-mole produced biomass we calculate that:

$$D_s^o / r_{Ax} = 266 \text{ kJ/C-mol}$$

If we compare this result for anaerobic biomass production with the aerobic biomass production (Table 10.3, glucose as substrate) two important facts emerge.

- the heat production per unit produced biomass is a factor of 3-4 times lower for the anaerobic process $83 \leftarrow \rightarrow 277$.

- the Gibbs energy dissipation per unit produced biomass is more or less the same $266 \leftarrow \rightarrow 310 \text{ kJ/C-mol}$.

energy dissipation per unit of biomass independent of electron acceptor

This indicates that the Gibbs energy dissipation per unit biomass appears to be a constant for a given C-source, irrespective of the applied electron acceptor. This is understandable because, depending on the C-source structure, a certain amount of effort is needed to modify this towards biomass precursors. This effort seems independent of the electron acceptor.

SAQ 10.6

What is the thermodynamic maximum value of Y_{sx} in the anaerobic ethanol fermentation described above? (Hint: we know that at equilibrium $D_s^o = 0$ and we know that $Y_{sx}^{max} = \dfrac{r_{Ax}}{-r_{As}}$).

10.4.4 Concluding remarks on the application of the 2nd law of the thermodynamics

In conclusion we have shown that the 2nd law of thermodynamics provides us with

- the maximum value of biomass yield Y_{sx} which can be obtained (Equation 10.20) for aerobic systems;

- the Gibbs free energy dissipation/unit biomass produced (D_s^o / r_{Ax}) depends very much on the type of C-source (Table 10.3, Figure 10.2);

- dissipation of Gibbs free energy/unit biomass for a specific C-source can be used to estimate Y_{sx} for different electron acceptors.

10.5 Application of metabolic modelling

10.5.1 Defining metabolism and the concept of maintenance

The usefulness of a metabolic description has been shown in the previous chapter.

By analogy to the conservation principles for aerobic growth one can generalise the metabolic modelling for aerobic growth on organic substrates. The procedure is completely identical to the example in Chapter 9 (see Section 9.8). Two modifications will however be introduced:

- in industrial practice a product is often formed in an aerobic fermentation;

- in microbial growth systems one has to take into consideration the so-called maintenance energy requirement. This follows from the observation that biomass yield from substrate decreases significantly if the growth rate decreases. A number of possible explanations have been given all of which are related to the understanding that living systems have to 'maintain' their complex and ordered structure against the natural tendency of erosion, degradation, dissipation etc.

For example:

- proteins denature at a given rate and become inactive;

- concentration and charge gradients over the various membranes have to be maintained against chemical and electrical leakage;

- many of the biochemical intermediates can become chemically degraded and need to be recycled.

The continuous repair and recycling effort costs Gibbs energy, which means ATP expenditure.

Based on these considerations one can design the following 6 reactions as a metabolic model for aerobic growth and product formation and including maintenance:

reaction 1	anabolism (synthesis) of biomass precursors
reaction 2	polymerisation of precursors to biomass
reaction 3	substrate catabolism (breakdown)
reaction 4	oxidative phosphorylation
reaction 5	product formation from substrate
reaction 6	maintenance

The relevant extracellular compounds for this microbial system are already defined in Table 10.1.

In addition, also defined are the intracellular intermediates ATP, $NADH_2$, biomass precursor by analogy to Chapter 9 (Section 9.8).

Based on the defined relevant compounds, we can draw up the 6 reaction equations; these we present in Figure 10.3.

We have used the same conventions we used in Chapter 9, except we have generalised the stoichiometric coefficients. Thus, '$C_{cx}H_{hx}O_{ox}N_{nx}$' represents biomass precursors and $C_{cx}H_{nx}O_{ox}N_{nx}$ is biomass.

∏ Read through Figure 10.3 and circle all of the intracellular intermediates. As you examine this Figure bear in mind that:

- in reactions 1 and 5, α_{11} and α_{51} depend upon the amount of decarboxylation reactions;

- in reaction 1, α_{12} is the amount of ATP needed for synthesis of biomass precursors;

- in reaction 2, α_{22} is the ATP needed for polymerisation of precursors to product biomass;

- in reaction 3, α_{32} is the ATP used or produced in catabolism. This includes ATP needed for substrate transport and ATP from substrate phosphorylation;

- in reaction 4, α_{42} is related to the so-called P/O ratio; $\alpha_{42} = 1/(P/O)$.

In biochemistry the P/O ratio is the amount of ATP produced by oxidative phosphorylation in relation to the amount of oxygen consumed by the oxidative phosphorylation process.

You should also note that:

- in reaction 5, α_{52} is the ATP needed for product formation

You should also note that:

- stoichiometric coefficients in each reaction follow from specific biochemical knowledge;

- the stoichiometric coefficients follow from application of elemental conservation to each reaction. In reactions r_1 to r_6, use has been made of the definition of degree of reduction γ_i. If, for example in reaction 1, γ_x and γ_s are replaced by their full expression in terms of the composition coefficients (Equation 10.2c) one can check that indeed the C, H, O, N-conservation in reaction 1 is satisfied;

- a product forming reaction r_5 is specified;

- the maintenance process is represented by reaction 6 which specifies the degradation of biomass to its precursors.

$- r_1$ **anabolism of biomass precursors**

$$- \alpha_{11} C_{cs} H_{hs} O_{os} - \alpha_{12} ATP - nxNH_3 + 0.5 (\gamma_s \alpha_{11} - \gamma_x) NADH_2 +$$
$$1 \; ' \underline{C_{cx} H_{hx} O_{ox} N_{nx}} \; ' + (cs\alpha_{11} - cx) CO_2 + (2cx - ox + os\,\alpha_{11} - 2cs\,\alpha_{11}) H_2O = 0$$

$- r_2$ **polymerisation of biomass precursors to biomass**

$$- 1 \; ' C_{cx} H_{hx} O_{ox} N_{nx} ' - \alpha_{22} ATP + \underline{1 \; C_{cx} H_{hx} O_{ox} N_{nx}} = 0$$

$- r_3$ **substrate catabolism**

$$- \frac{2}{\gamma_s} C_{cs} H_{hs} O_{os} + \alpha_{32} ATP + \underline{1NADH_2} - (2 - os) \frac{2}{\gamma_s} H_2O + \frac{2}{\gamma_s} CO_2 = 0$$

$- r_4$ **oxidative phosphorylation**

$$- \alpha_{42} NADH_2 - 0.5 \, \alpha_{42} O_2 + \alpha_{42} H_2O + \underline{1 ATP} = 0$$

$- r_5$ **product formation from substrate**

$$- \alpha_{51} C_{cs} H_{hs} O_{os} - \alpha_{52} ATP - npNH_3 + 0.5 (\gamma_s \alpha_{51} - \gamma_p) NADH_2 +$$
$$1 \; \underline{C_{cp} H_{hp} O_{op} N_{np}} + (cs\alpha_{11} - cp) CO_2 +$$
$$(2 cp - op + os\,\alpha_{51} - 2 cs\,\alpha_{51}) H_2O = 0$$

$- r_6$ **maintenance**

$$- C_{cx} H_{hx} O_{ox} N_{nx} + ' \underline{C_{cx} H_{hx} O_{ox} N_{nx}} ' = 0$$

Figure 10.3 Metabolic description of microbial aerobic growth and product formation (see text for details).

From this description we can see that there are:

- 10 chemical compounds (3 intracellular, 7 extracellular) for which one can formulate 10 reaction balances analogous to Equation 9.11a - i;

- 13 unknown rates (7 times r_{Ai} and 6 times r_i).

There are therefore $13-10 = 3$ degrees of freedom, for which we can choose r_{Ax}, r_{Ap} and r_6 as independent.

It is now possible to calculate in an analogous way to that described in Chapter 9 (Section 9.8), a relation between these 3 rates and the substrate consumption. This leads to the following relation:

$$(-r_{As}) = \frac{1}{Y_{sx}^{max}}\, r_{Ax} + \frac{1}{Y_{sp}^{max}}\, r_{Ap} + m_S\, C_x$$

$$(E-10.22)$$

This equation is well worth considering more closely. It describes the rate of substrate use $(-r_{As})$ in relation to the maximum of biomass yield (Y_{sx}^{max}); the rate of biomass production (r_{Ax}); the maximum product yield (Y_{sp}^{max}); the rate of product production (r_{Ap}) and the maintenance requirements $(m_s\, C_x)$.

m_s is the maintenance coefficient in terms of the rate of substrate consumption per unit of biomass for maintenance. It has the form of unit of substrate/unit of biomass x unit of time. C_x is the biomass concentration and $m_s = \dfrac{r_6}{C_x}$.

Note that if no product is made the relationship is simplified to $-r_{As} = \dfrac{1}{Y_{sx}^{max}} + m_s C_x$.

Herbert/Pirt relation

This relation, without r_{Ap}, is known as the Herbert/Pirt relation. It appears therefore that these types of relations can be 'derived' from a metabolic model description.

In general it is advisable to derive, for each case of microbial growth under consideration, the proper form of this linear relation. Especially in situations with multiple substrates and/or products it is not possible to predict intuitively the form of Equation 10.22. This can only be found from metabolic modelling.

In Equation 10.22 Y_{sx}^{max}, Y_{sp}^{max} and m_s are related to the metabolic model parameters of the 6 reactions in Figure 10.3 as follows:

$$Y_{sx}^{max} = \frac{\alpha_{32}\,\alpha_{42}\,\gamma_s + \gamma_s}{\alpha_{11}\,\alpha_{32}\,\alpha_{42}\,\gamma_s + 2\,\alpha_{42}\,(\alpha_{12} + \alpha_{22}) + \gamma_x}$$

$$(E-10.23)$$

$$Y_{sp}^{max} = \frac{\alpha_{32}\,\alpha_{42}\,\gamma_s + \gamma_s}{\alpha_{51}\,\alpha_{32}\,\alpha_{42}\,\gamma_s + 2\,\alpha_{42}\,(\alpha_{52}) + \gamma_p}$$

$$(E-10.24)$$

$$m_S = \frac{2\,\alpha_{42}\,\alpha_{22}}{\alpha_{32}\,\alpha_{42}\,\gamma_s + \gamma_s} \times \frac{r_6}{C_x}$$

$$(E-10.25)$$

Dividing Equation 10.22 by C_x leads to a popular version (Equation 10.26).

Note:

$$-q_s = \text{rate of substrate use per unit of biomass} = \frac{-r_{As}}{C_x}; \quad \mu = \text{specific growth rate} = \frac{r_{Ax}}{C_x}; \quad q_P$$

$$= \text{rate of product formation per unit of biomass} = \frac{r_{Ap}}{C_x}$$

m_s = maintenance coefficient = rate of substrate consumption to provide maintenance energy per unit of biomass

$$(-q_s) = \frac{1}{Y_{sx}^{max}}\,\mu \; + \; \frac{1}{Y_{sp}^{max}}\,q_p + m_s \qquad\qquad\qquad \text{(E - 10.26)}$$

This version is well worth remembering, it has broad application.

In this relation q_s, q_p, μ are the biomass specific conversion rates of substrate, product and biomass. These may be defined as:

$$(- q_s) = - r_{As}/C_x$$

q_s is also known as the metabolic quotient for the substrate and is the rate of substrate consumption per unit of biomass.

$$\mu = + r_{Ax}/C_x$$

μ is the rate of biomass production per unit of biomass and is known as the specific growth rate.

$$q_p = + r_{Ap}/C_x$$

q_p is the metabolic quotient for product synthesis and is the rate product is made per unit of biomass.

If we re-examine Equation 10.23. It provides us with the maximum yield (Y_{sx}^{max}) of biomass from a substrate. From this equation, it follows that maximum yield:

- increases if γ_s increases;

- increases if α_{42} decreases, which means a higher P/O ratio;

- increases if $\alpha_{12} + \alpha_{22}$ decreases, which means less ATP-need for biomass synthesis from substrate.

In other words:

- the more reduced the substrate the greater the biomass yield;

- the more efficient oxidative phosphorylation (the greater the P/O ratio) the greater the yield;

- the lower the energy demand for biomass production, the greater the biomass yield.

A moments thought will show that these are, in fact, common sense conclusions.

- Equation 10.24 provides the maximal yield Y_{sp}^{max} of product from substrate. The effects of various α's are analogous to Y_{sx}^{max}.

- Equation 10.25 provides the substrate need for maintenance. A popular assumption is that the rate r_6/C_x is constant. The product of $\alpha_{22} * r_6/C_x$ can then be identified as the rate of ATP needed for maintenance purposes m_{ATP} (mol ATP/C-mole biomass h). Elimination of r_6/C_x leads then to Equation 10.27, where m_{ATP} is a constant.

$$m_s = \frac{m_{ATP}}{\gamma_s/2\,(\alpha_{32} + 1/\alpha_{42})}$$

(E - 10.27)

From literature one can find typical m_{ATP} -values in the range of 0.04 - 0.1 mol ATP/C-mol biomass h.

10.5.2 Effect of different assumptions for the kinetics of the maintenance reaction rate r_6

For the derivation of Equation 10.26 it was assumed that the rate of reaction 6 was constant. However, it could also well be that the rate of reaction 6 is not constant but for example varies linearly according to the growth rate μ as described in Equation 10.28.

$$\frac{r_6}{C_x} = m_o + b\,\mu$$

(E - 10.28)

∏ Combine this with Equation 10.25 and 10.22 and make a comment on the merit of the proposal (10.28). Try this on a piece of paper before reading on.

If $\frac{r_6}{C_x}$ is eliminated from Equation 10.28 and Equation 10.25, one does obtain the same relation 10.22 but for Y_{sx}^{max} and m_s one obtains:

$$Y_{sx}^{max} = \frac{\alpha_{32}\,\alpha_{42}\,\gamma_s + \gamma_s}{\alpha_{11}\,\alpha_{32}\,\alpha_{42}\,\gamma_s + 2\,\alpha_{42}\,(\alpha_{12} + \alpha_{22}(1 + b)) + \gamma_x}$$

$$m_s = \frac{2\alpha_{42}\,\alpha_{22}}{\alpha_{22}\,\alpha_{42}\,\gamma_s + \gamma_s} * m_O$$

The parameter b enters in Y_{sx}^{max} and m_o in m_s.

Hence, the assumption of a constant rate (that is a linear increase with μ for m_s) leads to the same overall result of Equation 10.22. The difference resides in the interpretation of Y_{sx}^{max} in relation to the reaction stoichiometric coefficients.

In other words, if one finds a constant m_s from Equation 10.22 and Equation 10.28, this does not prove that the rate of the maintenance reaction is constant because in general it is not possible to determine b from Y_{sx}^{max}.

Equation 10.22 provides us, for the special case of no product formation ($r_{Ap} = 0$), with a relation between the *actual* growth yield Y_{sx} and the maximal yield Y_{sx}^{max}.

Y_{sx} is defined in Equation 10.3. Furthermore, the growth rate μ is defined as $\frac{r_{Ax}}{C_x}$

Elimination of r_{As} and r_{Ax} from Equations 10.22, 10.27 and 10.3, leads to:

$$\frac{1}{Y_{sx}} = \frac{1}{Y_{sx}^{max}} + \frac{m_s}{\mu}$$

(E - 10.29)

This is a very useful relation because, experimentally, it can be used to determine Y_{sx}^{max} and m_s. If we determine Y_{sx} values at several different values of μ, then we can determine Y_{sx}^{max} and m_s.

This is commonly done by plotting $1/Y_{sx}$ versus $1/\mu$. The intercept then provides $1/Y_{sx}^{max}$ and the slope provides m_s (Equation 10.29). An alternative is plotting $(-q_s)$ versus μ, the intercept is m_s and the slope is $1/Y_{sx}^{max}$ (Equation 10.26 a without product formation).

The effect of maintenance can be very dramatic with respect to yields, especially at low μ. A typical value of m_s for aerobic growth on organic substrates is 0.020 C-mol substrate/C-mol biomass h

Suppose that in a microbial system $Y_{sx}^{max} = 0.5$ C-mol biomass/C-mol substrate.

Using Equation 10.29 one can then calculate the actual value of Y_{sx} for various growth-rates μ. This Y_{sx} can then be used to calculate Y_{ox} (Equation 10.4) and $(Y_{Hx})^{-1}$, (Equation 10.12). These are the rates of biomass production/rate of oxygen consumed and rate of biomass production/rate of heat produced respectively.

The result shown in Table 10.4 were obtained for biomass ($\gamma_x = 4.2$) and glucose ($\gamma_s = 4$).

μ hr^{-1}	Y_{sx} (C-mol biomass/ C-mol substrate)	Y_{ox} (C-mol biomass/ mol O$_2$)	$(Y_{HR})^{-1}$ kJ/C-mol biomass
1	0.495	1.03	455
0.5	0.490	1.01	464
0.25	0.480	0.97	484
0.12	0.460	0.89	527
0.06	0.428	0.78	603
0.03	0.375	0.62	758
0.01	0.250	0.34	1382
0.005	0.167	0.20	2312
0.002	0.083	0.09	5148

Table 10.4 Relationship of growth rate to yield expressed in terms of substrate, oxygen consumption and heat production.

Considering the practical fact that industrial fermentations are often run as fed batch processes where the growth rate μ is low (0.01 - 0.05 h^{-1}) it appears that maintenance is a very important aspect. It leads to highly increased O$_2$ consumption and heat production per C-mol biomass produced at the lower growth rates.

Π One final point on maintenance, write down what you would anticipate the affect of mutating a bacterium from a motile (ie processing flagella) into a non-motile form on the value of m_s.

This was really a trick question. Motility of course expends energy. The energy the motile bacterium uses to drive its flagella means this energy is not available for synthesising new biomass. Thus this use of energy is similar to that expended in maintenance, it is energy consumption which does not lead to any net increase in biomass. For practical purposes energy of motility can be regarded as a component of the maintenance requirement. Thus we would expect the maintenance (m_s) value of motile organism to be greater than that of a similar, but non-motile one.

SAQ 10.7

Calculate the value of Y_{sx}^{max} (using Equation 10.23) for the aerobic growth on glucose. In this growth, there is no product other than biomass (with composition $CH_{1.8}O_{0.5}N_{0.2}$), CO_2 and H_2O.

Using a similar annotation as described in Figure 10.3, the following stoichiometric relationships have been established:

Anabolism of biomass precursors

- 1.095 glucose - 0.051 ATP - 0.2 NH_3 + 0.09 $NADH_2$ + 1 '$C_{1.8} H_{0.5} O_{0.2} N_{nx}$' + 0.095 CO_2 + 0.405 H_2O = 0

Polymerisation of biomass precursor to biomass

- '$C H_{1.8} O_{0.5} N_{0.2}$' - 1.7 ATP + $C H_{1.8} O_{0.5} N_{0.2}$ = 0

Substrate catabolism

- 0.5 CH_2O + 0.33 ATP + $NADH_2$ - + 0.5 H_2O + 0.5 CO_2 = 0

Oxidative phosphorylation

- 0.5 $NADH_2$ - 0.25O_2 + 0.5H_2O + ATP = 0

The degree of reduction of substrate and biomass (γ_s and γ_x are 4 and 4.2).

SAQ 10.8

The following results have been obtained from some studies on cultures of a micro-organism.

Specific growth rate μ (h^{-1})	Growth yield Y_{sx} (C-mol biomass/C-mol substrate)
1	0.495
0.5	0.490
0.25	0.480
0.12	0.460
0.06	0.428

From this data, determine the maintenance coefficient (m_s) and the maximum growth yield Y_{sx}^{max}.

Summary and objectives

In this chapter we have applied conservation principles (elements, energy) the second law of thermodynamics and the modelling of metabolism to establish a large number of extra relations between net-conversion rates and limits to maximal yields. Despite the number of relations that we can establish, there are always one or two net conversion rates which are unspecified. These are usually the rates of substrate conversion (r_{As}), growth (r_{Ax}) or the rate of product formation (r_{Ap}). These either have to be determined experimentally, or we need to use kinetic relations for these conversion rates. These are the topic of the next chapter.

Now that you have completed this chapter, you should be able to:

- calculate stoichiometric coefficients from supplied data;

- use the term growth yield and calculate growth yield values;

- calculate rates of heat production and substrate consumption from supplied metabolic data;

- explain why substrate chain length and state of reduction (γ_s) influences growth yield and rate of heat production;

- use metabolic modelling to calculate growth yields and maintenance from supplied data.

Kinetics of microbial processes

Kinetics of microbial processes

11.1 Introduction

In the preceding chapters it has been shown that microbial systems can be fully characterised but that there always remain 1 or 2 degrees of freedom. This means that at least for 1 or 2 net-conversion rates appropriate kinetic relations are needed. In general these will be a kinetic relation for:

- substrate uptake or biomass production;

- product formation.

Furthermore the kinetics of maintenance are also needed. It is well established that microbial growth and product formation is the result of long chains of sequential or branched enzyme catalysed reaction pathways. Often, one of the enzyme-catalysed steps is rate limiting. Hence it is logical to expect that growth and product formation can be based on single enzyme kinetics.

In Section 11.2 we will therefore present a concise treatment of single enzyme kinetics. In Section 11.3 we will extend this treatment to consider metabolic pathways. This chapter will be concluded with a brief examination of the kinetics of maintenance (Section 11.4).

11.2 Single enzyme kinetics

11.2.1 Introduction

In this section, several aspects of single enzyme kinetics will be examined. These include:

- a mechanistic view to obtain a rate equation based on elementary kinetics. (It is assumed that the reader is familiar with the transition state theory of reaction rates and catalysis - the BIOTOL texts of 'The Molecular Fabric of Cells' gives a descriptive account of these);

- enzyme inhibition;

- effect of temperature on enzyme kinetics.

11.2.2 Basic rate equation

In general it is assumed that a substrate S combines with the enzyme E to form an enzyme substrate complex ES, which subsequently is converted into the original enzyme E and product P. Here it is assumed that the breakdown of ES into E and P is the rate limiting step of the overall process and that it does not affect the equilibrium.

This is basically the Michaelis-Menten theory of enzyme action. This theory assumes that the enzyme forms a complex with the substrate and that the complex dissociates into the free enzyme and the end product of the enzymatic reaction. We can represent this diagrammatically as shown in Figure 11.1.

$$\text{Enzymatic mechanism} \qquad E + S \underset{k_{-1}}{\overset{k_1}{\rightleftharpoons}} ES \underset{k_{-2}}{\overset{k_2}{\rightleftharpoons}} E + P$$

Where S = substrate, P = product, E = enzyme, ES = enzyme substrate complex; k_1, k_{-1}, k_2 and k_{-2} are reaction constants

Figure 11.1 Mechanistic representation of a simple enzyme catalysed reaction.

In this theory of enzymatic reaction, it is assumed that the various reactions are reversible. For the chemical reaction system shown in Figure 11.1 in a batch situation, we can write the following 4 macroscopic balances for the 4 chemical species E, S, ES, P.

The transport terms in these macroscopic balances are absent. The kinetic equations are 1st or 2nd order. Thus:

$$\overset{o}{C}_E = -k_1 C_E C_S + k_{-1} C_{ES} + k_2 C_{ES} - k_{-2} C_E C_P \qquad \text{(E - 11.1a)}$$

$$\overset{o}{C}_{ES} = k_1 C_E C_S - k_{-1} C_{ES} - k_2 C_{ES} + k_{-2} C_E C_P \qquad \text{(E - 11.1b)}$$

$$\overset{o}{C}_S = -k_1 C_E C_S + k_{-1} C_{ES} \qquad \text{(E - 11.1c)}$$

$$\overset{o}{C}_P = k_2 C_{ES} - k_{-2} C_E C_P \qquad \text{(E - 11.1d)}$$

where C_E concentration of enzyme, C_{ES} concentration of enzyme substrate complex, C_S concentration of substrate, C_P concentration of product, $\overset{o}{C}_i$ signifies the change of component i with time.

From Equation 11.1a and 11.1b it appears that $\overset{o}{C}_E + C_{ES} = 0$ (try it).

This is correct because there is supposed to be available a total amount of enzyme C_{E0} which is equal to the free enzyme C_E and the complexed enzyme C_{ES}. Thus if $\overset{o}{C}_E$ changes $\overset{o}{C}_{ES}$ must change in an equal but opposite direction. We can write Equation 11.1e.

$$C_E + C_{ES} = C_{E0} \qquad \text{(E - 11.1e)}$$

where C_{E0} = original amount of enzyme

It is now possible to obtain a relation for the rate of substrate consumption by solving the above 4 relations for 4 unknown compounds. However there is no analytical solution because the set of differential equations is non-linear.

A good approximation can however be obtained by using again the pseudosteady state assumption. It is assumed that ES is formed as quickly as it breaks down.

Thus we may write for Equation 11.1b.

$$\overset{o}{C_{ES}} = 0 \qquad\qquad\qquad (E\text{-}11.2)$$

Equation 11.1b leads in combination with Equation 11.1e and Equation 11.2, to a relation for the concentration C_{ES} (Equation 11.3).

$$\frac{C_{ES}}{C_{E0}} = \frac{k_1 C_S + k_{-2} C_p}{k_1 C_S + k_{-1} + k_2 + k_{-2} C_p} \qquad\qquad (E\text{-}11.3)$$

Elimination of C_E and C_{ES} from 11.1c leads for r_{As} to

$$\frac{-r_{As}}{C_{E0}} = \frac{+k_1 k_2 C_S - k_{-1} k_{-2} C_p}{k_1 C_S + k_{-1} + k_2 + k_{-2} C_p} \qquad\qquad (E\text{-}11.4)$$

∏ It would be helpful for you to derive these for yourself. Try this on a piece of paper first then look at our solution below.

From Equations 11.1b and 11.2 we can write:

$$k_1 C_E C_S - k_{-1} C_{ES} - k_2 C_{ES} + k_{-2} C_E C_P = 0$$

$$11.1e \;\rightarrow\; C_E = C_{E0} - C_{ES}$$

$$k_1 (C_{E0} - C_{ES}) C_S - k_{-1} C_{ES} - k_2 C_{ES} + k_{-2} (C_{E0} - C_{ES}) C_P = 0$$

$$k_1 C_{E0} C_S - k_1 C_{ES} C_S - k_{-1} C_{ES} - k_2 C_{ES} + k_{-2} C_{E0} C_P - k_{-2} C_{ES} C_P = 0$$

$$(-k_1 C_S - k_{-1} - k_2 - k_{-2} C_P) C_{ES} = (-k_1 C_S - k_{-2} C_P) C_{E0}$$

$$\frac{C_{ES}}{C_{E0}} = \frac{k_1 C_S + k_{-2} C_P}{k_1 C_S + k_{-1} + k_2 + k_{-2} C_P}$$

$$11.1c \;\rightarrow\; r_{As} = -k_1 C_E C_S + k_{-1} C_{ES} = -k_1 (C_{E0} - C_{ES}) C_S + k_{-1} C_{ES}$$

$$\rightarrow \frac{r_{AS}}{C_{E0}} = \frac{-k_1 C_{E0} C_S + k_1 C_{ES} C_S + k_{-1} C_{ES}}{C_{E0}} = -k_1 C_S + (k_1 C_S + k_{-1}) \frac{C_{ES}}{C_{E0}}$$

$$= -k_1 C_S + (k_1 C_S + k_{-1}) \frac{k_1 C_S + k_{-2} C_P}{k_1 C_S + k_{-1} + k_2 + k_{-2} C_P}$$

$$= \frac{-k_1 k_1 C_S C_S - k_1 k_{-1} C_S - k_1 k_2 C_S - k_1 k_{-2} C_S C_P + k_1 k_1 C_S C_S + k_1 k_{-2} C_S C_P + k_{-1} k_1 C_S + k_{-1} k_{-2} C_P}{k_1 C_S + k_{-1} + k_2 + k_{-2} C_P}$$

$$\frac{-r_{As}}{C_{E0}} = \frac{+ k_1 k_2 C_S - k_{-1} k_{-2} C_P}{k_1 C_S + k_{-1} + k_2 + k_{-2} C_P}$$

Equation 11.4 is called the reversible Michaelis-Menten equation for enzyme kinetics.

$\left(\dfrac{-r_{As}}{C_{E0}}\right)$ is the enzyme specific substrate conversion rate (ie the rate of substrate used per specified amount of enzyme).

It should be noted that the reaction rate constants k_1, k_{-1}, k_2, k_{-2} are not totally independent. They are subject to the equilibrium condition.

Equilibrium is achieved when r_{As} and r_{Ap} are zero.

From Equation 11.1c and 11.1d it follows that:

$$\frac{C_P^*}{C_S^*} = K_{EQ} = \frac{k_1}{k_{-1}} * \frac{k_2}{k_{-2}}$$

(E - 11.5)

(superscript * for equilibrium)

Note that since at equilibrium $r_{As} = 0$ then from Equation 11.1c:

$$k_1 C_E C_S^* = k_{-1} C_{ES} \quad \text{and} \quad \frac{k_1}{k_{-1}} C_S^* = \frac{C_{ES}}{C_E}$$

and from Equation 11.1d when $r_{Ap} = 0$:

$$k_2 C_{ES} = k_{-2} C_E C_P^* \quad \text{and} \quad \frac{C_{ES}}{C_E} = \frac{k_{-2}}{k_2} C_P^*$$

Thus $\dfrac{k_1}{k_{-1}} C_S^* = \dfrac{k_{-2}}{k_2} C_P^*$

Thus $\dfrac{k_1}{k_{-1}} \cdot \dfrac{k_2}{k_{-2}} = \dfrac{C_P^*}{C_S^*}$ at equilibrium

Now for many enzyme catalysed reactions the reverse reaction (formation of ES from E and P) is very slow. Hence k_{-2} is close to zero. Equation 11.5 shows that this implies that K_{EQ} is large.

Michaelis-Menten relation

In this case Equation 11.4 can be written ($k_{-2} = 0$) as the well known Michaelis-Menten relation. Thus:

$$\frac{(-r_{As})}{C_{E0}} = \frac{V^{max} C_S}{K_M + C_S}$$

(E - 11.5a)

where V^{max} is maximal enzyme activity, K_M is the Michaelis Menten constant and C_S is the concentration of substrate.

Hence:

$$K_M = \frac{k_{-1} + k_2}{k_1} \qquad \text{(E - 11.5b)}$$

$$V^{max} = k_2 \qquad \text{(E - 11.5c)}$$

alternative expression of Michaelis-Menten relation

It is worth noting here that biochemists often use a different method for deriving the Michaelis-Menten relation. To add confusion, they also use a different nomenclature usually writing the Michaelis-Menten relationship in the form:

$$v_o = \frac{V_{max}[S]}{[S] + K_M}$$

where v_o is the initial velocity of the reaction; V_{max} is the maximum velocity;. K_M is the Michaelis constant; [S] is the substrate concentration.

To all intents, the relationship is the same as that described in Equation 11.5a except that v and V_{max} relate to an unspecified but fixed amount of enzyme (ie it could be 1 mg of enzyme or 1 mol of an enzyme solution).

From Equation 11.5a it appears that the enzyme kinetics follow a hyperbolic saturation curve (Figure 11.2).

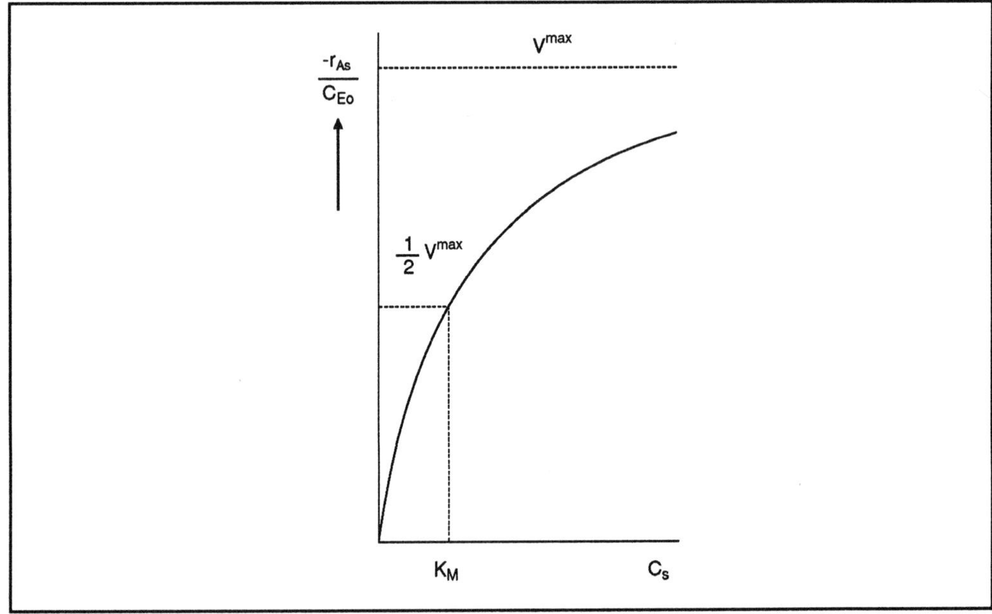

Figure 11.2 Plot of specific substrate conversion rate (ie velocity of the reaction S → P) against substrate concentration (C_s). See text for further details.

$\prod$ Using Equation 11.5a, calculate C_s when $\dfrac{-r_{As}}{C_{E0}} = \dfrac{1}{2} V^{max}$.

You should have come to the conclusion that under this condition $C_S = K_M$.

Since $\dfrac{-r_{As}}{C_{E0}} = \dfrac{1}{2} V^{max} = \dfrac{V^{max} C_S}{K_M + C_S}$

then $\dfrac{1}{2} = \dfrac{C_S}{K_M + C_S}$, thus $K_M = C_S$

We have shown this in Figure 11.2. In practice, it is experimentally difficult to determine V^{max}. Often the high concentrations of substrate that are needed cause inhibitory side effects. Apart from that, biochemical substrates are sometimes too expensive to use in high concentrations just for experimental purposes. An alternative way of determining K_M is to invert (ie take the reciprocals) of Equation 11.5a.

Thus $\dfrac{C_{E0}}{-r_{As}} = \dfrac{K_M}{V^{max}} \dfrac{1}{C_S} + \dfrac{1}{V^{max}}$

This has the form of a simple linear equation of the type $y = ax + b$ where $a = K_M/V^{max}$ and $b = \dfrac{1}{V^{max}}$

This has been plotted in Figure 11.3. Note that the slope $= K_M/V^{max}$ and the intercept with the vertical axis $= \dfrac{1}{V^{max}}$.

Lineweaver-Burke plot

We have also extrapollated the line to the $1/C_S$ axis where it intercepts at $-1/K_M$. This type of plot is known as a Lineweaver-Burke plot and is the usual method employed to determine V^{max} and K_M. For this suitable values of C_s are chosen and $\dfrac{C_{E0}}{-r_{As}}$ measured.

Biochemists regard the K_M value as reflecting the 'affinity' of the enzyme for its substrate. If the affinity of the enzyme is high (ie k_1 is high, k_{-1} is low) then the K_M value is low (see Equation 11.5b). If the binding of the substrate to the enzyme is weak (ie k_1 is low and k_{-1} relatively high) then the value of K_M will be high.

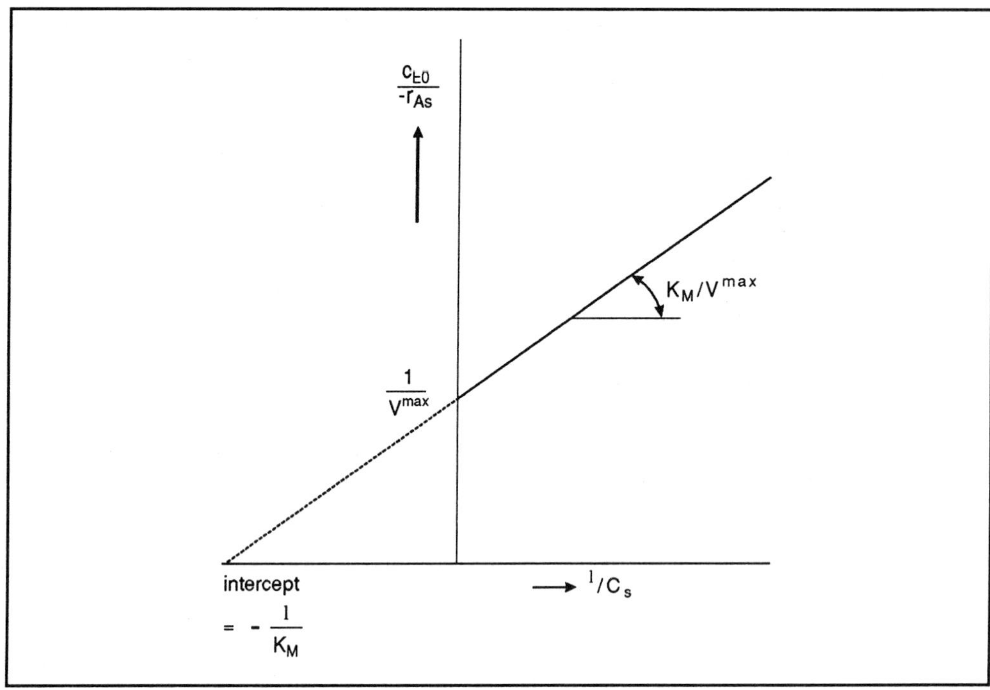

Figure 11.3 Lineweaver-Burke plot (see text for details)

11.2.3 Enzyme inhibition

active site   In general the catalytic action of an enzyme is explained by the lock and key or induced fit hypothesis (Figure 11.4). We do not intend to elaborate of these hypothesis here except to say that the substrate interacts with a specific site on the enzyme, the active site. It should be recalled that an enzyme has typically a MW of about $40000 - 10^6$ Daltons, while reactants have MW of 100-400 Daltons. The active site is therefore only a small part of the enzyme. (Fuller descriptions of the lock and key and induced fit hypothesis are provided in the Biotol text 'The Molecular Fabric of Cells').

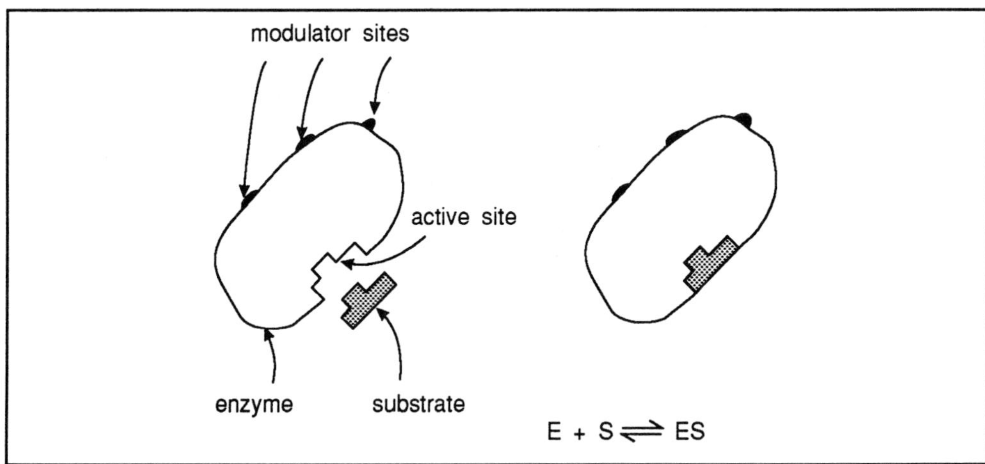

Figure 11.4 Enzyme-substrate interactions

modulator sites
allosteric
modification

Besides an active site, an enzyme often has modulator sites away from the active site. These allow the regulation of enzymatic activity. This is called allosteric modification which can be stimulating (+) or inhibiting (-).

This possibility is functional from a metabolic control point of view. For example, many products of metabolic pathways such as in the synthesis amino acids are involved in the control of enzyme activity. The reason is clear, if for example the production of a particular amino acid is too high the amino acid concentration will increase, which will then lead to a decrease of the too high enzyme activity (ie the amino acid regulates its own production). There are many examples of this kind of regulation. It is one reason why the complex metabolism of cells can proceed in a controlled and orderly manner.

Conventional inhibition

Quite another form of inhibition comes from effects of inhibition on the active site. This may be called conventional inhibition. Its importance lies more in the technical use of enzymes (or micro-organisms) at high substrate or product concentration. This form of inhibition bears no relation to control of enzyme activity from allosteric mechanisms.

There are several forms of conventional inhibition, which are all caused by effects of the inhibitor on the active site of the enzyme.

Three important cases are provided in Figure 11.5. There are other forms of inhibition which we will not deal with here. (The reader is referred to the Biotol texts 'The Molecular Fabric of Cells' or 'Technological Applications of Biocatalysts'.

∏ Examine Figure 11.5 and see if you can detect in what ways we have represented the overall reaction differently from the way we represented it earlier (Figure 11.1).

In our earlier description we represented the formation of product as being reversible ie:

$$ES \underset{k_{-2}}{\overset{k_2}{\rightleftharpoons}} P + E$$

whereas here we have represented this as irreversible (ie $ES \rightarrow E + P$).

You may recall that we showed earlier that k_{-2} was in most instance virtually zero. Thus the representation of this in Figure 11.5 is justified.

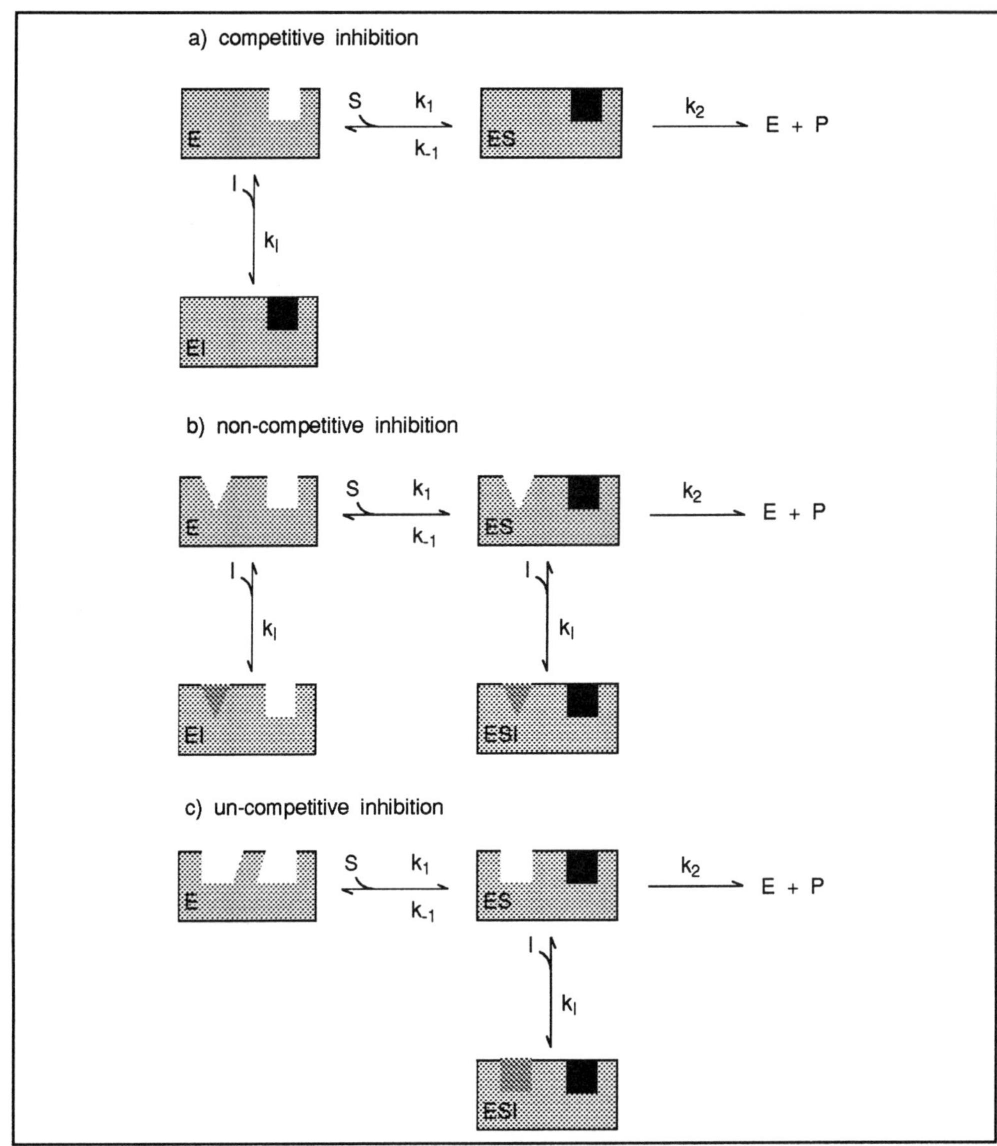

Figure 11.5 Diagrammatic representation of conventional forms of enzyme inhibition. a) competitive inhibition, b) non-competitive inhibition, c) un-competitive inhibition. I = inhibitor, E = enzyme, S = substrate, ES = enzyme-substrate complex, ESI = enzyme-substrate-inhibitor complex, EI = enzyme inhibitor complex. See text for further details.

Competitive inhibition

In the case of competitive inhibition, the inhibitor I blocks the active site of the enzyme E by forming an unproductive EI-complex (Figure 11.5). It is usual that the inhibitor I should somehow resemble the structure of the substrate S, in order to bind effectively to the active site (steric analogy). We might consider there to be a competition between the substrate and the inhibitor for binding to the enzyme. Thus we can represent this by:

$$E + S \rightleftharpoons ES \rightarrow E + P$$

$$K_I \updownarrow \quad I$$

$$E\,I$$

V^{max} is the same, K_M changes

Thus at high substrate concentration a lot of ES is formed and a little EI whilst at high inhibitor concentrations, EI is formed in greater quantities than ES. In principle, therefore, with this type of inhibition, the inhibition can be overcome by using a higher substrate concentration. With competitive inhibition we can achieve the same V^{max} but at a higher substrate concentration. Graphically, this is shown in Figure 11.6.

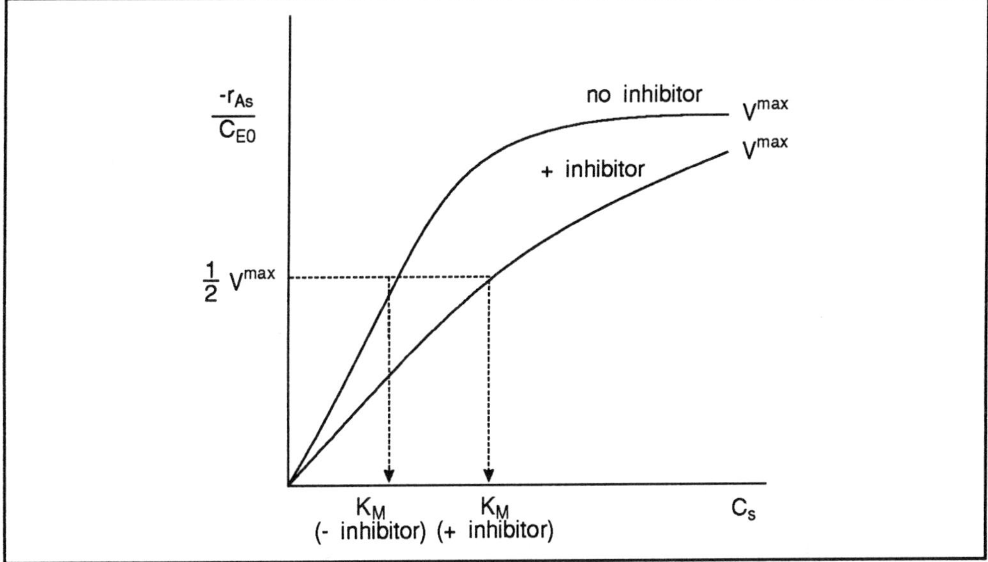

Figure 11.6 Effects of competitive inhibitor on reaction rate.

Using the same kind of reasoning as in the previous Section 11.2.2, (including the pseudo-steady state condition to ES and EI intermediate) leads to the Equation (11.6).

$$\frac{-r_{As}}{C_{Eo}} = \frac{V^{max}\,C_S}{C_s + K_M\,(1 + C_I/K_I)} \qquad\qquad (E\text{-}11.6)$$

where C_{E0} is the total concentration of the enzyme; C_S concentration of the substrate; K_M Michaelis-Menten constant; C_I concentration of inhibitor. K_I is the equilibrium binding constant of the EI complex.

From Equation 11.6 it appears that the kinetics are analogous to the Michaelis-Menten relationship.

Examine Figure 11.6 again. Note that although V^{max} remains the same, the C_s at which V^{max} is achieved has altered (ie the enzyme gives the appearance of having a lower affinity for its substrate). Thus there is an apparent shift in the K_M value. In fact $\frac{1}{2}V^{max}$ in the presence of a competitive inhibitor is achieved when:

$C_S = K_M (1 + C_I/K_I)$

The apparent K_M under these circumstances is $(K_M + K_M\,C_I/K_I)$ and is usually given the symbol K'_M.

If we invert Equation 11.6 and plot $\dfrac{C_{E0}}{-r_{As}}$ against $1/C_S$ (equivalent to a Lineweaver-Burke plot described in Figure 11.3) then the result we get is of the type shown in Figure 11.7.

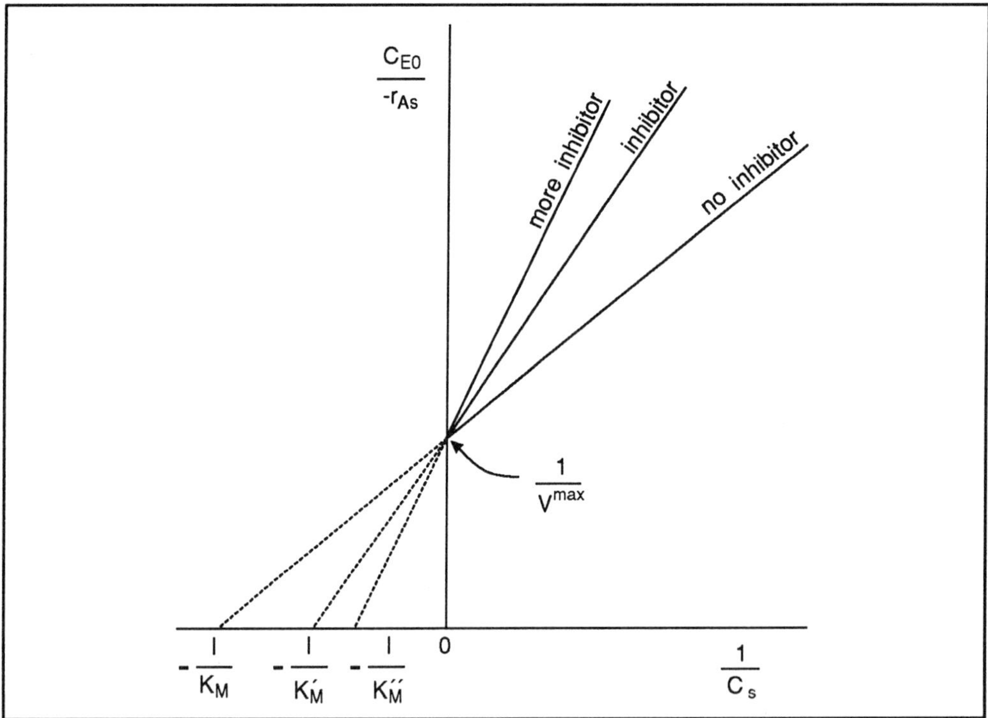

Figure 11.7 Lineweaver-Burke plot showing the effects of competitive inhibition on enzyme kinetics.

Examine Figure 11.7 carefully and note that V^{max} has not changed, but that K_M has. Note that the greater the quantity of inhibitor used, the greater the value of the apparent K_M.

Uncompetitive inhibition

In the case of uncompetitive inhibition the inhibitor I forms an unproductive complex ESI with the enzyme substrate complex ES.

We can represent this by:

$$E + S \rightleftharpoons ES \rightarrow E + P$$

$$\big\downarrow\!\!\uparrow\; I$$

$$E\,S\,I$$

When the inhibitor binds with enzyme-substrate to form an enzyme-substrate-inhibitor complex which is inactive, the inhibitor cannot be removed simply by adding more substrate. Again, by following the reasoning of the previous section of a pseudo-steady state hypothesis for ES and ESI complexes, leads to Equation 11.7.

$$\frac{-r_{As}}{C_{E0}} = \frac{V^{max}\,C_S}{K_M + C_S\,(1 + C_I/K_I)} \qquad\qquad (E\text{ - }11.7)$$

apparent K_M
and V_{max} are
reduced In this case, the inhibition cannot be overcome by increasing the substrate concentration. From Equation 11.7, it would appear that the maximum velocity (V^{max}) at ($C_s \rightarrow \infty$) is decreased by a factor of $1/(1 + C_I/K_I)$.

The effects of an uncompetitive inhibitor on enzyme reaction rates is shown in Figure 11.8.

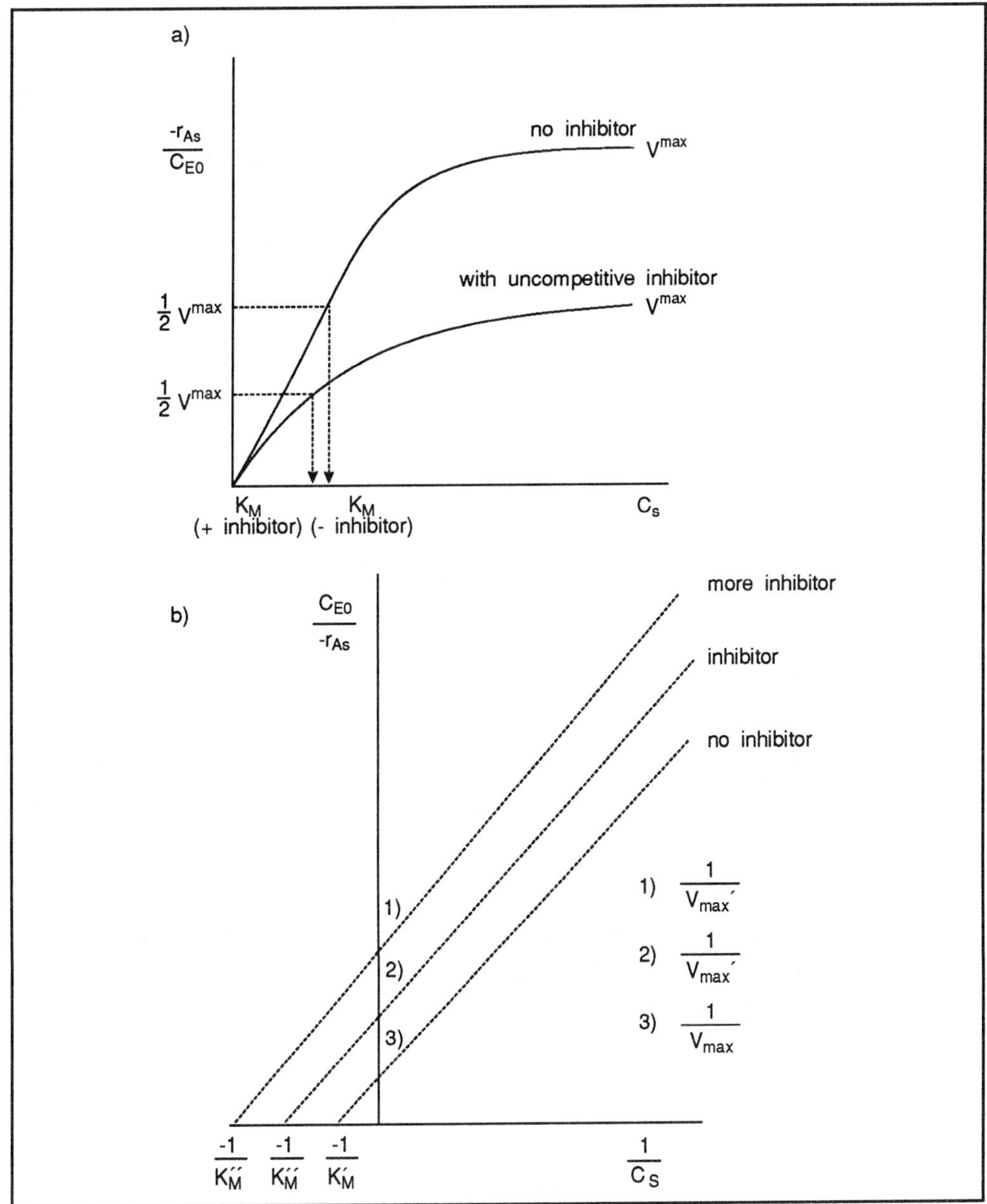

Figure 11.8 Effects of an uncompetitive inhibitor on reaction rates at various substrate concentration; a) direct plot b) Lineweaver-Burke plot.

Note $\dfrac{1}{K_M''} = \dfrac{1 + C_I/K_I}{K_M}$ and $\dfrac{1}{V^{max'}} = \dfrac{1 + C_I/K_I}{V^{max}}$

Non-competitive inhibition

A third type of inhibition is characterised by inhibitors which bind with an enzyme substrate complex with no effect on the K_M value but with a reduction in V^{max}. These so called non-competitive inhibitors effectively remove some enzyme from being available for catalysing the reaction. The unbound enzyme behaves identical to the enzyme in the absence of substrate. Thus they bind substrate with the same affinity (ie K_M remains the same), but because there is less enzyme available, V_{max} is lowered. This type of behaviour is shown graphically in Figure 11.9.

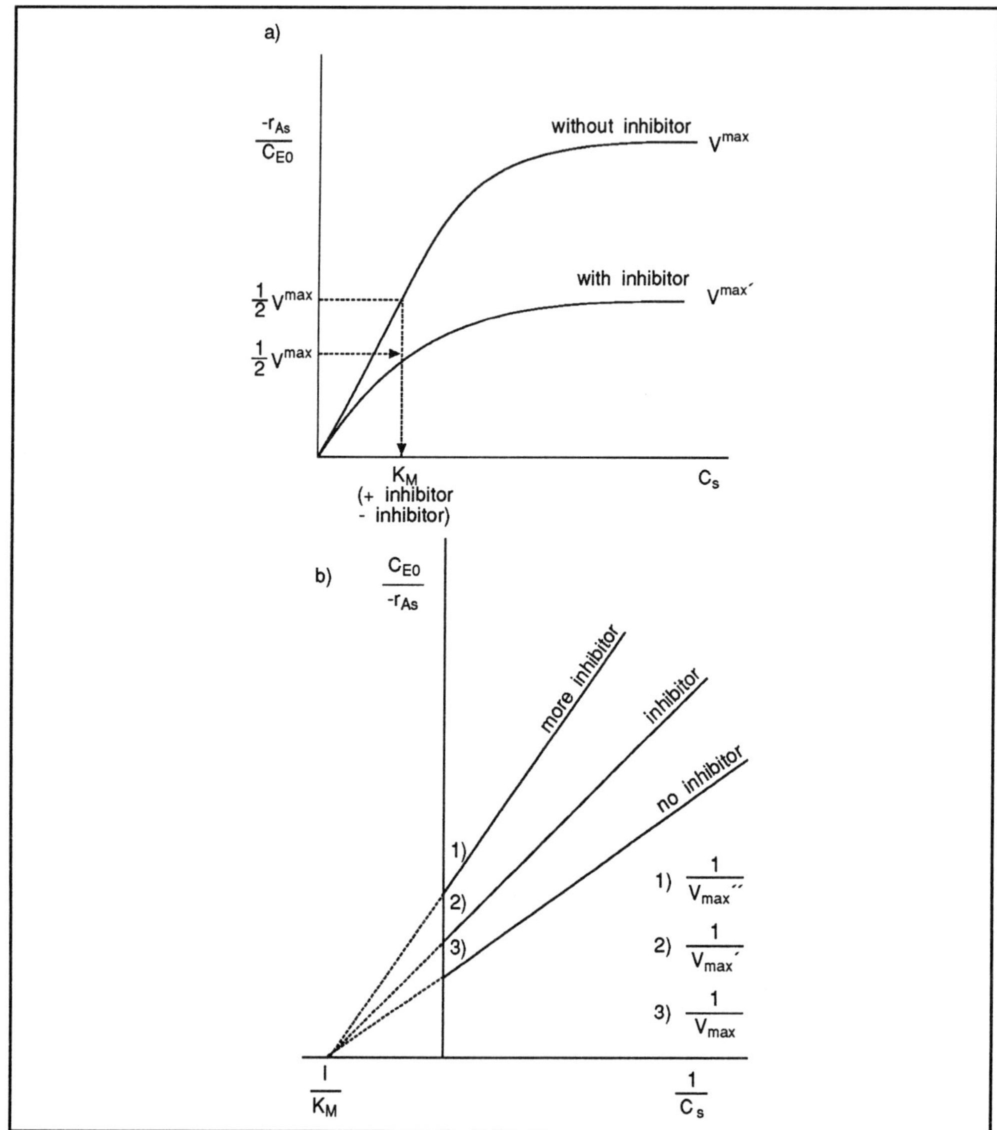

Figure 11.9 Effects of a non-competitive inhibitor on reaction rates at various substrate concentration. a) direct plot, b) Lineweaver-Burke plot (see text for discussion). V^{max} = maximum velocity in the absence of inhibitor, $V^{max'}$ = maximum velocity with inhibitor present.

$$\text{Note } \frac{1}{V^{max'}} = \frac{1 + \dfrac{C_I}{K_I}}{V^{max}}$$

For non-competitive inhibitions it can be shown that:

$$\frac{-r_{As}}{C_{E0}} = \frac{V^{max} C_S}{K_M + C_S(1 + C_I/K_I)} \tag{E-11.8}$$

Let us now summarise what we have learnt about enzyme inhibition in the form of a table (Table 11.1). It might be helpful to copy this out as a form of revision.

Type of inhibitor	Kinetic relationship	Effect on V^{max}	Effect on apparent K_M
None	$\dfrac{-r_{As}}{C_{E0}} = \dfrac{V^{max}C_S}{K_M + C_S}$	-	-
Competitive	$\dfrac{-r_{As}}{C_{E0}} = \dfrac{V^{max}C_S}{C_S + K_M(1 + C_I/K_I)}$	no effect	increased
Uncompetitive	$\dfrac{-r_{As}}{C_{E0}} = \dfrac{V^{max}C_S}{K_M + C_S(1 + C_I/K_I)}$	decreased	decreased
Non-competitive	$\dfrac{-r_{As}}{C_{E0}} = \dfrac{V^{max}C_S}{K_M + C_S(1 + C_I/K_I)}$	decreased	no effect

Table 11.1 Summary of kinetic expression for enzyme action in the presence and absence of various types of inhibition.

If we are studying inhibition, it is usual to try to examine the effect on V^{max} and K_M using a Lineweaver-Burke plot. In this way it is quite straightforward to determine what type of inhibition is taking place.

We can put this into practise by attempting SAQ 11.1.

In conclusion, it appears that we can always find a rate equation for enzyme catalysed processes based on:

- a mechanistic description of the catalysed reaction (analogous to Figure 11.1);

- application of the macroscopic balances and employing a pseudo-steady state assumption.

The result is invariably an equation of the hyperbolic type.

<table>
<tr><td>

SAQ 11.1

</td><td>

Enzyme kinetics and enzyme inhibition.

The following set of concentration-velocity data is obtained for an enzyme catalysed reaction.

</td></tr>
</table>

C_s mol l^{-1}	$-r_{As}/C_{E0}$ mol/mol enz min	$-r_{As}/C_{E0}$ (+ inhibitor) (mol/mol enzyme min)
2×10^{-5}	2.12×10^{-9}	1.01×10^{-9}
3×10^{-5}	2.70×10^{-9}	1.47×10^{-9}
4×10^{-5}	2.94×10^{-9}	1.73×10^{-9}
6×10^{-5}	3.33×10^{-9}	2.33×10^{-9}
10×10^{-5}	3.85×10^{-9}	2.78×10^{-9}

1) Does this reaction follow Michae lis-Menten kinetics?

2) Determine K_M and V^{max}.

3) An inhibitor is added at a concentration of 1.5×10^{5} mol l^{-1}, which changes the rates as indicated above.

Determine K_M and V^{max} and explain which type of inhibition is taking place. You may also attempt to calculate K_I.

11.2.4 Effect of temperature on the enzyme kinetics

Does the temperature have specific effects on an enzyme reaction?

Yes, an increase in temperature will increase the rate of an enzyme reaction but it will also increase the rate of inactivation of the enzyme. These two factors have opposite effects on the rate of the enzyme catalysed reaction.

For microbial cells the effect of temperature on the rates of reaction is very important because micro-organisms cannot maintain intracellularly a temperature different from their environment. This is contrary to pH, where micro-organisms can maintain a constant intracellular pH over a broad range of extracellular pH-values. Therefore in general temperature dependence of microbial conversion rates is strong, while pH dependence is much weaker.

There are 3 aspects of relevance for the influence of temperature on enzyme kinetics. These are:

- effect of temperature on the reaction rate constants (k values in Figure 11.1);

- effect of temperature on the fraction of active enzyme; this is called reversible denaturation;

- effect of temperature on destruction of the enzyme; this is called irreversible denaturation.

The first two aspects will be treated here.

The third aspect will be dealt with in the section considering maintenance (Section 11.4).

Effect of temperature on rate constants of enzymes

Temperature exerts its influence on the various kinetic rate coefficients. k_i and k_{-i} (note we have written k_i and k_{-i} to represent the more specific rate coefficients of k_1, k_{-1}, k_2 and k_{-2} used in Figure 11.1).

The effects of temperature on K_i is quantitatively described by the Arrhenius Equation 11.9.

$$k_i\,(T_1) = \; k_i\,(T_2)\; \exp\left[\frac{-\Delta H_i^{\#}}{R}\,(\frac{1}{T_1} - \frac{1}{T_2})\right]$$

$$(E - 11.9)$$

Here:

$k_i\,(T_1)$ is the rate constant k_i at the absolute temperature T_1

$k_i\,(T_2)$ is the rate constant k_i at the absolute temperature T_2

R is the gas constant ($=8.314 \times 10^{-3} kJmol^{-1}K^{-1}$).

$\Delta H_i^{\#}$ is the activation enthalpy of the reaction described by rate coefficient k_i ($kJmol^{-1}$).

For biochemical reactions one generally finds that at $T = 300$ K, the rate increases approximately twice with an increase of 10k. Hence one can calculate from Equation 11.9 that a typical value of $\Delta H_i^{\#} = 60$ $kJmol^{-1}$.

If we now inspect the Michaelis-Menten relation (Equation 11.5a) and how V^{max} and K_M depend on the rate coefficients (k_i see Equation 11.5b) it is clear:

- that V^{max} depends on one rate coefficient and hence the temperature dependence of V^{max} is according to Equation 11.9 with an enthalpy of activation $\Delta H^{\#}$ of about $60 kJmol^{-1}$.

- K_M depends on a ratio of rate constants, each of which depends on the temperature with similar values of $\Delta H_i^{\#}$. This means that their ratio depends only weakly on temperature.

Hence K_M is expected to show only weak temperature dependence. This is indeed often observed.

Effect of temperature on the fraction of active enzyme

reversible
denaturation

Enzymes are characterised by their conformation which is important for their activity. An enzyme is not a rigid structure, it continuously modifies its conformation due to thermal motion within the protein molecule and thermal motion of the surrounding solvent molecules. This raises the possibility that an enzyme can enter into a different conformation which is not active. This is called a denatured conformation. We can represent this by:

$$E_A \rightleftarrows E_D$$

$$(E - 11.10)$$

If we assume that such an interaction occurs continuously, we may assume that an enzyme preparation contains an active fraction E_A and denatured fraction E_D, which are in equilibrium (Equation 11.10).

If we assume an equilibrium constant K_D, we can write:

$$K_D = \frac{C_{E_D}}{C_{E_A}} \qquad\qquad (E - 11.11)$$

Because the total enzyme concentration is C_{E0} we can also write:

$$C_{EO} = C_{ED} + C_{EA} \qquad\qquad (E - 11.12)$$

We obtain then for C_{EA} from Equation 11.11 and Equation 11.12.

$$\frac{C_{EA}}{C_{EO}} = \frac{1}{1 + K_D} \qquad\qquad (E - 11.13)$$

The equilibrium constant K_D is related to the Gibbs free energy of denaturation ($\Delta G_D'$) of reaction Equation 11.10. Thus:

$$- \Delta G_D' = RT \ln (K_D) \qquad\qquad (E - 11.14)$$

This gives us for the temperature dependence of K_D.

$$K_D = \exp \frac{(- \Delta G_D')}{RT} = \exp \frac{\Delta S_D'}{R} \exp \frac{(- \Delta H_D')}{RT} \qquad\qquad (E - 11.15)$$

Since $\Delta G' = \Delta H' - T\Delta S'$

Where $\Delta S_D'$ is the reaction entropy and $\Delta H_D'$ is the reaction enthalpy. A typical value for $\Delta S_D'$ is 640 J $K^{-1}mol^{-1}$, while for $\Delta H_D'$ it is 200 kJmol^{-1}.

This leads to $\Delta G_D' = 200 - 0.64 \times T$. Hence at 312 K, $\Delta G_D' = 0$, which means that $K_D = 1$ at 39°C.

If one calculates the effect of temperature on the fraction of active enzyme (combination of Equation 11.13 and 11.15 then one will find an active fraction of 0.87 at a temperature of 305 K and a fraction of 0.14 at a temperature of 320 K.

Hence it appears that within a short temperature interval there is a near complete conversion of active to inactive enzyme conformation.

Π   Write down whether you believe that temperature has the same influence on the ratio of active to denatured (ie C_{EA}/C_{ED}) enzyme for all enzymes. Give reasons for your conclusion.

The answer is that temperature does not have exactly the same effect on the C_{EA}/C_{ED} ratio for all enzymes. Different enzymes are made up of different arrangements of amino acids. Some enzymes are therefore more stable than others. In other words $\Delta G_D'$

is not exactly the same for all enzymes. Thus K_D is not identical for all enzymes. Some enzymes remain active at higher temperatures than do others.

It will now be clear that for the effect of temperature on enzyme kinetics there operate two opposing effects at increasing temperature.

- increase of the specific rate of the active enzyme;

- decrease of the fraction active enzyme in the total enzyme preparation.

This means that one may expect a T-maximum in the activity of the enzyme (see Figure 11.10) .

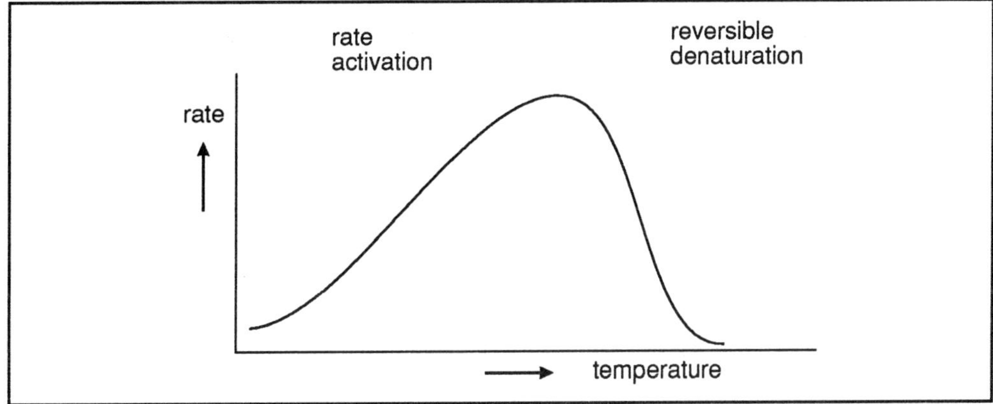

Figure 11.10 Effect of temperature on enzyme catalysed rates (stylised).

∏ Re-draw Figure 11.10 but this time put on two enzymes, one thermally relatively more stable than the other.

You should have drawn something similar to that shown below:

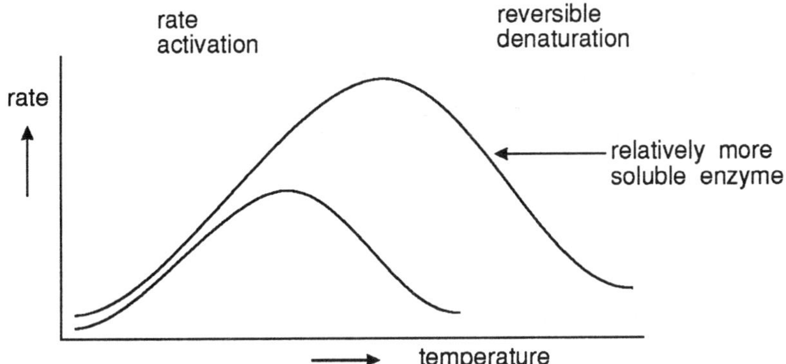

Mathematically we can treat this in the following way.

If we consider the Michaelis-Menten relation Equation 11.5a we can write for the maximal value of substrate conversion (K_M is assumed T-independent).

$$- r_{As}^{max} = V^{max} \times C_{E0} \tag{E - 11.16}$$

In this relation C_{E0} is in fact the concentration of active enzyme, which is generally not equal to the total enzyme concentration.

Introduction of the T-dependence of V^{max} and C_{EA}/C_{E0} leads to Equation 11.17.

$$\frac{r_{As}^{max}(T_1)}{r_{As}^{max}(T_0)} = \exp\left[\frac{-\Delta H_i^{\#}}{R}\left(\frac{1}{T_1} - \frac{1}{T_0}\right)\right] \times \frac{1 + \exp\left(\frac{\Delta S_D'}{R}\right)\exp\left(\frac{-\Delta H_D'}{RT_0}\right)}{1 + \exp\frac{(\Delta S_D')}{R}\exp\left(\frac{-\Delta H_D'}{RT_1}\right)} \tag{E - 11.17}$$

We have not shown all of the steps in this derivation, perhaps you would like to try it!

In principle what this equation tells us is that if we know $\Delta S_D'$ and $\Delta H_D'$ we can calculate the maximum velocity (V^{max}) of an enzyme catalysed reaction. Conversely, if we determine V^{max} experimentally at a variety of temperatures, we can determine $\Delta S_D'$ and $\Delta H_D'$.

Let us attempt to calculate the effects of temperature on V^{max} using some supplied data by attempting SAQ 11.2.

SAQ 11.2	For a specific enzyme, we are given that: $\Delta H^{\#} = 70$ kJ/mol, $\Delta S_D' = 640$ J/mol K, $\Delta H_D' = 200$ kJ/mol. Assume $T_0 = 300$ K. 1) Calculate the effect of temperature on $\dfrac{r_{As}^{max}(T)}{r_{As}^{max}(300)}$ in the range T = 290 - 340 K. We would suggest you use the following temperatures 290, 295, 300, 310, 320 and 330 K. 2) Approximately, what is the optimum temperature for this enzyme (the temperature which shows maximal activity, in terms of rate of substrate conversion).

The sort of enzyme denaturation we have been discussing in this section must not be confused with irreversible denaturation. Raising the temperature of an enzyme often leads to irreversible denaturation ie $E_{active} \rightarrow E_{denatured}$. We will examine this phenomenon in the final section of this chapter.

11.2.5 Enzymes in series

We will conclude this section dealing with the kinetics of single enzymes by briefly considering two or more enzymes working together. Usually in biotechnology, use is made either of a single (often purified) enzyme or a whole series of enzymes linked together into metabolic pathways. It is much more uncommon to use a short series of two (or more) enzymes. The reasons for this are quite straightforward. In many instances we wish to use an enzyme to carry out a single and specific reaction. As a result of the high cost involved in producing highly purified enzymes, industrially based processes based on enzyme catalysis is largely concentrated into the high priced commodity markets. Clearly, using two or three or more purified enzymes becomes even more expensive. The alternative is to make use of a whole series of enzymes in an unpurified state; namely the metabolic pathways found in cells.

We can represent such a metabolic pathway in the following way:

$$S \xrightarrow{\text{enzyme 1}} P_1 \xrightarrow{\text{enzyme 1}} P_2 \xrightarrow{\text{enzyme 1}} P_3 \xrightarrow{\text{enzyme 1}} P_4 \xrightarrow{\text{enzyme 1}} P_5$$

S = substrate, P = product

In this pathway, the product of the first enzyme (P_1) is the substrate of the second enzyme and so on. When whole metabolic pathways are used, the enzymes are contained within cells. We will be dealing with the kinetics of such pathways in the next section. Increasingly, however biotechnologists are examining ways of using what we might call partial pathways of 2 or 3 enzymes working together. For example we might visualise circumstances in which we could use two enzymes from the pathway above to convert $P_2 \rightarrow P4$. The overall kinetic of such a process are quite complex and depend upon:

- the distribution of the enzymes; are they still within cells, have they been purified and free in solution; have they been immobilised and fixed to a support; if so are they evenly distributed on the support or is each localised?

- the relative proportion of the two enzyme activities. If one of the enzymes is in vaste excess, then the kinetics are very similar to the kinetics of the single, rate-limiting enzyme. If the activities are very similar than in principle we apply Michaelis-Menten kinetics to both enzymes to derive overall kinetics.

- whether or not the two enzymes naturally associate with each other. This will, of course, influence mass transport between the two enzymes.

We can therefore visualise that the kinetic of double-enzyme systems will reflect the kinetics of the individual by mass transport between the enzymes. This area of enzyme technology is in its relative infancy and we anticipate the application of enzymes working in harness with each other will increase significantly during the next few years. Here we have chosen to focus onto the kinetics of whole cell catalysis in which multi-enzyme catalysis is employed to bring about desired chemical changes.

11.3 Kinetics of growth and product formation

In the preceding sections the kinetics of a single enzyme have been presented. We now turn our attention to substrate utilisation by whole cells. In metabolism the reactions, which were specified in a metabolic model, can be considered as the overall result of a number of pathways. Such pathways contain a large number of enzymes which perform consecutive reactions upon the substrate, and finally deliver the biomass precursors.

In principle one can model such a network completely using the Michaelis-Menten kinetic approach. However this would lead to very large numbers of parameters and a mathematical nightmare.

11.3.1 The kinetics of growth

rate limiting step

The first important feature is that simplifications must be made. For example it can be assumed that there does exist somewhere an enzyme which is the bottle neck of material conversion through the pathway. We should regard this bottleneck as the rate-limiting step in the pathway; it acts rather like a tap or valve in a pipeline. The rate at which material is transferred along the pipeline depends on how open the valve or tap is. Thus the rate limiting step in a metabolic pathway governs how fast material is passed along the pathway. This means that the kinetics of the pathway reduce to the kinetics of the bottleneck enzyme.

We will draw on another analogy, the daily phenomenon of traffic jams. The capacity of a system of motor highways is completely determined by the points where traffic jams occur.

This analogy also serves to illustrate the limitation of the bottleneck concept. If one would, for example by genetic manipulations, double the amount of a bottleneck enzyme in the organism, this will generally not lead to a double metabolic capacity, but perhaps to only a 10% increase. The reason is that there has arisen a new bottle neck enzyme somewhere else. Nevertheless, despite its short comings, it is generally accepted to model a pathway flux by a Michaelis-Menten type of relation. It is stressed that the V^{max} and K_M values are then not enzyme properties but pathway parameters.

allosteric regulation

A second important aspect of metabolic processes is that the metabolic network is controlled by the effect of allosteric modulators. This means that for example the material flux is controlled at the beginning of the pathway to avoid undue accumulation of metabolites in the cell.

This makes the substrate conversion step, located at the beginning of a pathway a likely candidate for a Michaelis-Menten type of kinetic pathway regulation. A biochemically based discussion of flux through a metabolic pathway is presented in the BIOTOL text 'Biosynthesis and the Integration of Cell Metabolism'.

A third important aspect is that growth and product formation are usually proportional to the biomass concentration C_x.

Let us remind you of some terms before we develop the kinetics for substrate uptake based on the above description:

μ: specific growth rate

- μ is the specific growth rate of an organism. It has analogy with the velocity of a reaction and has the units of time^{-1} (eg h^{-1}).

Y: growth yield coefficient

- Y is the growth yield coefficient (ie the amount of product produced for a known amount of substrate used). We can write Y_{sx} to represent the amount of biomass produced at the expense of a known amount of substrate(s).

$$Y_{sx} = \frac{\Delta C_x}{-\Delta C_s} = \frac{r_{Ax}}{-r_{As}}$$

q_s is the metabolic quotient which is the rate of substrate(s) consumed per unit of biomass.

$$\text{Thus } -q_s = \frac{-r_{As}}{C_x}$$

Since the uptake and use of a substrate in metabolism behaves like the substrate uptake by an enzyme, we can use Michaelis-Menten type kinetics. Remember also that substrate uptake will be proportional to the amount of biomass present. Using this type of argument, the rate of uptake of substrate is given by:

$$- r_{As} = \frac{q_s^{max} \, C_S \, C_x}{K_S + C_S} = (-q_s) \cdot C_x \qquad\qquad (E-11.18)$$

(Note the analogy with the Michaelis-Menten equation of $\dfrac{-r_{As}}{c_{E0}} = \dfrac{V^{max} C_S}{K_M + C_S}$).

K_S saturation constant The constant K_S is known as the saturation constant or Monod constant and is analogous to the Michaelis constant (K_M) used in single enzyme kinetics.

In the situations of high growth rates (μ) which occur for example in the growth phase in batch reactors, we can use the yield relationship between $-r_{As}$ and r_{Ax}.

$$-r_{As} = \frac{r_{Ax}}{Y_{Sx}}$$

where r_{Ax} is the rate of increase in biomass Y_{Sx} is the growth yield

$$-r_{As} = \frac{r_{Ax}}{Y_{Sx}} = (-q_s) \cdot C_x$$

Thus $\dfrac{r_{Ax}}{C_x} = (-q_s) \cdot Y_{Ax}$

$\dfrac{r_{Ax}}{C_x}$ is the rate of growth per unit of biomass concentration (ie the specific growth rate = μ).

Thus we can write $\mu = \dfrac{r_{Ax}}{C_x} = -q_s \cdot Y_{Sx}$

By analogy:

$$\mu^{max} = - q_s^{max} \cdot Y_{Sx} \qquad\qquad (E-11.19)$$

We can re-write Equation 11.18 in another form.

Since $\mu^{max} = q_s^{max} Y_{Sx}$ and $\mu = -q_s \cdot Y_{Sx}$ then $\dfrac{\dfrac{\mu^{max}}{Y_{Sx}} (C_s)(C_x)}{K_s + C_s} = \dfrac{\mu}{Y_{Sx}} (C_x)$

We can cancel C_x and Y_{Sx} $\qquad\qquad\qquad\qquad\qquad (E-11.20)$

Thus $\mu = \dfrac{\mu^{max} C_S}{K_S + C_S}$

This is the usual form in which this relationship is used. It is called the Monod growth relationship after the scientist who first deduced it.

SAQ 11.3

1) Under a particular set of conditions an organism grows with a specific growth rate of $1h^{-1}$. If the growth yield Y_{sx} of the organism on the substrate is $0.35\ g\ g^{-1}$, at what rate does 5 g of this organism consume substrate?

2) Given the following set of data, calculate the saturation constant (K_s) for the organism for the substrate.

What is the maximum specific growth rate (μ^{max}) of this organism growing on this substrate?

Substrate concentration C_s (mol l^{-1})	Specific growth rate μ (h^{-1})
2×10^{-4}	0.20
3×10^{-4}	0.27
4×10^{-4}	0.29
5×10^{-4}	0.33
10×10^{-4}	0.38

Since the kinetics of substrate uptake and utilisation are similar to those of single enzyme catalysed reactions, then we can use a similar approach to inhibition phenomena as has been described in Equation 11.6 and 11.7. We will not however explore this further here.

11.3.2 Kinetic relation for product formation

The kinetic relation for product formation is sometimes linked stoichiometrically to Equation 11.18 for substrate-uptake by the specified metabolic model. An example is ethanol fermentation where, by the metabolic model description, the net-conversion rate of ethanol is usually directly coupled to substrate uptake. It is desirable to provide a kinetic relation for r_{Ap} so let us see if we can develop a strategy for doing so.

A fundamental approach would be to determine the product forming pathway and its interactions with the growth pathway. As usual this leads to a large number of model parameters. Therefore simplifications are needed. A very simple approach has been given by Roels. Consider Figure 11.12, the product is considered to be a side branch of the pathway leading to biomass production. It is also assumed that intermediate I is the product of the limiting bottleneck (rate limiting step) which limits both growth and product formation.

Roels approach

For both branches one may assume a Michaelis-Menten/Monod type rate relation. This leads for μ (= r_{Ax} / C_x) and q_p (= r_{Ap} / C_x), to: (Note that q_p is the rate of formation of product per unit of biomass per unit of time. It is another metabolic quotient).

$$\mu = \mu^{max} \frac{C_i}{K_x + C_i} \qquad \text{(E - 11.21)}$$

where C_i = concentration of intermediate I, K_x = saturation constant

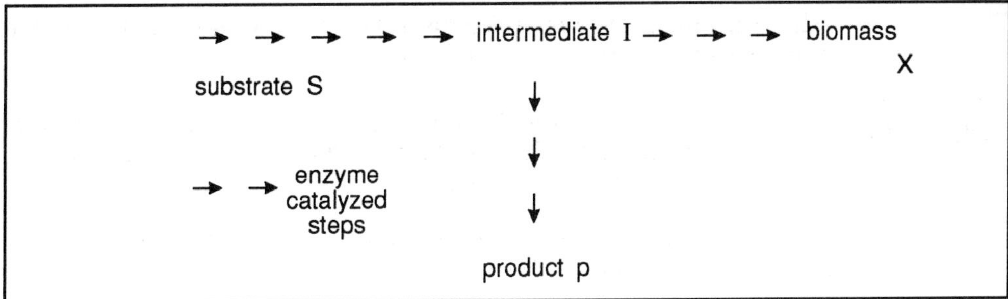

Figure 11.12 Product formation scheme used by Roels to model product formation (see text for details).

$$q_P = q_P^{max} \frac{C_i}{K_p + C_i} \tag{E - 11.22}$$

Elimination of C_i from Equations 11.21 and 11.22 gives Equation 11.23.

$$q_P = q_P^{max} * \frac{\mu}{\left(\dfrac{K_p}{K_x}\mu^{max}\right) + \mu\left(1 - \dfrac{K_p}{K_x}\right)} \tag{E - 11.23}$$

This hyperbolic relation shows in a quite general way that one can expect that:

$$q_P = q_P^{max} \times F(\mu) \tag{E - 11.24}$$

where F is a function related to K_p and K_x.

Equation 11.24 expresses the fact that the specific rate of product formation q_p is somehow correlated to μ. In practice however it is difficult to predict the exact form or the nature of F can seldom be predicted. Generally, its form is obtained from experiments in continuous culture. Figure 11.13 provides some observed forms.

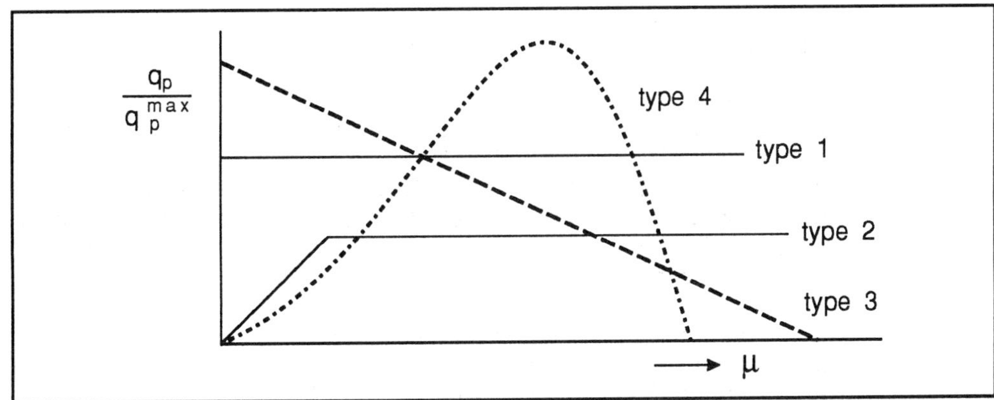

Figure 11.13 Some observed forms of product formation kinetics showing the relationship between rate of product formation and specific growth rate (see text).

For example the kinetics of penicillin production show that q_p increases linearly with μ, up to about $\mu = 0.02 \text{ h}^{-1}$ then remains constant at a value q_p^{max} over higher values of μ.

∏ Which curve in Figure 11.13 illustrates the relationship between penicillin production and μ?

You should have concluded that it is the curve labelled type 2.

∏ Which curves in Figure 11.13 show product formation independent of μ and which one shows an inverse relationship between product formation and μ?

Type 1 represents the case of product formation independent of μ, whilst type 3 shows an inverse relationship between product formation and μ. These two types of relationships are typical of what biologists called secondary metabolites; metabolites not linked to growth.

In conclusion it appears that product formation of kinetic relations can be represented by Equation 11.24 where $F(\mu)$ can take a variety of forms as in Figure 11.13. Often a linear or hyperbolic relation is found.

Substrate uptake is usually represented by a Michaelis-Menten type of relation, where, optionally, inhibition terms can be introduced.

SAQ 11.4

Calculate the efficiency of product yield on substrate in C-mol/C-mol biomass (Y_{Sp}) as a function of growth rate in the range $0 \rightarrow 0.02$ h^{-1} from the following relationship.

$$-q_s = \frac{1}{Y_{Sx}^{max}} \mu + \frac{1}{Y_{Sp}^{max}} q_P + m_s$$

(We derived this relationship in Chapter 10, here we will remind you of some of the terms).

Y_{Sx}^{max} is the maximal yield of biomass on substrates(s).

Y_{Sp}^{max} is the maximum yield of product on substrate(s).

q_P specific conversion rate of product formation

m_s is the substrate based maintenance coefficient (ie substrate consumed to maintain cells and which does not yield biomass).

You are supplied with the following data:

$Y_{Sx}^{max} = 0.5$, $Y_{Sp}^{max} = 0.66$,

$m_S = 0.02$ C-mol/C-mol hr

$q_P = 0.26\,\mu$ C-mol product/C-mol biomass h

We suggest you calculate the product yield when the specific growth rate, $\mu = 0$, 0.005, 0.01, 0.015 and 0.02 h^{-1}.

11.4 Kinetics of maintenance

It has been shown in the previous chapter that the maintenance reaction (r_6) in the metabolic model can be assigned a constant biomass specific rate (r_6/c_x = constant) or as a linear increase with μ. In many cases, the maintenance reaction appears to be constant when based on biomass and is described by the maintenance coefficient m_s.

The maintenance coefficient describes the rate of use of substrate per unit amount of biomass per unit time. In strict terms it should have the units of C-mol C-mol^{-1} s^{-1}. In practice other units are used. For example an m_s value for an organism of 1 gg^{-1}h^{-1} implies that 1g of substrate is consumed per gram of biomass per hour to fulfil its maintenance reaction requirements. Here we will extend the discussion of maintenance kinetics a little further.

The basic thought behind maintenance is that there occurs a constant and irreversible denaturation and degradation of complex biomass molecules into simple precursors and there is a dissipation of concentration gradients by leakage.

maintenance metabolism

maintenance energy

The maintenance of concentration gradients or the replacement of denatured or degraded biomass molecules requires energy. This energy is supplied by metabolism. Since this metabolism is directed towards maintaining cells, it is sometimes referred to as maintenance metabolism (the energy requirement is also called maintenance energy). We could also imagine that some of the ATP within the cell will be slowly chemically hydrolysed without being coupled to useful chemical work (ie the energy released is lost as heat). The replacement of the ATP would be an energy cost which will fall within the maintenance requirement of the cell.

A moments thought should lead you to the conclusion that all these processes (denaturation, gradient dissipation, hydrolysis) are all temperature dependent.

They all increase their rate if the temperature is increased. From literature data it has been found that this increase can indeed be described by an Arrhenius relation with an activation enthalpy $\Delta H^{\#}$ of about 40kJ mol^{-1}.

correlation between MATP and temperature

Furthermore at normal temperatures (25°C = 298 K) a typical m_{ATP} -value (mol ATP consumed per unit of biomass per unit of time to supply maintenance requirements) is about 0.04 mol ATP/C-mol biomass h.

Hence we can write Equation 11.25:

$$\frac{m_{ATP}^{T}}{m_{ATP}^{298}} = 0.04 \ \exp\left[-\frac{40}{8.314}\left(\frac{1000}{T}-\frac{1000}{300}\right)\right]$$

(E - 11.25)

<table>
<tr><td>

SAQ 11.5

</td><td>

Investigate the effect of a temperature rise from 300 K to 310 K on the yield of product formation in SAQ 11.4. Use the answer of SAQ 11.2 as the effect of temperature on production kinetics.

This is quite a difficult question so we will give you some guidelines.

First use the answer of SAQ 11.2 to determine how a rise of temperature of 10 K from 300 K to 310 K effects V_{Sp}^{max}. Then use this to adjust the relationship of q_p with μ described in SAQ 11.4 and to adjust the value of m_s. Then you can use these new values to calculate Y_{sp} at various μ values (0, 0.005, 0.010, 0.015 and 0.02 h^{-1}) using the relationship described in SAQ 11.4 (ie $-q_s = \dfrac{1}{Y_{Sx}^{max}}\, \mu + \dfrac{1}{Y_{Sp}^{max}}\, q_P + m_S$).

</td></tr>
</table>

11.5 Final comments

In the earlier chapters the basis of mathematical modelling was presented. Throughout this chapter we have used two principal assumptions:

- the macroscopic assumption;

- the pseudo-steady state assumption, leading to a constant biomass composition. (Black box assumption).

The limitations of the macroscopic assumption have already been dealt with. Until now, however little has been said about the limits of the pseudo-steady state assumption.

This approximation is only valid if the biomass composition remains constant or changes very fast compared to the changes outside the biomass. These conditions are not always adequately met under practical situations.

An example is the change from batch exponential growth to stationary growth or the lag phase before exponential growth. Here the biomass pseudo-steady state assumption is violated, because the changes outside the biomass are much faster than those within the biomass.

Another example is the formation of storage materials in biomass which are subsequently consumed. Furthermore, it is well known that protein, RNA, carbohydrate content of biomass does change significantly with culture conditions. For example, biomass with a high specific growth rate (μ) contains more RNA than do slower growing cells.

We must conclude that black box descriptions offers a valuable simple and first approach but their application fails in situations where the biomass composition changes extensively.

Therefore, a second level of mathematical modelling is needed where the variable biomass composition is taken into account. Such descriptions are called 'structured models' contrary to the 'unstructured models' of the present black box presentation. We will not, however, explore these further here.

Summary and objectives

In this chapter we have examined the kinetics of microbial processes. We began by examining the kinetics of single enzymes and went on to show that these kinetics could be applied to metabolic pathways and to metabolism and growth. We concluded the chapter with a brief examination of the kinetics of maintenance.

Now that you have completed this chapter you should be able to:

- calculate V^{max}, K_M, μ^{max} and K_S from supplied data;

- determine whether inhibition is competitive, non-competitive or uncompetitive from supplied data;

- calculate the effects of temperature on the rates of enzyme catalysed reactions from supplied data;

- use supplied production kinetic relationships to calculate the efficiency of product yield on substrate as a function of growth rate;

- determine the effects of maintenance on product yield.

Responses to SAQs

Responses to Chapter 2 SAQs

2.1 The ratio of coconut-oil to palm-oil in the out flow will be 1 : 0.65.

The way to calculate this is to use the relationship described in Equation 2.24.

$$C_{c,out}(t) = C_{c,0} \cdot \exp\left(\frac{-\phi_V}{V}t\right)$$

where $C_{c,out}(t)$ is the concentration of coconut-oil at time t; $C_{c,0}$ is the concentration of coconut-oil at a time $t = 0$; ϕ_V is the volumetric flow rate; V is the volume of the vessel; t is the time.

In the question set $C_{c,0} = 100\%$; $\phi_V = 2\,lh^{-1}$; $V = 4\,l$; $t = 1\,h$ (2pm $\rightarrow$ 3pm).

Therefore $C_{c,out}(t) = 100 \cdot \exp\left(-\frac{2}{4} \times 1\right) = 100 \cdot e^{-0.5} = 100 \times 0.607 = 60.7\%$.

Since the total flow makes up 100% then $C_{coconut-oil} + C_{palm-oil} = 100\%$.

Therefore $C_{palm-oil} = 100 - 60.7 = 39.3\%$.

So the ratio of coconut-oil to palm-oil in the out flow will be 60.7 : 39.3 or 1 : 0.65.

2.2 40 mol CO_2 would be produced per hour.

It is best to do this calculation by setting up a mole balance. We can represent the data provided in the question in the following way.

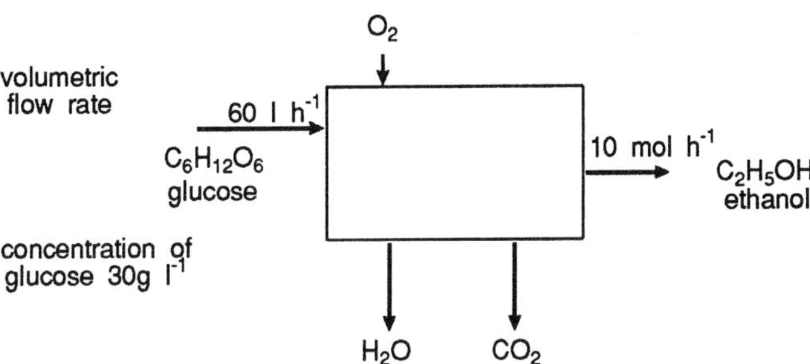

Concentration of glucose $= 30g\,l^{-1}$ or $\frac{30}{180}\,mol\,l^{-1}$. The molar flow of glucose into the tank (ϕ_{gluc}) = flow rate * concentration $= \frac{60 * 30}{180}\,mol\,h^{-1} = 10\,mol\,h^{-1}$.

From the reaction $C_6H_{12}O_6 \rightarrow 2C_2H_5OH + 2CO_2$ we know that the molar flow rate for ethanol $\phi_{eth} = 2\phi_{gluc}$ if all the glucose is converted to ethanol by the above reaction.

We were, however, told in the question that ethanol was produced at the rate of 10 mol h^{-1}. Therefore $\phi_{eth} = 10\,mol\,h^{-1}$.

From the reaction equation the rate of CO_2 production by this process would equal that of ethanol ie $\phi_{CO2} = 10\,mol\,h^{-1}$. Therefore, to support the production of ethanol we must have a flow rate of glucose of $\frac{\phi_{eth}}{2} = 5\,mol\,h^{-1}$.

But the total molar glucose flow rate = flow rate to support ethanol production + flow rate to support aerobic CO_2 production

ie $\phi_{gluc} = \phi_{gluc}^{eth} + \phi_{gluc}^{CO_2,aerobic}$.   Therefore, $10\,mol\,h^{-1} = 5\,mol\,h^{-1} + \phi_{gluc}^{CO_2,aerobic}$

Thus $5\,mol\,h^{-1}$ glucose is aerobically oxidised to produce CO_2. But from the reaction equation we see that 1 mole of glucose produces 6 moles of CO_2. So $5\,mol\,h^{-1}$ glucose flow would produce $5 \times 6 = 30\,mol\,h^{-1}$ CO_2.

Thus the combination of aerobic and anaerobic production of CO_2 in this system would yield $(10 + 30)\,mol\,h^{-1}$ of $CO_2 = 40\,mol\,h^{-1}$.

Diagrammatically:

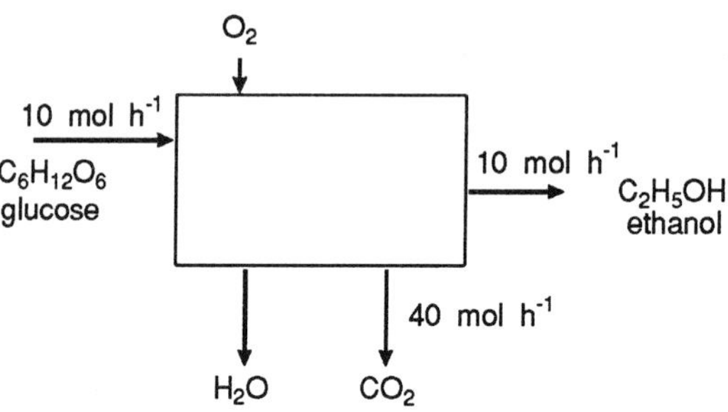

2.3

1) 3.93%

2) 6.32%

To calculate the sucrose concentration in the outflow, we can use the relationship

$$\frac{C_{out}}{C_0} = 1 - \exp\left(\frac{-\phi_v}{V}t\right) \quad (E-2.44)$$

In part 1; $C_0 = 10\%$ sucrose concentration in the inlet = 10% w/v; V = volume of vessel = 40 l; ϕ_v = flow rate = 40 l h^{-1}; t = 30 minutes = 0.5h.

Thus $\dfrac{C_{out}}{10} = [1 - \exp - \dfrac{40}{40}\cdot 0.5] = (1 - e^{-0.5}) = 1 - 0.607 = 0.393$

$C_{out} = 10 \times 0.393 = 3.93\%$

For part 2; $t = 1$h. Thus $\dfrac{C_{out}}{10} = [1 - \exp - \dfrac{40}{40} \times 1\ ] = (1 - e^{-1}) = 1 - 0.368 = 0.632$

$C_{out} = 10 \times 0.632 = 6.32\%$

2.4

39.1 litres

Since $C_1 = C_0\left(1 - \exp\left\{-\dfrac{\phi_v}{V}t\right\}\right)$ (see Equation 2.24 or 2.50) and $C_1 = 2g\,l^{-1}$; $C_0 = 5g\,l^{-1}$

Then $\dfrac{C_1}{C_0} = 0.4 = 1 - \exp\left\{-\dfrac{\phi_v}{V}t\right\}$

$$0.6 = \exp\left\{-\frac{\phi_v}{V}\,t\right\}, \text{ but } t = 1h \text{ and } \phi_V = 20\,l\,h^{-1}$$

$$\text{Thus } 0.6 = \exp\left\{\frac{-20}{V}\right\} \text{ Therefore } V = 39.1 \text{ litres}$$

2.5

1) $8\,l\,h^{-1}$

rate of change in mass of water in the vessel	=	rate of flow of water in via A	-	rate of flow of water out via B, C and D

$$\frac{dM_w}{dt} = \phi^A_{w,in} - \phi^B_{w,out} - \phi^C_{w,out} - \phi^D_{w,out}$$

where superscripts A, B, C and D refer to the pipes and subscript w = mass of water

If we keep the vessel full of water, then $\dfrac{dM_w}{dt} = 0$.

$$\text{Thus } \phi^A_{w,in} - \phi^B_{w,out} - \phi^C_{w,out} - \phi^D_{w,out} = 0$$

$$\phi^A_{w,in} = 2 + 1 + 5 = 8\,l\,h^{-1}$$

We have assumed that mass is proportional to volume. Thus we must pump in water at $8\,l\,h^{-1}$.

2) 6.25 h

$$\text{Since } \tau = \frac{V}{\phi_w} = \frac{50}{8} = 6.25 \text{ h}$$

2.6

1) Yes - the change in concentration of glucose in the outflow shows a sharp step up indicating that there has been no mixing in the tube.

2) $\tau = 20$ mins (or 0.33h). τ = the time it takes fluid entering the tube to exit the other end.

3) $\theta = 1$. Remember θ = the time (t) divided by the residence time thus $(\tau) = \dfrac{20}{20} = 1$.

4) $\theta = 2$ since $t = 40$min and $\tau = 20$min (see 3).

5) 0, since $F(\theta) = \dfrac{C_{out}}{C_0} = \dfrac{0g\,l^{-1}}{10g\,l^{-1}} = 0$.

6) 1 since $F(\theta) = \dfrac{C_{out}}{C_0} = \dfrac{10g\,l^{-1}}{10g\,l^{-1}} = 1$.

7) $0.06\,m^3\,h^{-1}$ (or $0.001\,m^3\,min^{-1}$ or $1\,min^{-1}$ or $60\,l\,h^{-1}$).

It takes 20 minutes for the fluid to flow through the pipe. The volume of liquid in the pipe = cross-sectional area x length.

$$= \frac{10}{10000}\,m^2 \text{ x } 20m \text{ (note Area} = 10cm^2 = \frac{10}{10000}\,m^2\text{). Thus the volume} = 0.02m^3.$$ But this volume is displaced in 20 minutes.

So the flow rate = $0.001\,m^3\,min^{-1}$ or $= 0.06\,m^3\,h^{-1}$.

You may have calculated this in a different way using different units. It is traditional to use m^3 and hours for most calculations of this type, especially for processes operating on an industrial scale; m^3 are, however, inappropriate for small scale processes, where it is more usual to use litres (= dm^3 or 10^{-3} m^3).

2.7

1) $10\,l\,h^{-1}$

Since $\tau = \dfrac{\text{total volume}}{\text{flow rate}} = \dfrac{2\,V}{\phi_{v,A}}$ where V is the volume of a single tank.

$2 = \dfrac{2\,x\,10}{\phi_{v,A}}$ Thus $\phi_{v,A} = 10\,l\,h^{-1}$

2) Since $F(\theta) = 1 - (1 + 2\theta)\,\exp\,(-\,2\theta)$ (see E - 2.51) and $\theta = \dfrac{t}{\tau}$

After $1\,h\,\theta = 0.5$ since $\tau = 2h$

Thus $F(\theta)$

$= 1 - (1 + 2 * 0.5)\,(\exp - 2 * 0.5) = 1 - 2\,\exp^{-1} = 1 - 0.736 = 0.264$

2.8

1) It is a wise plan to begin tackling this question by making a drawing to represent the system.

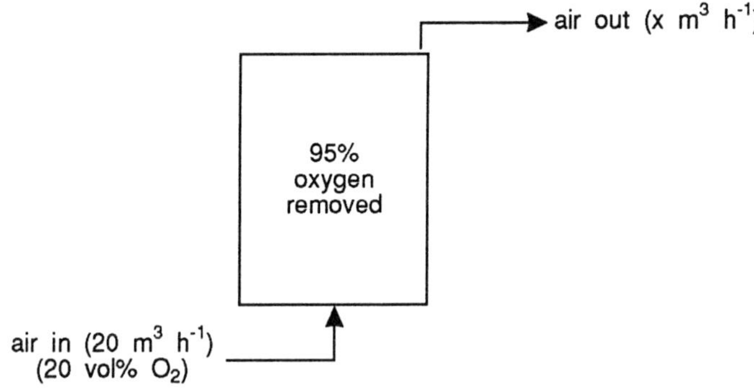

Oxygen balance

change of O_2 in the reactor	=	rate of flow of O_2 into the vessel	-	rate of flow of O_2 out of the vessel	-	rate of consumption in vessel

$\dfrac{dM_{O_2}}{dt} = \phi_{M_{O_2},in} - \phi_{M_{O_2},out} - R_{O_2}$; where R_{O_2} is oxygen consumption in the vessel

2) $16.2\,m^3\,h^{-1}$

We can split the volumetric gas low into two parts $\phi_{air,in} = \phi_{O_2,in} + \phi_{x,in}$

where $x = air - O_2$

But the question gave us that $\phi_{air,in} = 20\,m^3\,h^{-1}$ and $\phi_{O_2,in} = 20\%$ of $\phi_{air,in}$

Thus $\phi_{O_2,in} = 4\,m^3\,h^{-1}$ and $\phi_{x,in} = 16\,m^3\,h^{-1}$

The volumetric outflow rate of air can also be split into two parts $\phi_{air,out} = \phi_{x,out} + \phi_{O_2,out}$

But, since the non-oxygen components of air are not consumed, then $\phi_{x,in} = \phi_{x,out} = 16\,m^3\,h^{-1}$

We also know from the question that $\phi_{O_2,out} = 5\%$ of $\phi_{O_2,in}$

Thus $\phi_{O_2,out} = \dfrac{5}{100} \times 4 \text{ m}^3 \text{ h}^{-1} = 0.2 \text{ m}^3 \text{ h}^{-1}$

Thus $\phi_{air,out} = 16 + 0.2 \text{ m}^3 \text{ h}^{-1} = 16.2 \text{ m}^3 \text{ h}^{-1}$

3) 158 mol h^{-1}

Since the oxygen balance can be written as:

$$\frac{dM_{O_2}}{dt} = \phi_{M_{O_2},in} - \phi_{M_{O_2},out} - R_{O_2}$$

At steady-state $\dfrac{dM_{O_2}}{dt} = 0$. Thus $R_{O_2} = \phi_{M_{O_2},in} - \phi_{M_{O_2},out}$

If we use volumetric flows

$$R_{O_2} = \phi_{O_2,in} - \phi_{O_2,out} = 4 - 0.2 = 3.8 \text{ m}^3 \text{ h}^{-1}$$

or in mol h^{-1} $R_{O_2} = \dfrac{3.8 \times 1000}{24}$ mol h^{-1} (1 mol occupies 24 l) $= 158 \text{ mol h}^{-1}$

2.9 $300.9\,^{\circ}K$

We use the relationship $T_2 - T_1 = \dfrac{p_1 - p_2}{\rho c_v}$ (E - 2.59)

Therefore $T_2 - 300 = \dfrac{10^6 - 10^5}{10^3 \times 1 \times 10^3} = \dfrac{9 \times 10^5}{1 \times 10^6} = 0.9$. Thus $T_2 = 300.9\,^{\circ}K$.

2.10 1) $0.05\,^{\circ}C$ (approximately)

Again we can use the relationship $T_2 - T_1 = \dfrac{p_1 - p_2}{\rho c_v}$

Thus $T_2 - T_1 = \dfrac{2 \times 10^5}{10^3 \times 4.2 \times 10^3} = 0.0476 = 0.05\,^{\circ}C$

2) $0.5\,^{\circ}C$ (approx)

The total energy charge change over the tube is:

$$\frac{dE}{dt} = 0 = \phi_M \left\{ u_1 + \frac{p_1}{\rho_1} + \frac{1}{2} v_1^2 + g\, z_1 \right\} - \phi_M \left\{ u_2 + \frac{p_2}{\rho_2} + \frac{1}{2} v_2^2 + g\, z_2 \right\}$$

Again $v_1 = v_2$ and $\rho_1 = \rho_2$ but in this case $z_1 = z_2 = -200m$

Thus $u_1 + \dfrac{p_1}{\rho_1} + g\, z_1 = u_2 + \dfrac{p_2}{\rho_2} g\, z_2$

$u_1 - u_2 = \dfrac{p_2 - p_1}{\rho} + g\,(z_2 - z_1)$

But $u_1 - u_2 = c_v\,(T_1 - T_2)$

$(T_1 - T_2) = \dfrac{p_1 - p_2}{c_v\,\rho} + \dfrac{g\,(z_2 - z_1)}{c_v} = \dfrac{2 \times 10^5}{4.2 \times 10^3 \times 10^3} + \dfrac{9.81 \times 200}{4.2 \times 10^3} = \text{approximately } 0.51\,^{\circ}C$

2.11

We use the relationship $T(t) = T_0 + \dfrac{\phi_q}{\rho c_V V} t$ (see Equation 2.65)

$$T(t) - T_0 = 0.1 = \frac{\phi_q}{10^3 \times 4.2 \times 10^3 \times 50} \times 1$$

(NB temperature rise $(T(t) - T_0) = 0.1$ in time t where t = 1 sec).

Thus $\phi_q = 0.1 \times 10^3 \times 10^3 \times 50 \times 4.2 = 21.0 \times 10^6 = 2.1 \times 10^7$ J s^{-1}

2.12

1) $0.05°C$ s^{-1}

We can use the relationship

$$T(t) = T_0 + \frac{\phi_q}{\rho c_V V} t \text{ (Equation 2.65)}$$

$$T(t) - T_0 = \frac{2 \times 10^6}{10^3 \times 2 \times 10^3 \times 20} \text{ (where t = 1 second)} = 0.05°C$$

Thus the temperature rises by $0.05°C$ s^{-1}

2) 0.2 kg s^{-1}

We use the relationship

$$\phi_{M,ev} = \frac{\phi_q}{\Delta h_{ev}} = \frac{2 \times 10^6}{10^7} = 0.2 \text{ kg s}^{-1}$$

2.13

2.10^8 J h^{-1}

Again we can use Equation 2.65.

$$T(t) - T_0 = \frac{\phi_q}{\rho c_V V} t$$

where in this case $T(t) - T_0 = -10°C$ and t = 1 h = 3600 s

Thus $-10 = \dfrac{\phi_q}{10^3 \times 2 \times 10^3 \times 10} . 3600$; $\phi_q = -5.5 . 10^4$ W $= -5.5 . 10^4$ J s^{-1}. Thus $\phi_q = -2 . 10^8$ J h^{-1}

Notice it is a negative value since we are removing heat.

2.14

1)

| change in thermal energy in the reactor | = | thermal energy input from outside reactor | − | thermal energy loss to the outside | + | thermal energy produced inside the reactor | − | loss of thermal energy by conversion to other energy forms within the reactor |

2) It is best to begin by making certain all figures are in the same dimensions. In the question set we have two different time values (minutes and seconds). We have converted each to a second basis. Thus heat generated = 6×10^4 J min^{-1} = 1×10^3 J s^{-1}.

Now we substitute our values into the balance given in 1).

$$\boxed{\begin{array}{c}\text{change in}\\ \text{thermal energy}\\ \text{in the reactor}\end{array}} = 0 - 2 \times 10^3 + 1 \times 10^3 - \frac{0.01 \times 10^6}{60} = -1{,}167 \text{ J s}^{-1}$$

Note that the loss of thermal energy by conversion to other energy forms is thermal energy of evaporation =

$$\phi_{M,ev} \times \Delta h_{ev} = \frac{0.01 \times 10^6}{60} \text{ J s}^{-1}$$

The net overall thermal change in the reactor is a negative value, thus the culture will cool down.

3)   We require a heater with a capacity to supply $1{,}167 \text{ J s}^{-1}$ in order to balance the heat loss.

4) $1.12 . 10^5 \text{ J s}^{-1}$

The difference in the thermal energy of the inflow and outflow can be regarded as a heat loss from the system.

The difference is given by the difference in temperature multiplied by the specific heat per unit mass.

ie $(T_{in} - T_{out})$ cv = difference in thermal energy per unit mass = $-4.10^4 \text{ J kg}^{-1}$

Thus if the flow rate is $10 \text{ m}^3 \text{ h}^{-1}$ and the density = 10^3 kg m^{-3} then the flow rate is:

$= 10^4 \text{ kg h}^{-1}$

Thus the thermal energy loss as a result of the flow of media through the system.

$= -4 . 10^4 . 10^4 \text{ J h}^{-1} = -1.11 . 10^5 \text{ J s}^{-1}$

Thus the heater would require a total capacity.

$= 1{,}167 \text{ (from question 3)} + 1.11 . 10^5 \text{ J s}^{-1} = 1.12 . 10^5 \text{ J s}^{-1}$

These calculations should show you how energy balances can be used. In principle you could use the same approach to work out thermal energy balances for such devices as heat exchanges and radiators.

Responses to Chapter 3 SAQs

3.1

1)   363 K

We can calculate the temperature from the relationship

$$T_x = \frac{T_1 - T_0}{L} x + T_0 \quad \text{(see Equation 3.8)}$$

Substituting in the values given in the question

$$T_x = \frac{300 - 390}{1} . 0.3 + 390 \text{ K}; \quad T_x = -\frac{90 \times 0.3}{1} + 390 = 363 \text{ K}$$

(Note we have converted x into meters)

2) 9 J s^{-1} or 9 W

The heat flow can be calculated from the equation

$$\phi_q = -\frac{\lambda}{L} A (T_1 - T_0) \quad \text{(see Equation 3.9)}$$

Substituting in the values given in the question

$$\phi_q = -\frac{1}{1} \times 0.1\,(300 - 390) = -0.1\,(-90)\ \text{J s}^{-1} = 9\ \text{J s}^{-1}$$

3) $0.5 \cdot 10^{-6}\,\text{m}^2\,\text{s}^{-1}$

Since $\alpha = \dfrac{\lambda}{\rho c_p}$ (see Equation 3.3) $= \dfrac{1}{2 \cdot 10^3 \cdot 1 \cdot 10^3} = 0.5 \times 10^{-6}\,\text{m}^2\,\text{s}^{-1}$

3.2 $v = 1.4 \times 10^{-4}\,\text{m s}^{-1}$ (approx)

We obtained this value from Equation (3.18)

$$C_B\,(x) = C_0 \exp\left(\frac{-vx}{D}\right)$$

where: $C_B = 10^{-6}C_0$; $x = 10\ \text{cm} = 0.1\text{m}$; $D = 10^{-6}\,\text{m}^2\,\text{s}^{-1}$.

Thus $10^{-6}\,C_0 = C_0 \exp\left(-\dfrac{v \times 0.1}{10^{-6}}\right)$. Thus $v = 1.4 \times 10^{-4}\,\text{m s}^{-1}$

3.3 1) $100\ \text{J s}^{-1}\,\text{m}^{-2}$ or $100\ \text{Wm}^{-2}$

Since $\phi_q^{''} = -\dfrac{\lambda dT}{dx}$ (Equation 3.2) then $\phi_q^{''} = -1 \times 100\ \text{J s}^{-1}\,\text{m}^{-2}$

2) $10^{-6}\,\text{m}^2\,\text{s}^{-1}$

Since $\phi_{M,A}^{''} = -D\dfrac{dC_A}{dx}$ then $10^{-6}\,\text{kg s}^{-1} = -D \cdot 1\ \text{kg s}^{-1}\,\text{m}^{-2}$

$D = 10^{-6}\,\text{m}^2\,\text{s}^{-1}$

3) $10^{-2}\,\text{kg m}^{-1}\,\text{s}^{-2}$

We can use Equation 3.19. $\phi_{p,xy}^{''} = -\eta\dfrac{dv_y}{dx}$ but $\eta = v\rho t$ herefore $\phi_{p,xy}^{''} = -v\rho\dfrac{dv_y}{dx}$

$\phi_{p,xy}^{''} = -10^{-5} \times 1000 \times 1 = -10^{-2}\,\text{kg m}^{-2}\,\text{s}^{-1}$

4) Schmidt number $Sc = \dfrac{v}{D}$

5) Decrease - since Prandtl number $= \dfrac{v}{\alpha}$

3.4 1) Since the Schmidt number $= \dfrac{v}{D}$ and the Prandtl number $= \dfrac{v}{\alpha}$, then the Lewis number (Le) $= \dfrac{Sc}{Pr} = \dfrac{\alpha}{D}$. The Lewis number is therefore the ratio of the thermal diffusion coefficient to the mass diffusion coefficient. In effect, therefore, it gives a ratio of the transport of heat and the transport of mass.

2) Lewis number $= 1$. Diffusion coefficient $= 10^{-7}\,\text{m}^2\,\text{s}^{-1}$

Since $Le = \dfrac{Sc}{Pr} = \dfrac{10^2}{10^2} = 1$

Also, since $Le = \dfrac{\alpha}{D}$ and $\alpha = 10^{-7}\,\text{m}^2\,\text{s}^{-1}$ and $Le = 1$ then $D = 10^{-7}\,\text{m}^2\,\text{s}^{-1}$.

3) A high Prandtl number is indicative of either a high kinematic (kinetic) viscosity coefficient or a low thermal diffusion coefficient or a combination of both of these. Such a system does not dissipate heat very

effectively. Thus gradients of localised heat may be established. Such systems are prone to localised over-heating.

4) A high Schmidt number is indicative of either a high kinematic (kinetic) viscosity coefficient or a low diffusion coefficient or a combination of both of these. In other words, such a system will result in only a slow movement of solute from areas of high concentration to areas of low concentration and thus be prone to localised concentration gradients.

5) In biotechnology, it is usually not desirable to have localised high and low concentrations of solute since the rate at which cells (or enzymes) can convert substrate into products is dependent upon concentration. Likewise it is usually not desirable to have the build up of localised high concentration of toxic products. Similarly since biological systems are sensitive to temperature, each displaying an optimum temperature, then it is undesirable to have localised temperature variations within a bioreactor. For these reasons, systems with low Schmidt and Prandtl numbers are preferred. Of course it is not always possible to use a system in which α, D and v can be controlled. But, at least by determining Prandtl and Schmidt numbers, the process technologist is alerted to the likelihood of localised gradients being established in the system and that special measures may need to be taken to generate homogeneity.

Responses to Chapter 4 SAQs

4.1

Answer $= 0.5 \text{ cms}^{-1}$

We worked it out in the following way.

$$\frac{C_y^B}{C_0^B} = \frac{C_y^A}{C_0^A} \text{ and } \frac{C_y^B}{C_0^B} = f\,(Pe^B) \text{ and } \frac{C_y^A}{C_0^A} = f\,(Pe^A)$$

then $Pe^B = Pe^A$

But $Pe^B = \dfrac{v^B L^B}{D^B}$ and $Pe^A = \dfrac{v^A L^A}{D^A}$ (see Equation 4.3)

Thus $\dfrac{v^B L^B}{D^B} = \dfrac{v^A L^A}{D^A}$

But we were given in the question that: $D^B = 2D^A$; $L^A = 45$ cm; $L^B = 180$ cm; $v^A = 1$ cm s^{-1}.

Thus $\dfrac{v^B \times 180}{2D^A} = \dfrac{1 \times 45}{D^A}$. Therefore $v^B = \dfrac{1 \times 45 \times 2}{180} = 0.5$ cm s^{-1}

This is clearly much simpler than working out the value of: $\dfrac{C_y^A}{C_0^A} = \exp\left(-\dfrac{vL}{D}\right)$

4.2

Since $C_p = f\,(C_r, t, L, x)$ we can write

$C_p \sim C_r^\alpha . t^\beta . L^\gamma . x^\delta$

Replacing these by units

$$\frac{kg}{m^3} \sim \left(\frac{kg}{m^3}\right)^\alpha . s^\beta . m^\gamma . m^\delta$$

Collecting up dimensions

kg: $1 = \alpha$

m: $-3 = -3\alpha + \gamma + \delta$

Thus $\gamma = 3\alpha - \delta - 3$ and $\alpha = 1$. Thus $\gamma = -\delta$

s: $0 = \beta$

Thus we can write

$$C_p \sim C_r \, t^0 \, L^{-\delta} \, x^{\delta}$$

Therefore C_p does not depend upon t (since C_p is related to t^0)

$$C_p \sim C_r \left(\frac{x}{L}\right)^{\delta}$$

Thus $\dfrac{C_p}{C_r} \sim \left(\dfrac{x}{L}\right)^{\delta}$ or $\dfrac{C_p}{C_r} = f\left(\dfrac{x}{L}\right)$

This final equation tells us that the ratio of the concentration of the product (C_p) to that of the reactant (C_r) is related to the ratio of depth of the liquid (x) and the length of the catalyst (L) by some function (f).

This example illustrates the power of dimensional analysis. Without knowing much abouth the process under investigation, by considering the dimensions you have come to a simple conclusion relating the various parameters described in the question. This example was purely hypothetical but it will have enabled you to test your ability to carry out dimensional analysis.

The answer above has a suggestion that it is nonsense. In the question we declared that the concentration of the product was dependent upon time, yet the dimensional analysis indicates that this is not so. You will need to return to the text for an explanation.

4.3

1) We use p = n - m

   n = number of parameters = 3

   m = number of units = 3

   p = 3 - 3 = 0

   This means that the set of parameters we have selected is contradictory. Thus the answers are:

   a) We have chosen an incorrect or incomplete set of variables which influence M

   b) This question is meaningless in view of the result we obtained for a).

2) Using Buckingham π theorem

   p = n - m = 5 - 3 = 2 thus:

   a) it is possible that the parameters are correct and b) that two dimensionless groups might be obtained. (You might like to work these out for yourself). We came to the conclusion that:

   $$\frac{C_x}{C_0} = f\left[\frac{vL}{D}\right]$$

   where $\dfrac{vL}{D}$ is a dimensionless group (in fact the Péclet number).

   $\dfrac{C_x}{C_0}$ is a ratio of two concentrations and is therefore dimensionless.

Responses to Chapter 5 SAQs

5.1

2.6 bar

The mean water velocity can be calculated from

$$v = \frac{\phi_M}{\rho \frac{\pi}{4} D^2} = \frac{1}{10^3 \times \frac{\pi}{4} \times (5 \times 10^{-2})^2} = 0.51 \text{ m s}^{-1}$$

Using $Re = \frac{\rho.v.D}{\eta}$ then $Re = 2.55 \times 10^4$

The flow is therefore turbulent at this high Reynolds number.

We can now obtain 4f from Figure (5.4)

4f = 0.025 in a smooth pipe when Reynolds number = 2.55×10^4

Substituting this and the given values into Equation (5.9)

$$\Delta p = 4f \frac{L}{D} \frac{1}{2} \rho \, v^2 = 2.6 \text{ bar}$$

5.2

1) 10 000

2) 100

These have been calculated from

$$Re = \frac{\rho N D^2}{\eta} \quad \text{(see Equation 5.11)}$$

Thus for 1)

$$Re = \frac{1000 \times 100 \times (1)^2}{10} = 10\,000$$

For 2)

$$Re = \frac{1000 \times 1 \times (1)^2}{10} = 100$$

From this data, the high Reynold's number for condition 1) suggest that the culture would be well mixed by turbulent flow. On the other hand, condition 2) represents non-turbulent conditions. In practice with biological systems (cells, micro-organisms in culture) it may not be possible to create really turbulent mixing without damaging the cells as a result of the shear forces set up.

5.3

1) When a body is placed in a liquid flow, it is subject to two types of drag forces called **form** and **frictional** drag.

2) When a liquid flows such that the layers of the liquid are moving in parallel, the liquid is said to be showing **laminar** flow.

3) Flat bladed paddle types of impellers produce **radial** flow in stirred tanks.

4) Inclined blade turbine impellers produce **axial** flow in stirred tanks.

5) When the Reynolds number exceeds 10 000 in a stirred tank, the liquid in the tank will show **turbulent** flow characteristics.

6) The smallest eddies in the liquid in a stirred tank are called **Kolmogorov** eddies.

7) Liquids in which shear stress τ_{xy} is proportional to the velocity gradient $\frac{dv_y}{dx}$ are said to be **Newtonian** liquids.

8) Liquids in which the velocity gradient $\dfrac{dv_y}{dx} = 0$ providing the shear stress (τ_{xy}) is smaller than the stress yield (τ_0) are called **Bingham plastics**.

5.4

1) $C_w = 491$

First calculate the density of the sphere

Volume of sphere $= \dfrac{\pi}{6} d^3 = \dfrac{\pi}{6} \dfrac{125}{10^9} \, m^3$

Mass $= 0.07 \, g = 7 \times 10^{-5} \, kg$

Density $= \dfrac{7 \times 10^{-5}}{\left[\dfrac{\pi}{6} \cdot \dfrac{125}{10^9}\right]} \, kg\,m^{-3} = \dfrac{42 \times 10^4}{\pi \times 125} = 1069 \, kg\,m^{-3}$

Now we can use:

$C_w \, A \, \dfrac{1}{2} \, \rho_{liquid} \, v_2 = $ force on the sphere (see Equation 5.4)

But in steady fall, this equals the effects of buoyancy

Buoyancy $= \dfrac{\pi}{6} d^3 \, (\rho_{sphere} - \rho_{liquid}) \, g$

But A (= Area of the sphere $= \dfrac{\pi}{4} d^2$)

Thus $C_w \, \dfrac{\pi}{4} d^2 \, \dfrac{1}{2} \, \rho_{liquid} \, v^2 = \dfrac{\pi}{6} d^3 \, (\rho_{sphere} - \rho_{liquid}) \, g$

and $C_w = \dfrac{4}{3} d \, \dfrac{(\rho_{sphere} - \rho_{liquid}) \, g}{v^2 \, \rho_{liquid}}$

$= \dfrac{4}{3} \cdot \dfrac{5}{10^3} \cdot \dfrac{(1069 - 900)}{(5 \times 10^{-3})^2 \cdot 900} \, (9.81)$

$= \dfrac{20}{3 \cdot 10^3} \cdot \dfrac{169}{900} \cdot \dfrac{9.81}{(5 \times 10^{-3})^2} = 491$

2) $0.46 \, Nsm^{-2}$

There are several ways to calculate this. We have choosen to use Stoke's law where $F = 3\pi \, \eta \, d \, v$ (see Equation 5.6).

At steady fall the effects of buoyancy and gravity are counteracted by friction, thus:

$\dfrac{\pi}{6} d^3 \, (\rho_{sphere} - \rho_{liquid}) \, g = 3\pi \, \eta \, d \, v$

Thus $\dfrac{d^2}{18} \, \dfrac{(\rho_{sphere} - \rho_{liquid}) \, g}{v} = \eta$

Thus $\eta = \dfrac{1}{18} \left(\dfrac{5}{10^3}\right)^2 \cdot \dfrac{(1069 - 900)}{5 \times 10^{-3}} \cdot 9.81 \, Nsm^{-2} = 0.46 \, Nsm^{-2}$

5.5

$50 \, rev\,min^{-1}$

Let us call the critical Re number when laminar flow becomes turbulent flow Re^c.

Thus from Equation 5.11 $Re^c = \dfrac{\rho N_1^c D_1^2}{\eta_1}$ where N_1^c is the number of revolution per minute at critical Re.

Thus for the first vessel

$$Re^c = \frac{\rho_1 \, 400 \, D_1^2}{\eta_1}$$

In the second vessel ρ is the same, $\eta_2 = 2\,\eta_1$ and $D_2 = 4D_1$

Thus in this case $Re^c = \dfrac{\rho_2 \, N_2^c \, D_2^2}{\eta_2} = \dfrac{\rho_1 \, N_2^c \, (4D_1)^2}{2\eta_1}$

The liquid in the second vessel will become turbulent at the same Re value

Thus $\dfrac{\rho_1 \, 400 \, D_1^2}{\eta_1} = \dfrac{\rho_1 \, N_2^c \, (4D_1)^2}{2\,\eta_1}$

and $N_2^c = \dfrac{800}{16} = 50$ rev min^{-1}

Responses to Chapter 6 SAQs

6.1 Conduction is the transfer of heat from one part of the body to another or from one body to another in physical contact with each other, without appreciable displacement of the particles of the body.

6.2 $100 \, \text{J m}^{-2}\,\text{s}^{-1}$

You should have been able to work this out very quickly. Since $\Delta T = \phi_q'' \dfrac{D}{\lambda}$ (Equation 6.6) and D and λ are constant then ϕ_q'' is proportional to ΔT.

When $\Delta T = 10$ K; $\phi_q'' = 50 \, \text{J m}^{-2}\,\text{s}^{-1}$; thus when $\Delta T = 20$ K; $\phi_q'' = \dfrac{50 \times 20}{10} = 100 \, \text{J m}^{-2}\,\text{s}^{-1}$

6.3 296.7 K

We can write for layer A

$$T_1 - T_0 = \phi_{q,A}'' \cdot \frac{D}{\lambda_A} \qquad\qquad \text{(Equation A)}$$

and for layer B

$$T_0 - T_2 = \phi_{q,B}'' \cdot \frac{D}{\lambda_B} \qquad\qquad \text{(Equation B)}$$

but $\lambda_A = 2\lambda_B$ and $\phi_{q,A}'' = \phi_{q,B}''$

Thus substituting in and dividing A by B.

$\left(\dfrac{T_1 - T_0}{T_0 - T_2}\right) = \dfrac{\phi_{q,A}''}{\phi_{q,B}''} \, \dfrac{D}{2\lambda_B} \, \dfrac{\lambda_B}{D} = \dfrac{1}{2};$ Thus $2\,(T_1 - T_0) = (T_0 - T_2)$.

But the total temperature difference $(T_1 - T_0) + (T_0 - T_2) = 10$ K

thus $(T_1 - T_0) + 2(T_1 - T_0) = 10$ and $T_1 - T_0 = 3.3$ K

Thus $T_0 = 300 - 3.3 \,°K = 296.7 \, K$

6.4

1) Your sketch should have looked something like this:

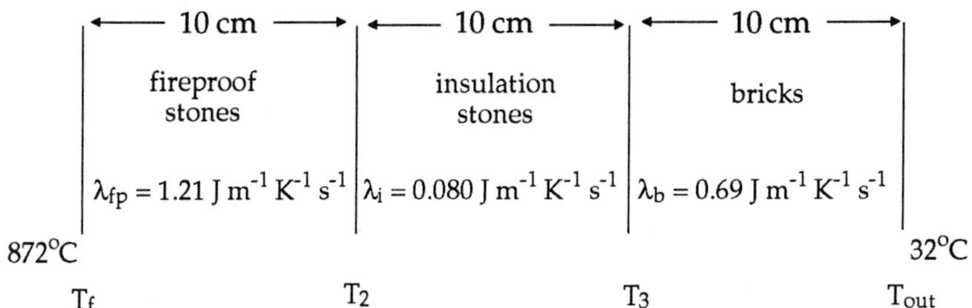

You may have used different symbols to represent the temperature at each of the interfaces.

2) The heat flow in the fire proof stone layer is given by: $T_f - T_2 = \phi_q'' \left(\dfrac{D}{\lambda_{fp}} \right)$

For the insulation stone layer: $T_2 - T_3 = \phi_q'' \left(\dfrac{D}{\lambda_i} \right)$

For the bricks: $T_3 - T_{out} = \phi_q'' \left(\dfrac{D}{\lambda_b} \right)$

3) $T_f = 872°C \; T_2 = 820°C \; T_3 = 39°C \; T_{out} = 32°C$. We carried out the calculation in the following way.

For the fire proof stones layer

$$T_f - T_2 = \phi_q'' \left(\frac{10 \times 10^{-2}}{1.21} \right) \tag{E-3a}$$

Thus $\phi_q'' = (T_f - T_2) \, 12.1$

Similarly for the insulation stone layer

$\phi_q'' = (T_2 - T_3) \, 0.8$

Thus $(T_f - T_2) \, 12.1 = (T_2 - T_3) \, 0.8$

Thus $(T_f - T_2) \, \dfrac{12.1}{0.8} = (T_2 - T_3)$

From the brick layer $\phi_q'' = (T_3 - T_{out}) \, 6.9$

Thus $(T_3 - T_{out}) \, 6.9 = (T_2 - T_3) \, 0.8$

Thus $(T_3 - T_{out}) = (T_f - T_2) \, \dfrac{12.1}{6.9}$

But $(T_f - T_2) + (T_2 - T_3) + (T_3 - T_{out}) = 840° = $ Total temperature differences.

Thus $(T_f - T_2) + (T_f - T_2) \, \dfrac{12.1}{0.8} + (T_f - T_2) \, \dfrac{12.1}{6.9} = 840$

Thus $T_f - T_2 = 47°C$

and $(T_2 - T_3) = 710.6°C$

$(T_3 - T_{out}) = 82.4°$

Thus temperatures at the interfaces are (approximately):

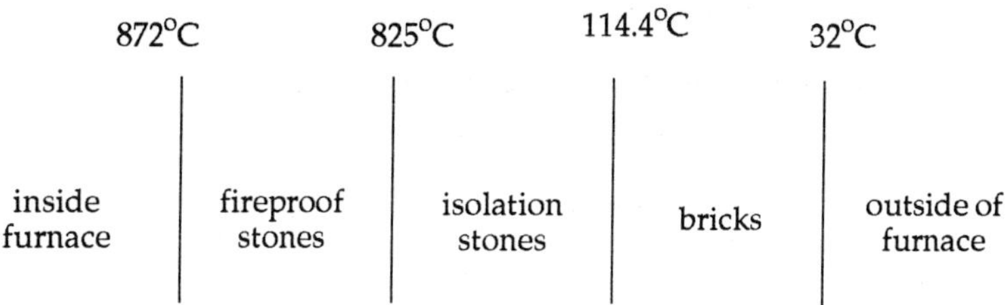

4) In the calculation carried out in 3, we assumed that the heat flow per unit area (ϕ_q'') was equal across all layers. This is, of course, only true if the layers are flat and in parallel. If the layers were curved then the surface areas of the layers would not be equal, (the outer layer would have a larger surface area). Thus in these cases ϕ_q'' would not be the same in each layer.

6.5

1) Approximately 91°C

2) Approximately 98.5°C

The way we did this calculation is as follows.

Since the temperature at the centre had risen to 60°C after 10 minutes, then the value of $\dfrac{T_1 - T_c}{T_1 - T_0} = \dfrac{100 - 60}{100 - 0}$ = 0.4.

From Figure 6.11 we can see that the value of Fo ($= \alpha t / D^2$) when $\dfrac{T_1 - T_c}{T_1 - T_0}$ is 0.4.

= 0.045 (you may not have been able to read this very accurately from the figure).

Since Fo $= \alpha t / D^2$, and α and D are constant, then if Fo = 0.045 when t = 10 minutes, Fo will = 0.09 when t = 20 minutes.

From Figure 6.11, when Fo = 0.09 then, for a cylinder

$$\dfrac{T_1 - T_c}{T_1 - T_0} = 0.09 \text{ (approx)}$$

Thus $\dfrac{100 - T_c}{100 - 0} = 0.09$

Therefore T_c = 91°C (approximately)

Similarly, if Fo = 0.045 when t = 10 minutes, then Fo = 0.135 when t = 30 minutes.

From Figure 6.12, when Fo = 0.135, then $\dfrac{T_1 - T_c}{T_1 - T_0} = 0.015$ (approximately).

Thus $\dfrac{100 - T_c}{100 - 0} = 0.015$

and T_c = 98.5°C

6.6

1) $12 \text{kJm}^{-2} \text{K}^{-1} \text{s}^{-1}$, $8 \text{kJm}^{-2} \text{K}^{-1} \text{s}^{-1}$, $2 \text{kJm}^{-2} \text{K}^{-1} \text{s}^{-1}$.

2) $1.412 \text{ kJm}^{-2} \text{K}^{-1} \text{s}^{-1}$.

3) 564.8 kJ s^{-1}.

Part 1 is answered by using the equation $h = \dfrac{\lambda}{D}$

Thus $h_A = \dfrac{0.6}{0.05}\,kJm^{-2}\,K^{-1}\,s^{-1}$ (with D in m) $= 12\,kJm^{-2}\,K^{-1}\,s^{-1}$

$h_B = \dfrac{0.4}{0.05}\,kJm^{-2}\,K^{-1}\,s^{-1} = 8\,kJm^{-2}\,K^{-1}\,s^{-1}$

$h_C = \dfrac{0.2}{0.1}\,kJm^{-2}\,K^{-1}\,s^{-1} = 2kJm^{-2}\,K^{-1}\,s^{-1}$

Part 2

We use the relationship

$$\frac{1}{U} = \frac{1}{h_1} + \frac{1}{h_2} + \frac{1}{h_3} = 0.083 + 0.125 + 0.5 = 0.708$$

Thus $U = 1.412\,kJm^{-2}\,K^{-1}\,s^{-1}$

Part 3

We use the relationship

$$\phi_q = U \cdot A \cdot \Delta T$$

$$= 1.412 \times 40 \times 10\,kJ\,s^{-1}$$

$$= 564.8\,kJ\,s^{-1}$$

6.7

1)

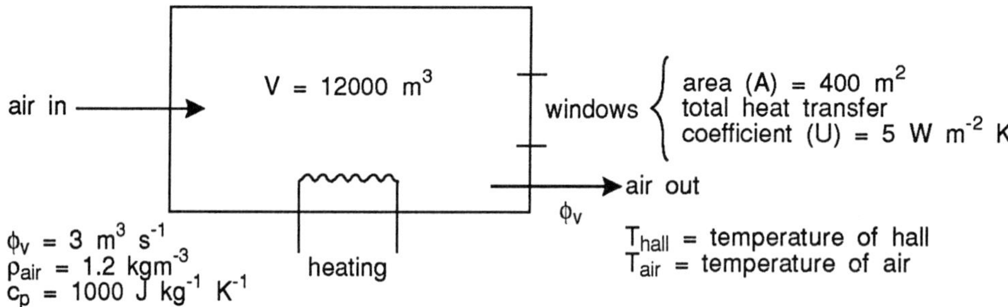

2) The rate of change of heat in the hall is given by:

$$\frac{d}{dt}(\rho_{air} \cdot V \cdot c_p \cdot T_{hall}) = \text{heat input in incoming air - heat output by outflowing air - heat loss through the windows + production of heat via the heater}$$

$$\frac{d}{dt}(\rho_{air} \cdot V \cdot c_p \cdot T_{hall}) = \phi_v\,\rho_{air}\,c_p\,T_{air} - \phi_v\,\rho_{air}\,c_p\,T_{hall} - U \cdot A \cdot (T_{hall} - T_{air}) + P_H$$

3) $T_{hall} - T_{air} = 30°C$ has to be reached. The power of the heater (P_H) to reach this is found from a steady-state balance with $T_{hall} - T_{air} = 30°C$ since in this case the heater just compensates for the heat losses.

$$0 = -\phi_v\,\rho_{air}\,c_p\,(T_{hall} - T_{air}) - U \cdot A \cdot (T_{hall} - T_{air}) + P_H$$

$$P_H = \phi_v\,\rho_{air}\,c_p\,(T_{hall} - T_{air}) + U \cdot A \cdot (T_{hall} - T_{air})$$

$$= (3 \cdot 1.2 \cdot 1000 + 5 \cdot 400) \cdot 30 = 1.68 \cdot 10^5\,W$$

4) To solve this we need to use the non steady-state relationship established in 2).

Since $\dfrac{d\,(\rho_{air}\,V\,c_p\,T_{hall})}{dt} = \phi_v\,\rho_{air}\,c_p\,T_{air} - \phi_v\,\rho_{air}\,c_p\,T_{hall} - U.A.\,(T_{hall} - T_{air}) + P_H$

it follows that:

$$\frac{d\,T_{hall}}{dt} = -\frac{\phi_v\,\rho_{air}\,c_p + U.A.}{\rho_{air}\,c_p\,V}\,(T_{hall} - T_{air}) + \frac{\tfrac{1}{2}P_H}{\rho_{air}\,c_p\,V}$$

since we are using only half the power of the heater.

This equation can be written in a more convenient form.

$$\frac{d\,T_{hall}}{dt} = -\left(\frac{\phi_v\,\rho_{air}\,c_p + U.A.}{\rho_{air}\,c_p\,V}\right)\left\{T_{hall} - T_{air} + \frac{1}{2}\frac{P_H}{\phi_v\,\rho_{air}\,c_p + U.A.}\right\}$$

Let $\alpha = \dfrac{\phi_v\,\rho_{air}\,c_p + U.A.}{\rho_{air}\,c_p\,V}$ and $T_s = T_{air} + \dfrac{1}{2}\dfrac{P_H}{\phi_v\,\rho_{air}\,c_p + U.A.}$

then $\dfrac{d\,T_{hall}}{dt} = -\alpha\,(T_{hall} - T_s)$

Thus $\dfrac{1}{T_{hall} - T_s}\,d\,T_{hall} = -\alpha\,dt$

The general solution: $\ln (T_{hall} - T_s) = -\alpha t + K_1$; the boundary condition is at $t = 0$ $T_{hall} = T_{air} = 5^{\circ}C$

Thus $K_1 = \ln (T_{air} - T_s)$

$$\Rightarrow \ln \frac{T_{hall} - T_s}{T_{air} - T_s} = -\alpha t$$

thus $t = -\dfrac{1}{\alpha}\ln\dfrac{T_{hall} - T_s}{T_{air} - T_s}$. If $T_{hall} = 15^{\circ}C$ then $t = 47$ min.

6.8   1) $0.34\ ms^{-1}$

The mean velocity $= <v> = \dfrac{\phi_v}{\tfrac{\pi}{4}D^2}$.

Remember $40\,l = 0.04\ m^3$, $5\ cm = 0.05\ m$ and our flow was described in the question in terms of $l\ min^{-1}$

$$<v> = \frac{0.040}{60 \times \frac{\pi}{4} \times (0.05)^2} = 0.34\ ms^{-1}$$

2) Turbulent. We first determine Reynold's number

$$Re = \frac{\rho<v>D}{\eta} = \frac{1000 \times 0.34 \times 0.05}{1 \times 10^{-3}} = 1.7 \times 10^4$$

with this high Reynold's number the flow would be turbulent.

3) We have used the same nomenclature as that associated with Figure 6.14. For the turbulent flow the transfer of heat is governed by the relationship.

$Nu = 0.027\ Re^{0.8}\ Pr^{0.33}$ (see Equation 6.53). We can construct (for the steady-state) a heat balance over a small slab of water flowing from x to x + dx in the tube (as shown in Figure 6.14).

(As the flow is turbulent we can work with the mean temperature of the water at position x).

The heat gained by the liquid will be equal to the heat transfered via the wall.

Thus $\phi_v\,\rho c_p\,[T]_x - \phi_v\,Pc_p\,[T]_{x+dx} = h\,\pi D\,dx\,(T_w - T)$

Thus $\phi_v\,\rho c_p\,\dfrac{dT}{dx} = h\,\pi\,D\,(T_w - T)$

Integrating $\dfrac{T_w - T}{T_w - T_{in}} = \exp\left(-\dfrac{h\pi D}{\phi_v \rho c_p}\,x\right)$

4) In order to calculate T at $x = L$, we would need to know T_w, T_{in}, the specific heat (c_p) of the liquid and the heat transfer coefficient (h). We have been given D, ρ and L (the length of the tube).

Responses to Chapter 7 SAQs

7.1

$0.05\ \mathrm{mol\ s^{-1}}$

We calculate this from $\phi_{M,A} = k\,A\,\Delta C_A$. (E - 7.12)

Thus $\phi_{M,A} = 0.1 \times 1 \times 0.5\ \mathrm{mol\ s^{-1}} = 0.05\ \mathrm{mol\ s^{-1}}$

7.2

There are several ways of solving this. Perhaps the most straight forward is as follows.

For a flat plane $Sh = 1$ but $Sh = \dfrac{kD}{D}$

Thus $\dfrac{k \times 1 \times 10^{-3}}{2 \times 10^{-3}} = 1$ and, therefore, $k = 2\ \mathrm{m\ s^{-1}}$

7.3

1) $4.5\ \mathrm{cm}$

2) $0.286\ \mathrm{kg\ m^{-2}}$

To calculate the depth the dye has penetrated we can use the penetration theory.

$x_e = \sqrt{\pi D\,t_e} = \sqrt{\pi \times 6.31 \times 10^{-5} \times 10}\,\mathrm{m} = 4.5\ \mathrm{cm}$ (E - 7.27)

The mass of dye diffusing into the layer is calculated in the following way.

The mean flux into the layer during the 10s is calculated using the mean value for the mass transfer coefficient ($<k>$) (see Equation 7.26).

$\phi_M'' = <k>\,(C^* - 0) = 2\,\sqrt{\dfrac{D}{\pi\,t_e}}\,.\,C^*$

From which we can calculate the mass per unit area diffusing into the layer

$M'' = <\phi_M''> = \dfrac{2}{\pi}\,\sqrt{\pi D\,t_e}\,.\,C^*$ (see equation 7.29)

Since $\sqrt{\pi D\,t_e} = 4.5\ \mathrm{cm} = 0.045\ \mathrm{m}$ and $C^* = 10\ \mathrm{kg\ m^{-3}}$ then $M'' = 0.286\ \mathrm{kg\ m^{-2}}$.

7.4

This problem is analogous to example 4 described in the text.

Since the column is 100 m, it takes the drop $\dfrac{100}{10}$ s to fall from the top to the bottom, ie $t_e = 10$ s.

Thus we can calculate the penetration depth of SO_2 into the drop by the time the drop reaches the bottom of the column from

$x_e = \sqrt{\pi D\,t} = 2.5 \times 10^{-4}\ \mathrm{m}$

Thus x_e is significantly smaller than the radius (2 cm) of the droplet so we can approximate the surface of the sphere to be a flat plate (see example 4 in the text). Thus we can use equations 7.25 and 7.26 to establish that

$$\Delta M = 2\, D^2 \cdot C^* \cdot x_e \quad \text{(see Equation 7.31)}$$

Thus $<C> = \dfrac{12\, C^*\, x_e}{\pi\, D}$ (see Equation 7.32)

where D = diameter of the droplet.

$$= \frac{12}{\pi} \times \frac{2.5 \times 10^{-4}}{0.04} \times 0.4 = 0.0095 \text{ kg m}^{-3}$$

7.5

1) 6.9×10^{-3} mol s^{-1}

2) 6.9×10^{-7} m s^{-1}

We can use Stefan's equations. Thus the molar flux of the vapour through the vertical tube

$$\phi_A'' = \frac{D\, C}{L} \ln\left[\frac{C - C_{AL}}{C - C_{A0}}\right] \quad \text{(E - 7.45)}$$

(we have denoted the solvent by A).

The total molar concentration C at the top of the tube is calculated from the ideal gas law.

$$C = \frac{P}{RT} = 37.5 \text{ mol m}^{-3}$$

The concentration of the solvent vapour C_{A0} at the liquid vapour interface is obtained $C_{A0} = \dfrac{Pv}{RT} = 13.9$ mol m^{-3}.

The concentration $C_{AL} = 0$.

Thus $\phi_A = \dfrac{2.0 \times 10^{-5} \times 37.5}{0.05} \ln\left[\dfrac{37.5}{23.6}\right] = 1.5 \times 10^{-2} \ln 1.589 = 6.9 \times 10^{-3}$ mol s^{-1}.

But this mole flux must all come from the liquid.

Thus $\dfrac{\rho v}{M} = \phi_A'' = 6.9 \times 10^{-3}$ mol s^{-1}

where v is the velocity by which the liquid level falls.

Thus $v = \dfrac{\phi_A'' \cdot M}{\rho} = \dfrac{6.9 \times 10^{-3} \times 180 \times 10^{-3}}{1800} = 6.9 \times 10^{-7}$ m s^{-1}.

7.6

1) $m = 30$

We were given that $C_\infty = 8$ mg l^{-1} = 0.008 kg m^{-3}.

We define the partition coefficient as $m = \dfrac{C_{air}}{C_\infty}$. The O_2-concentration in the air is $C_{air} = 0.2 \times 1.2 = 0.24$ kg m^{-3}, hence $m = 30$.

2) a) O_2 will flow from water to air (see notes for an explanation).

    b) 15 (since the pressure, and thus CO_2 in air, is reduced by 50%)

    c) 30

    d) 0.004g m^{-3} (4 mg l^{-1}).

The O_2-concentration in the air when the pressure is reduced is now $C_{new\ air} = 0.12$ kg m^{-3}, hence the quotient of the concentrations in the bulk of the liquid and the air is 15 and therefore much smaller than then the equilibrium value m $= 30$. Thus there is too much O_2 in the water and in order to approach equilibrium (where the quotient is m $= 30$) there will be a flow of O_2 out of the water into the air! Of course the concentrations at the interface immediately jump to new values so that there quotient equals m.

After a long time equilibrium will be restored and $m = \dfrac{C_{air}}{C_\infty}$ will again $= 30$. Thus the new O_2 concentration in water will be $c_\infty = \dfrac{0.12}{30} = 0.004$ kg m$^{-3} = 4$ mg l^{-1}.

7.7

1) False. The concentration profile is not continuous (see example Figure 7.5)

2) True. This is derived from the relationship

$$\frac{1}{K_1} = \frac{m}{k_2} + \frac{1}{k}$$ (equation 7.64) established in Section 7.6.2.

3) True. a = specific area = area per unit volume. Thus total area = volume x a

$$= 25\ m^2.$$

7.8

152 s

Here is our solution.

From Equation 7.87 we know that:

$$k = \frac{h}{\rho c_p}\left(\frac{D}{\alpha}\right)^{\frac{2}{3}} \quad \text{and} \quad \alpha = \frac{\lambda}{\rho c_p}$$

$$\text{Thus } k = \frac{h}{\rho c_p}\left(\frac{\rho c_p D}{\lambda}\right)^{\frac{2}{3}} = \frac{10^4}{10^3 \cdot 4.2 \cdot 10^3} \cdot \left(\frac{1}{100}\right)^{\frac{2}{3}} = 1.1 \cdot 10^{-4}\ ms^{-1}$$

Now we can examine the mass balance for the salt layer. We can represent this in the following way.

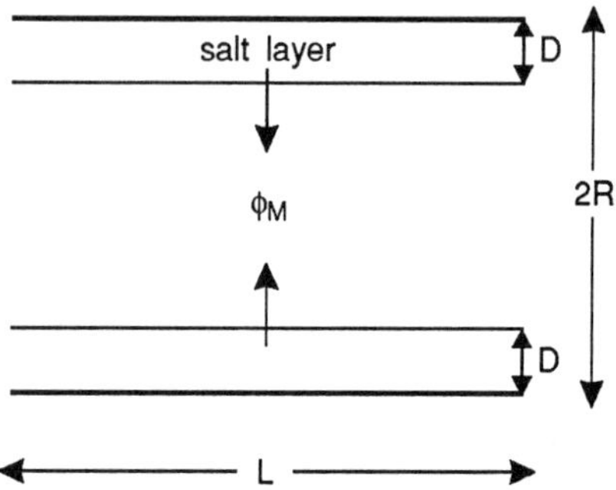

Mass balance for the salt layer:

$$\frac{d\ (\rho_{salt} \cdot 2\pi R L D)}{dt} = -\phi_M = -2\pi R L \cdot K (C^* - C_w)$$

where C^* solubility of salt in distilled water and C_w = concentration of salt in the water in the pipe. Given that $C^* = 300$ kg m^{-3} and $C_w = 0$.

Thus: $\rho_{salt}\dfrac{dD}{dt} = -kC^{*}$

Solution:

$$D = -\frac{KC^{*}}{\rho_{salt}}.t + K_1$$

The boundary conditions are $t = 0 \rightarrow D = D_0 = 2mm$

Hence $K_1 = D_0$

Thus $D - D_0 = -\dfrac{kC^{*}}{\rho_{salt}}.t$ and $D = 0$

Therefore $t = \dfrac{\rho_{salt}\,D_0}{kC^{*}} = \dfrac{2500.2.10^{-3}}{1.1.10^{-4}.300} = 152\,s$

7.9   1)   44 h (approximately $1.6 . 10^5$ s)

2)   1100 s

Our solution to these problems are as follows.

Part 1)

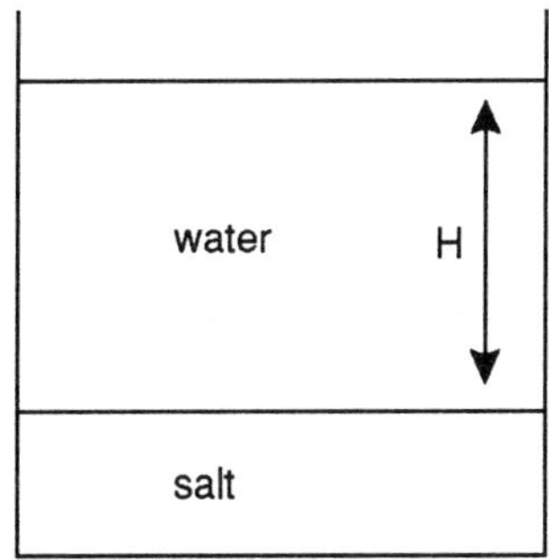

When all the salt is dissolved, the mean salt concentration in the water is:

$$<C> = \frac{200.10^{-3}\,kg}{10^3\,m^3} = 200\,kg\,m^{-3}$$

If we rotate the figure above by 90°, we can represent the salt concentration profile in the following way.

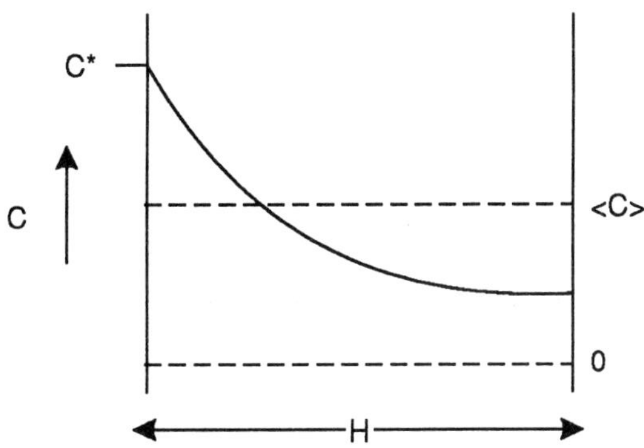

where C^{*} = solubility of salt in water = 300 kg m^{-3}

This is analogous to the half the picture used in discussing the temperature profile during heat transfer (see Figure 6.10).

If we double the figure; we obtain the analogy of heat penetrating into a body (see Figure 6.10 and associated text).

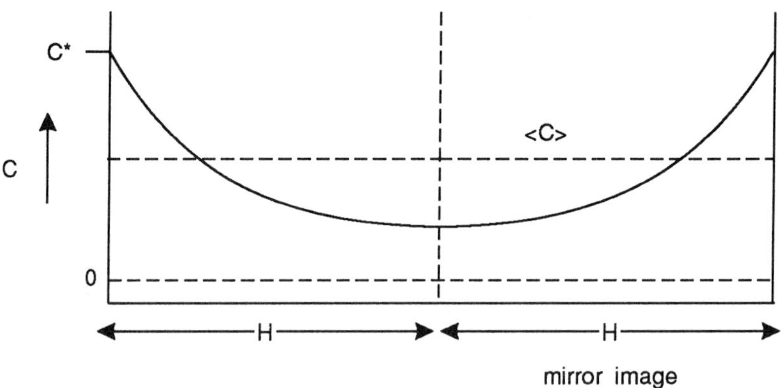

and $\dfrac{T_1 - <T>}{T_1 - T_0} \leftrightarrow F_0$ (see inset in Figure 6.11)

We can replace $\dfrac{T_1 - <T>}{T_1 - T_0}$ by $\dfrac{C^* - <C>}{C^* - C_0}$

and $D \to 2H$

$\alpha \to D$

Thus $F_0 = \dfrac{D\,t}{(2H)^2}$ (see Equation 6.27)

$\dfrac{C^* - <C>}{C^* - C_0} = \dfrac{300 - 200}{300} = \dfrac{1}{3}$

From Figure 6.10, $F_0 \approx 0.1$ when $\dfrac{C^* - <C>}{C^* - C_0} = \dfrac{1}{3}$ for a plate geometry.

Hence $t = \dfrac{4H^2}{D} F_0 = 4 \dfrac{(2 \cdot 10^{-2})}{10^{-9}} \cdot 0.1 = 1.6 \cdot 10^5 \, s = 44h.$

Part 2)

We can write mass balance

$\dfrac{d\,(V.C)}{dt} = + Ak\,(C^* - C)$

where V = volume of water, C = concentration of salt in the water, A = area of salt - water interface.

Given the boundary condition $t = 0 \to C = 0$.

We obtain the solution $\ln \dfrac{C^* - C}{C^*} = -\dfrac{A}{V} k\,t$

Using $V = A.H$, we can work out t when $C = 200 \, kg \, m^{-3}$ (the time when all of the salt is dissolved).

Thus, from the data given $v = 10^{-3} \, m^3$, $H = 2.10^{-2} \, m^2$ and $A = 5.10^{-2} \, m^{-2}$

Thus $t = 1100 \, s$

7.10   We can represent the situation by:

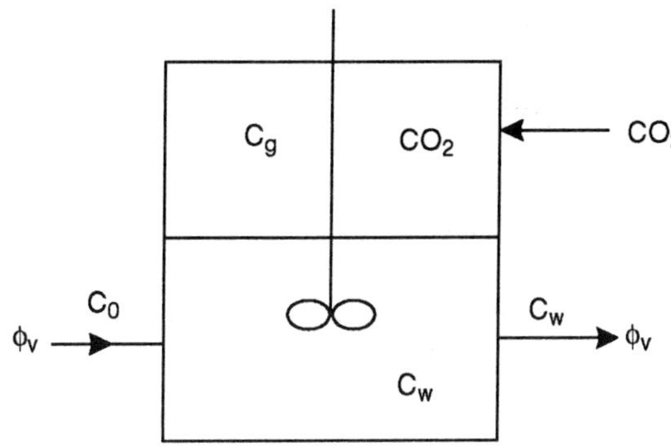

At the interface, there is an equilibrium.

$C_{wi} = \dfrac{C_{gi}}{m}$ where i signifies interface, m = partition coefficient

In the pure CO_2 gas phase, there is no resistance against CO_2 transport, hence:

$C_g = C_{gi}$

The flow of CO_2 from the water side of the interface into the bulk of the water can be written as:

$\phi_M = k. \, A \, (C_{wi} - C_w)$ where A = interfacial area

Thus $\phi_M = k. \, A \left(\dfrac{C_{gi}}{m} - C_w \right) = k. \, A \left(\dfrac{C_g}{m} - C_w \right)$

Thus a steady-state mass balance for CO_2 reads as:

$0 = \phi_v \, C_0 - \phi_v \, C_w + k. \, A \left(\dfrac{C_g}{m} - C_w \right)$

But $C_0 = 0$. Thus: $k = \dfrac{\phi_v \, C_w}{\left(\dfrac{C_g}{m} - C_w \right) A}$

7.11

1) Your sketch should be similar to this.

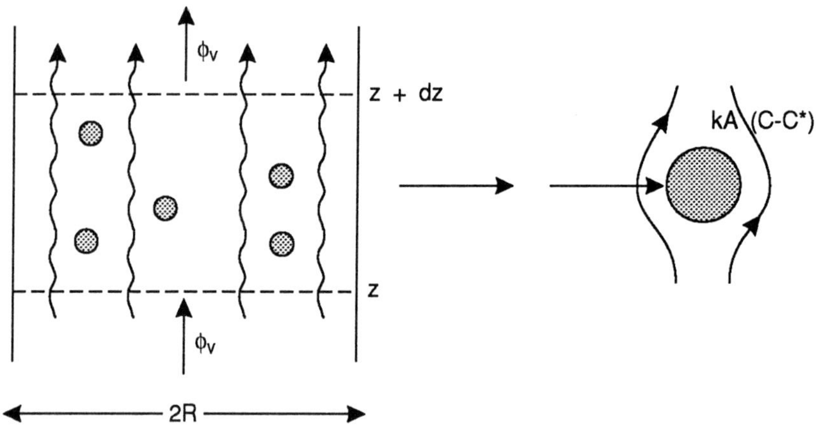

2) Steady-state Ca^{++} mass balance over the slab $\{z, z + dz\}$

$$0 = \phi_v\, C|_z - \phi_v\, C|_{z+dz} - k\,(S\,\pi\,R^2\,dz)\,.\,(C|_z - C^*)$$

The first two terms are the convective flows into and out of the slab.

The third term is the transfer to the ion-exchange, which are excluded from the 'control-volume'. The balance is over the **liquid** in the slab. The term $S\,\pi\,R^2\,dz$ is the total active surface of the exchange in the slab (S = active surface per $m^3 = 1.5\,.\,10^2\,m^{-3}$).

The balance can be simplified to:

$$+\,\phi_v\,.\,\frac{dC}{dz} = -\,k\,(S\,\pi\,R^2)\,(C - C^*)$$

Since $C^* \approx 0$ this becomes: $\dfrac{dC}{dz} = -\,\dfrac{k\,S\,\pi\,R^2}{\phi_v}\,.\,C = -\,A.C$ with $A = \dfrac{k\,S\,\pi\,R^2}{\phi_v}$

3) Since $\dfrac{dC}{dz} = -\,A.C$

$\ln C = A.z. + K_1$

The boundary conditions are:

$z = 0 \rightarrow C = C_i\ (= 500\ mg\ l^{-1})$

$\Rightarrow\ \ln\dfrac{C}{C_i} = -\,A.z.$

Let the concentration at the exit be C_f. Thus $\dfrac{C_f}{C_i} = \exp\,(-A.L)$ where L = column length.

4) Since $A = k\,\dfrac{S\,\pi\,R^2}{\phi_v}$ (see 2)) and the velocity $v_0 = \dfrac{\phi_v}{\pi\,R^2}$

But we gave the relationship $\dfrac{K\,d_0}{D} = 1.4\left(\dfrac{v_0\,d_0}{v}\right)^{\frac{1}{2}}\left(\dfrac{v}{D}\right)^{\frac{1}{3}}$

substituting in $V_0 = \dfrac{\phi_v}{\pi\,R^2}$

$$\frac{K\,d_0}{D} = 1.4\left(\frac{v}{D}\right)^{\frac{1}{3}} \cdot \left(\frac{d_0}{v}\right)^{\frac{1}{2}} \cdot \left(\frac{\phi_v}{\pi}\right)^{\frac{1}{2}} \cdot \frac{1}{R} \Rightarrow \frac{k\,S\,\pi\,R^2\,L}{\phi_v} = 1.4\,S\left(\frac{\pi}{\phi_v}\right)^{\frac{1}{2}} \frac{D}{d_0^{\frac{1}{2}}\,v^{\frac{1}{6}}}\,RL = 157\,RL$$

where 157 is in m^{-2}

Hence $C_f = C_i \exp(-157\,RL)$ Equation A

constraint $Re > 10$ (for k-relation) $\Rightarrow \dfrac{\phi_v}{\pi R^2}\dfrac{d_0}{v} > 0 \Rightarrow R < 4.21$ cm

Choose $R = 3.5$ cm $\rightarrow$ diameter $= 7$ cm

Take C_f to be $= 9\ \text{mg l}^{-1} < 10\ \text{mg l}^{-1} \Rightarrow RL = 2.56 \cdot 10^{-6}\ m^2$

Thus $L = 73$ cm.

7.12   1)   We can represent the system in the following way.

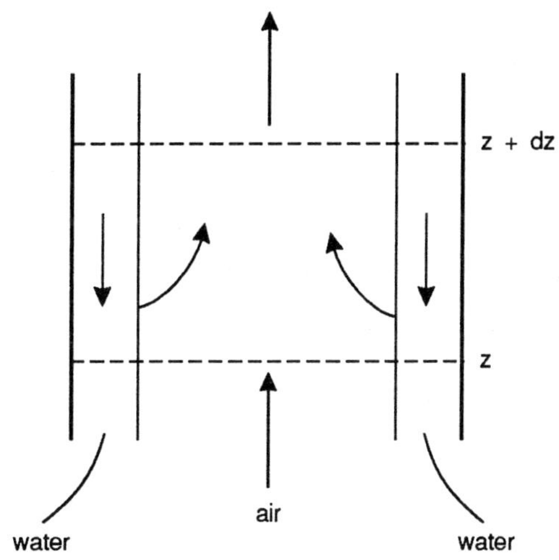

Steady-state mass balance for vapour.

$$0 = \phi_{v,\text{air}} \cdot C|_z - \phi_{v,\text{air}} \cdot C|_{z+dz} + k\,(\pi\,D\,dz)\,(C^* - C|_z)$$

$$= \phi_{v,\text{air}}\frac{dc}{dz} = k\,(C^* - C)\,\pi\,D.$$

2)   Boundary condition $z = 0 \rightarrow C = 0$

Solution: $-\ln(C^* - C) = \dfrac{\pi\,D\,K}{\phi_{v,\text{air}}} + K_1$

Boundary condition $\Rightarrow K_1 = -\ln C^*$

Thus:

$$\ln\frac{C^* - C}{C^*} = -\frac{\pi\,D\,k\cdot z}{\phi_{v,\text{air}}}\quad\text{Equation 2A}$$

k is calculated from the Sh relation.

$$v = \frac{\phi_v}{\frac{\pi}{4} D^2} = 3.5 \text{ ms}^{-1} \rightarrow Re = \frac{\rho\, v\, D}{\eta} = 1.03 \cdot 10^4$$

$$Sc = \frac{v}{D} = \frac{\eta}{\rho\, D} = 0.68$$

Hence $Sh = 0.023\, Re^{0.83}\, Sc^{0.44} = 41.6$

$$= k = \frac{Sh\, D}{D} = 41.6 \cdot \frac{2.5 \cdot 10^{-5}}{5 \cdot 10^{-2}} = 2.1 \cdot 10^{-2} \text{ ms}^{-1}$$

At the end of the column (of length L) we must have $\dfrac{C}{C^*} = 0.99$

Using Equation 2A derived above, yields:

$$\ln\left(\frac{C^* - 0.99\, C^*}{C^*}\right) = -\frac{\pi \cdot 5 \cdot 10^{-2} \cdot 2.1 \cdot 10^{-2}}{6.9 \cdot 10^{-3}} L$$

$L = 9.7$ m.

7.13

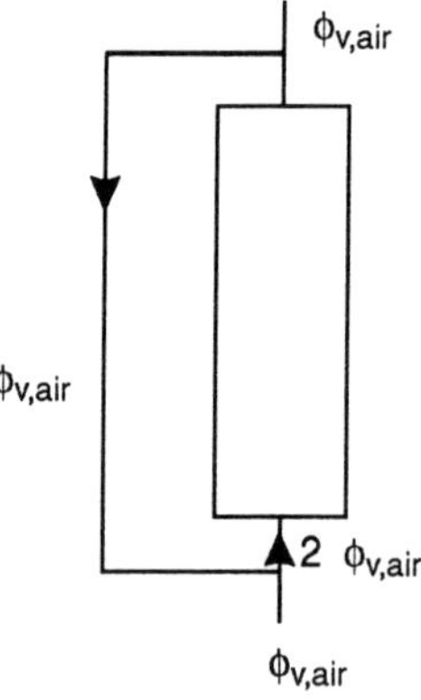

Since $\phi_{v,air}$ circulates, we can deduce that $2\,\phi_{v,air}$ flows through the column.

Thus we can use the same balance as in SAQ 7.11 if we substitute $2\,\phi_{v,air}$ instead of $\phi_{v,air}$.

Thus $- \ln (C^* - C) = \dfrac{\pi\, D\, k}{2\, \phi_{v,air}} z + K_1$

Now the boundary condition is:

at $z = 0 \rightarrow C = \dfrac{1}{2} C_L$ where C_L = concentration leaving the column (since half the gas leaving the column is returned to the column)

Thus $K_1 = - \ln C^* - \dfrac{1}{2} C_L$

Thus: $\ln \dfrac{C^* - C}{C^* - \dfrac{1}{2} C_L} = -\dfrac{\pi\, D\, k \cdot z}{2\, \phi_{v,air}}$

Since we have $2\,\phi_{v,air}$ flowing through the column, the velocity is doubled, hence Re is doubled.

Thus k increases by a factor $(2)^{0.83}$. Thus k is now $3.7 \cdot 10^{-2}$ ms^{-1}

Again at $z = L$; $C = C_L = 0.99\ C^*$ then

$$\ln \frac{C^* - 0.99\ C^*}{C^* - \frac{1}{2} \cdot 0.99\ C^*} = - \frac{\pi D k}{2\ \phi_{v,air}} \cdot L. \text{ Thus } L = 9.3m$$

Thus by recirculating half the gas, we can only reduce the column length from 9.7m to 9.3m to achieve the same result.

Responses to Chapter 8 SAQs

8.1

1) Earth is a closed system by approximation. There is significant input of free enthalpy by the sunlight. There is a negligible input of matter from meteorites etc.

2) The galaxy is probably isolated although of course it might receive very small quantities of light from other galaxies.

3) An isothermal anaerobic fermentor is closed system because of heat-transfer to or from the constant temperature room.

4) This anaerobic fermenter is open, there is the flow of heat (energy) into or out of the fermentor in order to maintain a constant temperature and also matter (CO_2) passes the system boundary.

8.2

As long as the number of cells is large, eg 10^6-10^9 ml^{-1} a macroscopic description is allowed. However certainly in the range of 10 cells ml^{-1}; down to 0 cells ml^{-1} one is dealing with a particle problem where the macroscopic approach is not allowed, but a stochastic description is needed.

The risks are that certain elements of the total volume will contain contaminating micro-organisms while others do not. As sterilisation proceeds the average number of micro-organisms present per element would approach 0 ml^{-1}. If the vessel was large (eg 50m^3) then there would be many elements of 1 ml (50 x 10^6). If a single organism was to survive the average number per element would be 0.2×10^{-8}. From a macroscopic stand point, this may be regarded as 0. Nevertheless, the system cannot be regarded as sterile. The macroscopic approach is only appropriate where we need to know an average value. For example if we were aiming to reduce the population of cells from say 10^7 ml^{-1} to 10^4 ml^{-1}.

8.3

Intensive	**Extensive**
- colour	- particle number
- shape	- system length
- particle diameter	- system area
- concentration of biomass	- amount of biomass and glucose
- concentration of ammonia	

This result can easily be understood by using the property that intensive properties do not change if a system size is changed. The extensive properties change proportional to system enlargement. Thus for example doubling the size of a system does not change the concentration of a component, it does however double the amount of the component.

8.4

1) A system in which only energy is exchanged with the surrounding is a closed system.

2) No. Intensive properties are not additive.

3) Yes. The reaction volume is 1 ml. If we are taking 5 µl portions and we regard each as an elementary volume, the system contains 200 elementary volumes. Each elementary volume contains 10^{-1} x Avagadros

number $x \dfrac{5}{10^6}$ molecules of each reactant (ie mol l^{-1} x Av no x volume in litres). Thus each elementary volume contains approximately 3×10^{17} molecules of each reactant. Thus the three conditions for the macroscopic approach (elementary volume is small compared to the system but is large compared to the particles and contain many particles) are fulfilled. A macroscopic approach is justified. Whether or not the small volume of samples would enable accurate measurement of reactants/products is however another matter!

4) Volume, energy and mass - these are extensive properties and we can therefore formulate macroscopic balances for them. Temperature is an intensive property and we cannot produce a macrobalance for such properties.

5) Accumulation = net conversion + net transfer. In a closed system, there is no net tranfer of mass. So, for a closed system: accumulation = net conversion.

8.5 A macroscopic balance must be made for each extensive quantity. The amount of each chemical compound is such an extensive quantity.

Hence for ethanol: $\overset{o}{E} = P_E + \phi_E$

production $P_E = + 1$ kg h^{-1} and transfer $\phi_E = - 0.2$ kg h^{-1}

The accumulation rate of ethanol is therefore:

$\overset{o}{E} = + 1 - 0.2 = + 0.8$ kg h^{-1}

Hence after 10 hrs, $\Delta E = 8$kg ethanol.

Responses to Chapter 9 SAQs

9.1 The substrate glucose is consumed, and hence its net-conversion rate is negative leading to $r_{As} = -4/4 = -1$ g l^{-1} h^{-1} ie 1 g of glucose is consumed per litre per hour.

Analogously $r_{Ax} = +2/4 = +0.5$ g l^{-1}h^{-1} ie 0.5 g of biomass are produced per litre per hour.

9.2 All systems contain biomass, H_2O and CO_2 as relevant compounds. Furthermore the following applies:

	1	2	3
C-source	glucose	CO_2	glucose
N-source	NH_3	HNO_3	NH_3
electron donor	glucose	H_2S	glucose
conjugate electron acceptor	CO_2	H_2SO_4	CO_2
electron acceptor	O_2	O_2	HNO_3
conjugate electron donor	H_2O	H_2O	N_2

To these we must add biomass, since this is a product in each case, H_2O must also be added as a relevant compound. Thus the total number of compounds in each case are 6; 7; 7 respectively.

System 1 is a typical heterotroph using glucose as both a source of energy (electron donor) and of carbon. Ammonia serves as a nitrogen source. Oxygen is the electron acceptor. CO_2 and H_2O are the products of glucose oxidation and oxygen reduction (ie the conjugated electron acceptor and donor).

System 2 is a typical autotroph, using the oxidation of sulphide as the energy source and carbon dioxide as a source of carbon. Sulphuric acid is the product of sulphide oxidation.

System 3 - denitrifying organisms are anaerobes which use NO_3^- ions as electron acceptors (ie these replace oxygen). The nitrate is reduced to dinitrogen gas.

9.3

1) If glucose is replaced by acetic acid, nothing changes because the 1-C-mole composition of both compounds is equal (CH_2O).

2) If glucose is replaced by ethanol then the composition of the substrate is changed from CH_2O to C_2H_6O (ie $CH_3O_{0.5}$), so for C conservation

$1r_{Ax} + 1r_{As} + 1r_{Ac} = 0$ thus $+r_{Ac} = -r_{Ax} - r_{As}$

For H conservation

$1.8\,r_{Ax} + 3\,r_{As} + 2r_{Aw} + 3\,r_{An} = 0$

For N conservation this remains unaltered

ie $-r_{An} = +0.2r_{Ax}$

For O conservation

$0.5r_{Ax} + 0.5\,r_{As} + 2r_{Ao} + 2r_{Ac} + 1r_{Aw} = 0$

(NB the substrate contains 0.5 mol of O for each mol of C, hence $0.5\,r_{As}$)

Substituting in for r_{Ac} we get $0.5r_{Ax} + 0.5r_{As} + 2r_{Ao} - 2r_{Ax} - 2r_{As} + 1r_{Aw} = 0$

Thus $-1.5r_{Ax} - 1.5r_{As} + 2r_{Ao} + 1r_{Aw} = 0$. (We will call this equation SAQ 9.3a).

We can eliminate the term r_{Aw} using the H-conservation.

Thus: $1.8r_{Ax} + 3r_{As} + 2r_{Aw} + 3r_{An} = 0$

Thus: $1.8r_{Ax} + 3r_{As} + 2r_{Aw} - 0.6r_{Ax} = 0$ (since $r_{An} = -0.2r_{Ax}$)

Thus: $1.2r_{Ax} + 3r_{As} = -2r_{Aw}$, $r_{Aw} = -0.6r_{Ax} - 1.5r_{As}$

We can now substitute this into Equation SAQ 9.3a.

Thus $-1.5r_{Ax} - 1.5r_{As} + 2r_{Ao} - 0.6r_{Ax} - 1.5r_{As} = 0$

$-r_{Ao} = -1.05r_{Ax} - 1.5r_{As}$

Thus we can write the relationships for carbon, nitrogen, oxygen and hydrogen as:

$+r_{Ac} = -r_{Ax} - r_{As}$

$-r_{An} = +0.2\,r_{Ax}$

$-r_{Ao} = -1.05\,r_{Ax} - 1.5\,r_{As}$

$+r_{Aw} = -0.6\,r_{Ax} - 1.5\,r_{As}$

3) This is a similar manipulation to that of part 2.

Using a parallel manipulation you should find the net-conversion rates of the elements when NH_3 is replaced by HNO_3 are:

$+r_{Ac} = -r_{Ax} - r_{As}$

$-r_{An} = +0.2\,r_{Ax}$

$-r_{Ao} = -1.45\,r_{Ax} - r_{As}$

$+r_{Aw} = -0.8\, r_{Ax} - 1.0\, r_{As}$

We will do part of the derivation in full.

Biomass: $CH_{1.8}O_{0.5}N_{0.2}$; Glucose: CH_2O; O_2: O_2; CO_2: CO_2; H_2O: H_2O; HNO_3: HNO_3

C - conservation: $r_{Ax} + r_{As} + r_{Ac} = 0$

H - conservation: $1.8\, r_{Ax} + 2\, r_{As} + 2\, r_{Aw} + r_{An} = 0$

O - conservation: $0.5\, r_{Ax} + r_{As} + 2\, r_{Ao} + 2\, r_{Ac} + r_{Aw} + 3\, r_{An} = 0$

N - conservation: $0.2\, r_{Ax} + r_{An} = 0$

C - conservation $\rightarrow r_{Ac} = -r_{Ax} - r_{As}$

N - conservation $\rightarrow r_{An} = -0.2\, r_{Ax}$

H - conservation $\rightarrow 2\, r_{Aw} = -1.8\, r_{Ax} - 2r_{As} - r_{An} = -1.8\, r_{Ax} - 2\, r_{As} - (-0.2\, r_{Ax})$

$= -1.6\, r_{Ax} - 2\, r_{As}$

thus $r_{Aw} = -0.8 r_{Ax} - r_{As}$

O - conservation: $2\, r_{Ao} = -0.5\, r_{Ax} - r_{As} - 2\, r_{Ac} - r_{Aw} - 3\, r_{An}$

$= -0.5\, r_{Ax} - r_{As} - 2\, (-r_{Ax} - r_{As}) - (-0.8\, r_{Ax} - r_{As}) - 3\, (-0.2\, r_{Ax})$

$= -0.5\, r_{Ax} - r_{As} + 2\, r_{Ax} + 2\, r_{As} + 0.8\, r_{Ax} + r_{As} + 0.6\, r_{Ax} = 2.9\, r_{Ax} + 2\, r_{As}$

thus: $r_{Ao} = 1.45\, r_{Ax} + r_{As}$

4) By changing the biomass composition changing from $CH_{1.8}O_{0.5}N_{0.2}$ to $CH_{1.7}O_{0.8}N_{0.1}$ then the net rate conversion rates are also changed. You should have calculated that the rates are:

$+r_{Ac} = -r_{Ax} - r_{As}$

$-r_{An} = +0.1\, r_{Ax}$

$-r_{Ao} = -0.95\, r_{Ax} - r_{As}$

$+r_{Aw} = -0.7\, r_{Ax} - r_{As}$

Here is our solution:

C - conservation: $r_{Ax} + r_{As} + r_{Ac} = 0$

H -conservation: $1.7\, r_{Ax} + 2\, r_{As} + 2\, r_{Aw} + 3\, r_{An} = 0$

O - conservation: $0.8\, r_{Ax} + r_{As} + 2\, r_{Ao} + 2\, r_{Ac} + r_{Aw} = 0$

N - conservation: $0.1\, r_{Ax} + r_{An} = 0$

C - conservation $\rightarrow r_{Ac} = -r_{Ax} - r_{As}$

N - conservation: $r_{An} = -0.1\, r_{Ax}$

H - conservation: $2\, r_{Aw} = -1.7\, r_{Ax} - 2\, r_{As} - 3r_{An}$

$= -1.7\, r_{Ax} - 2\, r_{As} - 3\, (-0.1\, r_{Ax}) = -1.7\, r_{Ax} - 2\, r_{As} + 0.3\, r_{Ax}$

$= -1.4\, r_{Ax} - 2\, r_{As}$

thus: $r_{Aw} = -0.7\, r_{Ax} - r_{As}$

O - conservation: $2\, r_{Ao} = -0.8\, r_{Ax} - r_{As} - 2\, r_{Ac} - r_{Aw}$

$$= -0.8\, r_{Ax} - r_{As} - 2\,(-r_{Ax} - r_{As}) - (-0.7\, r_{Ax} - r_{As})$$

$$= -0.8\, r_{Ax} - r_{As} + 2\, r_{Ax} + 2\, r_{As} + 0.7\, r_{Ax} + r_{As}$$

$$= 1.9\, r_{Ax} + 2\, r_{As}$$

thus: $r_{Ao} = 0.95\, r_{Ax} + 1\, r_{As}$

9.4

The system definition which includes S and P includes H_2SO_4 and H_3PO_4 as relevant component as S and P source. Therefore the system definition includes now 8 compounds (biomass, glucose, O_2, CO_2, H_2O, NH_3, H_2SO_4, H_3PO_4) and 6 elements (C, H, O, N, S, P). Hence there are still two degrees of freedom.

This means that it is still possible to choose r_{Ax} and r_{As} as independent rates and calculate the relations for r_{Ac}, r_{An}, r_{Ao}, r_{Aw} as function of r_{Ax} and r_{As}.

If this is done one will find the relation 9.10c and 9.10d only change with respect to the coefficient of r_{Ax}. This is understandable because S and P only occur in the biomass. By performing the calculations as before one finds that the coefficient of r_{Ax} in Equation 9.10c changes from (1.05) to (1.09) which is indeed a small change.

Here is our solution:

$CH_{1.8}O_{0.5}N_{0.2}S_{0.01}P_{0.02}$	r_{Ax}
CH_2O	r_{As}
O_2	r_{Ao}
CO_2	r_{Ac}
H_2O	r_{Aw}
NH_3	r_{An}
H_3PO_4	r_{Ap}
H_2SO_4	r_{Az}

C - conservation: $r_{Ax} + r_{As} + r_{Ac} = 0 \rightarrow r_{Ac} = -r_{Ax} - r_{As}$

H - conservation: $1.8\, r_{Ax} + 2\, r_{As} + 2\, r_{Aw} + 3\, r_{An} + 3\, r_{Ap} + 2\, r_{Az}$

P - conservation: $0.02\, r_{Ax} + r_{Ap} = 0 \rightarrow r_{Ap} = -0.02\, r_{Ax}$

S - conservation: $0.01\, r_{Ax} + r_{Az} = 0 \rightarrow r_{Az} = -0.01\, r_{Ax}$

N - conservation: $0.2\, r_{Ax} + r_{An} = 0 \rightarrow r_{An} = -0.2\, r_{Ax}$

from H - conservation: $2\, r_{Aw} = -1.8\, r_{Ax} - 2\, r_{As} - 3\,(-0.2\, r_{Ax}) - 3\,(-0.02\, r_{Ax}) - 2\,(-0.01\, r_{Ax})$

$$= 1.8\, r_{Ax} - 2\, r_{As} + 0.6\, r_{Ax} + 0.06\, r_{Ax} + 0.02\, r_{Ax}$$

$$= -1.12\, r_{Ax} - 2\, r_{As}$$

$r_{Aw} = -0.56\, r_{Ax} - r_{As}$

O - conversion: $0.5\, r_{Ax} + r_{As} + 2\, r_{Ao} + 2\, r_{Ac} + r_{Aw} + 4\, r_{Ap} + 4\, r_{Az} = 0$

$2\, r_{Ao} = -0.5\, r_{Ax} - r_{As} - 2\,(-r_{Ax} - r_{As}) - (0.56\, r_{Ax} - r_{As}) - 4\,(-0.02\, r_{Ax}) - 4\,(-0.01\, r_{Ax})$

$$= -0.5\, r_{Ax} - r_{As} + 2\, r_{Ax} + 2\, r_{As} + 0.56\, r_{Ax} + r_{As} + 0.08\, r_{Ax} + 0.04\, r_{Ax}$$

$r_{Ao} = 1.09\, r_{Ax} + r_{As}$

9.5

1) The system contains 7 compounds and 4 elements, hence there are $7 - 4 = \underline{3}$ degrees of freedom.

2) For the manipulation of the elemental conservation relations r_{Ax}, r_{As} and r_{Ap} are chosen as independent.

The 1-C-mole formula of ethanol (C_2H_5OH) is $CH_3O_{0.5}$. This leads to the following relations:

C-conservation

$$1\,r_{Ax} + 1\,r_{As} + 1\,r_{Ac} + 1\,r_{Ap} = 0$$

H-conservation

$$1.8\,r_{Ax} + 2\,r_{As} + 2\,r_{Aw} + 3\,r_{An} + 3\,r_{Ap} = 0$$

O-conservation

$$0.5\,r_{Ax} + 1\,r_{As} + 2\,r_{Ao} + 2\,r_{Ac} + 1\,r_{Aw} + 0.5\,r_{Ap} = 0$$

N-conservation

$$0.2\,r_{Ax} + 1\,r_{An} = 0$$

There are 4 relations between 7 rates (ie 3 degrees of freedom); hence we choose r_{Ax}, r_{As} and r_{Ap} as independent. This leads for r_{Ac}, r_{Ao}, r_{Aw}, r_{An} to the following equations:

$$+\,r_{Ac} = -\,r_{Ax} - r_{As} - r_{Ap}$$

$$-\,r_{An} = 0.2\,r_{Ax}$$

$$-\,r_{Ao} = -1.05\,r_{Ax} - r_{As} - 1.5\,r_{Ap}$$

$$+\,r_{Aw} = -\,0.6\,r_{Ax} - r_{As} - 1.5\,r_{Ap}$$

9.6

Table 9.2 is defined in C-mol units. This means that we need to recalculate the various rates, keeping also the sign convention in mind.

Biomass: the molar weight of 1 C-mole is $12 + 1.8 + 0.5 \times 16 + 0.2 \times 14 = 24.6$ g

Hence:

$$r_{Ax} = 1000/24.6 = \underline{+40.6}\ \text{C-mol h}^{-1}$$

Substrate: the molar weight of 1-C-mole glucose is $180/6 = 30$ g

Hence $r_{As} = -\,3000/30 = \underline{-100}\ \text{C-mol h}^{-1}$

$$O_2 : r_{Ao} = \frac{-1000}{32} = \underline{-31.2}\ \text{mol h}^{-1}$$

$$CO_2 : r_{Ac} = \frac{+1850}{44} = \underline{+41.9}\ \text{mol h}^{-1}$$

We can use initially Equation 9.10 a-d, because product formation is not taken into account. Out of 6 net-conversion rates r_{Aw} and r_{An} are not measured. Hence we can reduce the set of 4 relations (Equation 9.10a-d) between 6 rates to a set of 2 relations between the 4 measured rates by elimination of r_{Aw} and r_{An}.

This leads to:

$$r_{Ac} = -r_{Ax} - r_{As}$$

$$-r_{Ao} = -1.05\,r_{Ax} - r_{As}$$

The first equation leads, after substitution of the above values to a predicted carbon dioxide production of $-40.6 + 100 = +59.4$ mol h^{-1}, which is higher than the observed carbon dioxide production.

Apparently: $59.4 - 41.9 = 17.5$ C-mol h^{-1} are not converted to CO_2 but to a product.

The second equation leads to predicted O_2-consumption of:

$$(-r_{Ao}) = -\,1.05 \times 40.6 - (-100) = 57.4\ \text{mol h}^{-1}\ \text{which is again much too high.}$$

This also indicates that less glucose is oxidised to CO_2, and that some other product is formed.

In other words, some oxygen and glucose is consumed that does not end up in CO_2 or biomass.

(NB After extra analysis it is found that there is ethanol production at a rate of 17.5 C-mole h^{-1}).

The relations derived in SAQ 9.5 can be used to check this conclusion.

9.7

1) Degrees of freedom

There are 7 extra- and 3 intra-cellular compounds, which lead to 10 relations analogous to equations 9.11 a-f.

The extracellular compounds are glucose (CH_2O); ammonia (NH_3); CO_2; H_2O; ethanol ($CH_3O_{0.5}$) and glycerol ($CH_{2.66}O$) and biomass ($CH_{1.8}O_{0.6}N_{0.2}$)

The intracellular compounds are ATP, precursors ($CH_{1.8}O_{0.6}N_{0.2}$) and $NADH_2$.

For the extracellular compounds there are 7 net-conversion rates of the extracellular compounds and 4 unknown reaction rates, a total of 11.

Hence there are 10 equations and 11 unknown rates. Hence the degree of freedom is 1.

2) Derivation of relations

The relations are: $r_{Ax} = 0.133 \,(-r_{As})$

$r_{Ae} = 0.516 \,(-r_{As})$

$r_{Ag} = 0.077 \,(-r_{As})$

These can be obtained in the following way.

The following relations are obtained from the specified reactions.

$r_{As} = -\,1.095\, r_1 - 1.5\, r_3 - r_4$

$r_{An} = -\,0.2\, r_1$

$r_{Ac} = +0.095\, r_1 + 0.5\, r_3$

$r_{Aw} = +0.305\, r_1$

$r_{Ax} = r_2$

$r_{Ae} = r_3$

$r_{Ag} = r_4$

ATP: $0 = -\,0.051\, r_1 - 1.7\, r_2 + 0.5\, r_3 - 0.33\, r_4$

Precursors: $0 = r_1 - r_2$

$NADH_2$: $0 = 0.19\, r_1 - 0.33\, r_4$

After some algebraic manipulation one obtains for the reaction rates $r_1 \rightarrow r_4$:

$r_1 = r_{Ax}$

$r_2 = r_{Ax}$

$r_3 = 3.88\, r_{Ax}$

$r_4 = 0.58\, r_{Ax}$

(Since $r_1 = r_2$ and $r_2 = r_{Ax}$ and 0.19 $r_1 = 0.33r_4$ and $r_1 = 1.736r_4$ thus $r_4 = 0.58\ r_{Ax}$. Also since $0 = - 0.051\ r_1 - 1.7\ r_2 + 0.5\ r_3 - 0.33\ r_4$ thus $0 = - 0.051\ r_{Ax} - 1.7\ r_{Ax} + 0.5\ r_3 - 0.19\ r_{Ax}$. Thus $r_3 = 3.88\ r_{Ax}$)

We can now substitute these into the equations relating r_{Ax}, r_{Ae} and r_{Ag} to r_2, r_3 and r_4.

This leads then to:

$r_{Ax} = 0.133\ (-r_{As})$

$r_{Ae} = 0.516\ (-r_{As})$

$r_{Ag} = 0.077\ (-r_{As})$

3)   Yields. From the above relations it appears that: biomass yield is 0.133 C-mole biomass/C-mole glucose, ethanol yield is 0.516 C-mole ethanol/C-mole glucose, glycerol yield is 0.077 C-mole glycerol/C-mole glucose.

We remind you that by convention r_{As} is negative.

Note that the production of glycerol is a necessity and no accident. If you inspect the reactions of SAQ 9.7 it appears that the produced reducing equivalents in reaction 1 are only removed because of the glycerol production.

9.8   There is some flexibility in the way in which you could have answered this question. Nevertheless, you have learnt so far that the required number of model parameters can be reduced by using pseudo-steady state approximations, by using the principles of the conservation of elements and finally using metabolic information. Thus we would expect your answer to basically say that the number of required conversion kinetic relations can be minimized, and hence the number of required model parameters, by:

- using the pseudo-steady state approximation;

- using the conservation principles of elements;

- using metabolic information.

We will learn in the next section that a more general approach may be used to reduce the number of model parameters.

9.9   1)   In an anaerobic closed bioreactor only heat is transferred, hence no chemical transfer kinetics are required.

2)   In an aerobic batch fermentation O_2 is transferred into and CO_2 is transferred out of the system. Hence one needs 2 transfer kinetic relations.

3)   Here the transfer of acid/alkali needs to be specified.

4)   Transfer occurs for acid/alkali/glucose/NH_3/O_2/CO_2.

Responses to Chapter 10 SAQs

10.1   With HNO_3 as N-source $\gamma_x = 6.15$; with N_2 as N-source $\gamma_x = 4.9$; with aspartic acid as N-source $\gamma_x = 1.15$

We use equation 10.2 c to calculate these values.

Thus:

HNO_3 as N-source

$cn = 0$; $hn = 1$; $on = 3$; $nn = 1$;

$$\gamma_x = 4 \times 1 + 1.7 - 2 \times 0.4 - \frac{0.25}{1}\ (0 + 1 - 2 \times 3) = \underline{6.15}$$

N$_2$ as N-source

cn = 0, hn = 0; on = 0; nn = 2

$$\gamma_x = 4 \times 1 + 1.7 - 2 \times 0.4 - \frac{0.25}{2}(0 + 0 - 0) = \underline{4.9}$$

Aspartic acid as N-source

cn = 1; hn = 1.75; nn = 0.25; on = 1.0

$$\gamma_x = 4 \times 1 + 1.7 - 2 \times 0.4 - \frac{0.25}{0.25}(4 \times 1 + 1.75 - 2 \times 1.0) = \underline{1.15}$$

10.2   If we combine the defined biomass yields for CO_2, N-source and water with Equations 10.2 a, b, c and d, and the definition of Y_{sx} (Equation 10.3) one can derive a series of relationships:

Equation 10.1d gives us:

$$\delta_c\, r_{Ac} + \delta_x\, r_{Ax} + \delta_s\, r_{As} + \delta_p\, r_{Ap} = 0$$

But in this case $r_{Ap} = 0$.

We also know that $Y_{sx} = \dfrac{r_{Ax}}{-r_{As}}$ and $Y_{cx} = \dfrac{r_{Ax}}{r_{Ac}}$

Thus we can convert r_{Ac}, r_{As} and r_{Ax} into terms of Y_{sx} and Y_{cx}. Thus:

$$\delta_c\, \frac{r_{Ax}}{Y_{cx}} + \delta_x\, r_{Ax} + \delta_s\, \frac{r_{Ax}}{-Y_{sx}} = 0$$

$$\text{or } \frac{\delta_c}{Y_{cx}} + \delta_x + \frac{\delta_s}{-Y_{sx}} = 0$$

Rearranging gives:

$$Y_{cx} = \frac{\delta_c}{\delta_x} * \frac{Y_{sx}\, \delta_x/\delta_s}{(1 - Y_{sx}\, \delta_x/\delta_s)}$$

Similarly we can use 10.1b to establish

$$Y_{wx} = \frac{\beta_w}{\beta_x} * \frac{Y_{sx}\, \beta_x/\beta_s}{(1 - Y_{sx}\, \beta_x/\beta_s)}$$

and Equation 10.1a to establish

$$Y_{nx} = \frac{-\alpha_n}{\alpha_x} * \frac{Y_{sx}\, \alpha_x/\alpha_n}{(1 - Y_{sx}\, \alpha_x/\alpha_n)}$$

In principle therefore by determining the appropriate coefficients (ie α_i, β_i etc) we can calculate the biomass yield on the nitrogen source (Y_{nx}) and the carbon dioxide and water produced per unit of biomass produced (Y_{wx} and Y_{cx}) from the Y_{sx} value.

10.3   1)   9.59 mol.h^{-1}

since $\phi_H + 460\, r_{Ao} = 0$

4410 + 460 r_{Ao} = 0 thus: r_{Ao} = -9.59 mol h^{-1}

2)   46000 kJh^{-1}

Since 2 mol O_2 are consumed h^{-1} m^{-3} of vessel then 2 x 50 mol O_2 will be consumed h^{-1} for the total vessel.

$$\text{Again using } \phi_H + 460\, r_{Ao} = 0 \text{ then } \phi_H = 460 \times (2 \times 50) = 46000 \text{ kJh}^{-1}$$

10.4 Since $Y_H \left(= \dfrac{r_{Ax}}{\phi_H} \right)$ is the reciprocal of $\dfrac{\phi_m}{r_{Ax}}$ in Table 10.3, we can have all the terms of Equation 10.13 except γ_x.

Thus for methane: $\dfrac{1}{1330} = \dfrac{0.49}{115(8 - 0.49\gamma_x)}$ Therefore $\gamma_x = 4.76$

Using a similar calculation for methanol $\gamma_x = 4.36$

(since $\dfrac{\phi_H}{r_{Ax}} = 776$, $\gamma_s = 6$ and $Y_{sx} = 0.54$)

For lactic acid $\gamma_x = 4.33$

(since $\dfrac{\phi_H}{r_{Ax}} = 404$, $\gamma_s = 4.0$ and $Y_{sx} = 0.51$)

Thus your calculations γ_x values are 4.76, 4.36 and 4.33 (ie they are all similar).

10.5 Glucose produces biomass at the lowest cost price.

Our reasoning is as follows:

From Table 10.3 we find for glucose $Y_{sx} = 0.61$ C-mole biomass/C-mole. The heat production is 277 kJ/C-mole biomass.

Due to the limited cooling capacity the maximum biomass production is: $10^6/277 = 3610$ C-mole h^{-1}. This leads to fixed costs of $\dfrac{100}{3610} = 0.0277$ \$/C-mole biomass.

The variable (substrate) costs are calculated as follows:

1 kg of glucose costs \$0.60. Therefore 1C-mole of glucose costs \$ $\dfrac{0.60 \times 30}{1000}$ (Note 30g of glucose contains 1C-mole).

But we require $\dfrac{1}{0.61}$ C-mole glucose to produce 1 C-mole biomass. Therefore substrate costs are $\dfrac{1}{0.61} \times 0.60 \times \dfrac{30}{1000} = 0.0295$ \$/C-mole biomass. The total cost for biomass from glucose is, therefore, 0.0572 \$/C-mole biomass.

For methanol the heat production is 776 kJ/C-mole and the $Y_{sx} = 0.54$. This leads to a maximal biomass production of $10^6/776 = 1289$ C-mole h^{-1}, which is nearly 3 times less than for glucose. This then leads analogously to a total cost of 0.0901 \$/C-mole biomass.

We conclude that, despite the higher substrate price, the glucose is overall cheaper because it allows a much higher maximal biomass production.

10.6 From the derived relation

$$D^o_s + 10.8\, r_{Ax} + 36.8\, r_{As} = 0$$

and the 2nd law limit, $D^o_s = 0$, it appears that for the yield Y_{sx} at $D^o_s = 0$ (equilibrium).

$$Y^{max}_{sx} = \dfrac{r_{Ax}}{-r_{As}} = 3.4 \text{ since } 10.8\, r_{Ax} = 36.8\, r_{As}$$

10.7 $Y^{max}_{sx} = 0.70$

By comparing the coefficients described in the question with Figure 10.3, it follows that $\alpha_{11} = 1.095$; $\alpha_{12} = 0.051$; $\alpha_{22} = 1.7$; $\alpha_{32} = 0.33$; $\alpha_{42} = 0.5$; $\gamma_s = 4$; $\gamma_x = 4.2$

Substitution in Equation 10.23 leads to:

$$Y_{sx}^m = \frac{0.33 * 0.5 * 4 + 4}{1.095 * 0.33 * 0.5 * 4 + 2 * 0.5 (0.051 + 1.7) + 4.2}$$

$$= \frac{0.66 + 4}{0.72 + 1.751 + 4.2} = \underline{0.70}$$

10.8 $m_S = 0.02$ C-mol substrate/C-mol biomass h

$Y_{sx}^{max} = 2.5$ C-mol biomass/C-mol substrate

These may be determined graphically:

μ (h^{-1})	$\dfrac{1}{\mu}$ (h)	Y_{sx}	$\dfrac{1}{Y_{sx}}$ (C-mol sub/ C-mol biomass)
1	1	0.495	2.02
0.5	2	0.490	2.04
0.25	4	0.480	2.08
0.12	8.3	0.460	2.17
0.06	16.7	0.428	2.34

We now plot $\dfrac{1}{Y_{sx}}$ against $\dfrac{1}{\mu}$.

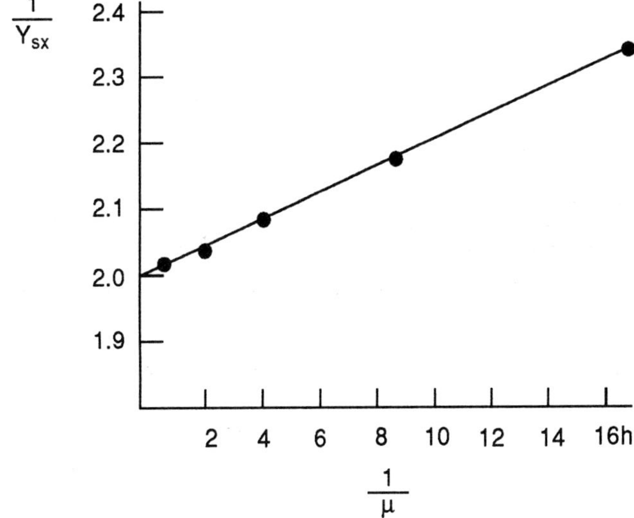

Slope = 0.02 C-mol substrate/C-mol biomass h = m_s

Intercept = 2 mol C-mol substrate/C-mol biomass = $\dfrac{1}{Y_{sx}^{max}}$. Therefore $Y_{sx}^{max} = 0.5$ C-mol biomass/C-mol substrate

Responses to Chapter 11 SAQs

11.1

1) The enzyme appears to follow Michaelis-Menton kinetics. The best way to show this is to use a Lineweaver Burke plot. First we calculate $C_{Eo}/-r_{As}$ and $\frac{1}{C_s}$ and then plot $C_{Eo}/-r_{AS}$ against $\frac{1}{C_s}$. Thus from the data given in the question we calculate:

$\frac{1}{C_s}$ (l mol^{-1})	$C_{Eo}/-r_{As}$
5×10^4	4.7×10^8
3.3×10^4	3.7×10^8
2.5×10^4	3.4×10^8
1.67×10^4	3.0×10^8
1×10^4	2.6×10^8

Now we now plot the graph using these values:

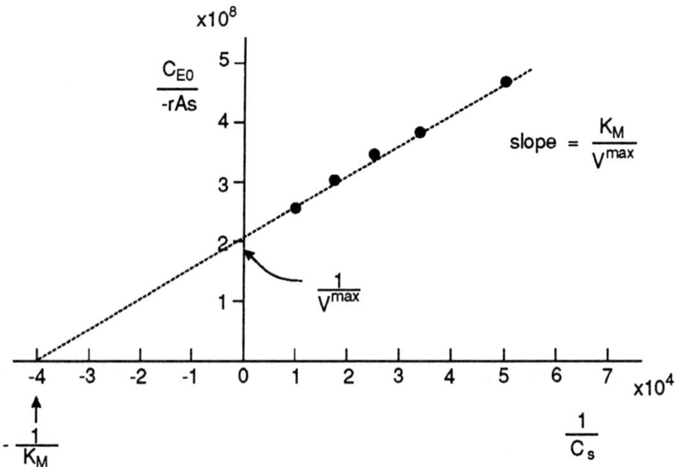

The straight line indicates that Michaelis-Menten kinetics are being followed.

2) From intercept and slope of the graph plotted above we can determine $V^{max} = 5 \times 10^9$ (approx) mol/mol min and $K_M = 2.5 \times 10^{-5}$ mol l^{-1}.

3) Competitive inhibitor. We can determine this by calculating $\frac{1}{C_s}$ and $C_{Eo}/-r_{as}$.

We can plot these graphically as in question 1) and 2).

Thus:

$\dfrac{1}{C_s}$ (l mol^{-1})	$C_{Eo}/-r_{As}$ (mol enzy m mol^{-1})
5×10^4	10.0×10^8
3.3×10^4	6.8×10^8
2.5×10^4	5.8×10^8
1.67×10^4	4.3×10^8
1.0×10^4	3.6×10^8

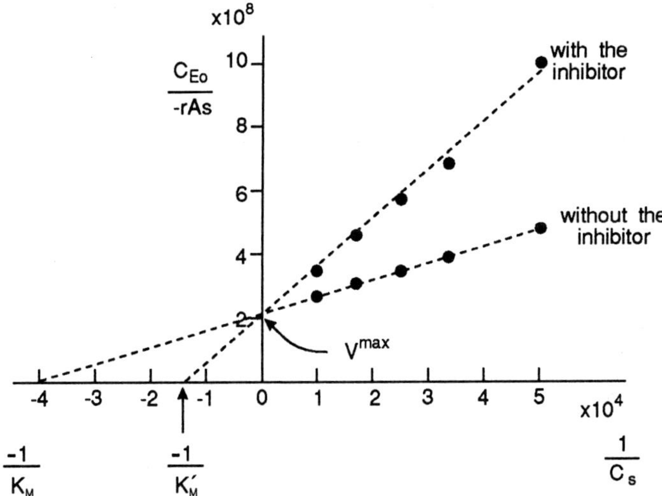

In this case we can determine $\dfrac{1}{V^{max}}$ and $\dfrac{1}{K_M}$ from the intercepts of this graph.

$$\frac{1}{V^{max}} = 2 \times 10^8 \text{ thus } V^{max} = 5 \times 10^{-9} \text{ mol/mol enz min}$$

and $K_M' = 6.7 \times 10^{-5}$ mol l^{-1} (approx)

Thus there is competitive inhibition (V^{max} unchanged; K_M is increased)

In this case $-\dfrac{1}{K_M''} = \dfrac{1 + C_I/K_I}{K_M}$ (see Figure 11.7 and associated text)

or $K_M' = K_M (1 + C_I / K_I)$

Thus using the figures determined for K_M' and K_M and $C_I = 1.5 \times 10^{-5}$ mol l^{-1}

then $6.7 \times 10^{-5} = 2.5 \times 10^{-5} \left(1 + \dfrac{1.5 \times 10^{-5}}{K_I} \right)$

$2.68 = 1 + \dfrac{1.5 \times 10^{-5}}{K_I}$. Therefore $K_I = 8.92 \times 10^{-6}$ mol l^{-1}

11.2

1) The values you should have obtained are:

Temperature K	$r_{As}^{max}(T) / r_{As}^{max}(300\ K)$
290	0.40
295	0.65
300	1.00
310	1.67
320	0.85
330	0.22

If you failed to get these then either you were using the wrong equation or you have made an arithmetic error.

The equation you need is Equation 11.17.

$$\frac{r_{As}^{max}(T_1)}{r_{As}^{max}(T_0)} = \exp\left[\frac{-\Delta H_i^{\#}}{R}\left(\frac{1}{T_1} - \frac{1}{T_0}\right)\right] \times \frac{1 + \exp\left(\frac{\Delta S_D'}{R}\right)\exp\left(\frac{-\Delta H_D'}{RT_0}\right)}{1 + \exp\left(\frac{\Delta S_D'}{R}\right)\exp\left(\frac{-\Delta H_D'}{RT_1}\right)}$$

2) The enzyme appears to show maximum activity at about 310 K since it is at this temperature that the ratio of $r_{As}^{max}(T)/r_{As}^{max}(300K)$ is highest.

11.3

1) $14.29\ g\ h^{-1}$

Since $\mu = -q_s\ Y_{sx}$ and $\mu = 1h^{-1}$ and $Y_{sx} = 0.35gg^{-1}$ then $-q_s = 2.86\ gg^{-1}h^{-1}$ (that is, each g of biomass uses 2.86 g of substrate each hour)

Thus 5g of cells will use $2.86 \times 5\ g\ h^{-1} = 14.29\ gh^{-1}$.

2) $K_s = 3 \times 10^{-4}\ mol\ l^{-1}$ (approx)

$\mu^{max} = 0.5\ h^{-1}$ (approx)

These can be calculated in an analogous way to the K_M and V^{max} of an enzyme.

We use the relationship;

$$\mu = \frac{\mu^{max}C_s}{K_s + C_s} \qquad\qquad\qquad (Eq.\ 11.20)$$

in its inverted form

Thus $\dfrac{1}{\mu} = \dfrac{K_s}{\mu^{max}}\dfrac{1}{C_s} + \dfrac{1}{\mu^{max}}$

A plot of $\dfrac{1}{\mu}$ against $\dfrac{1}{C_s}$ gives a straight line.

Thus:

$\dfrac{1}{C_s}$ (l mol^{-1})	$\dfrac{1}{\mu}$(h)
5×10^3	5
3.3×10^3	3.7
2.5×10^3	3.4
2.0×10^3	3.0
1.0×10^3	2.6

Graphically the intercept on the $\dfrac{1}{\mu}$ axis $= \dfrac{1}{\mu^{max}}$ and as the $\dfrac{1}{C_s}$ axis as $= -\dfrac{1}{K_s}$.

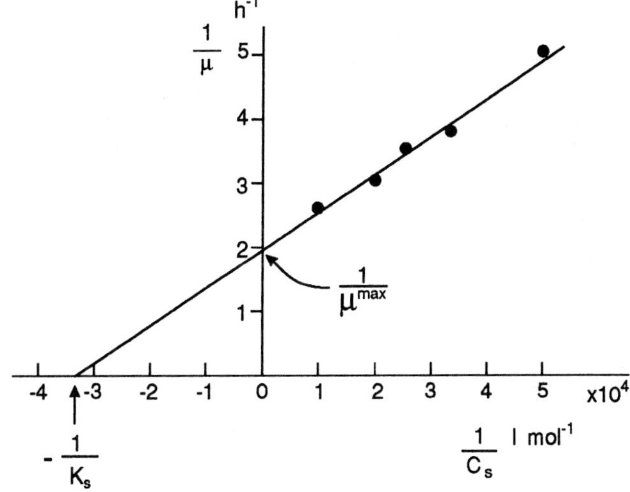

Thus $\bar{\mu}_{max} = 0.5$ h^{-1} approx. and $K_s = 3 \times 10^{-5}$ mol l^{-1} (approx).

11.4   The equation we are using is:

$$-q_s = \frac{1}{Y_{Sx}^{max}}\mu + \frac{1}{Y_{Sp}^{max}} q_p + m_S$$

You were required to calculate the yield of product formation on substrate (Y_{Sp}). This is by definition the ratio of q_p and $-q_s$ ie $Y_{Sp} = \dfrac{q_p}{-q_s}$. Hence it should be noted that $Y_{Sp} \neq Y_{Sp}^{max}$.

$$Y_{Sp} = \frac{q_p}{\dfrac{1}{Y_{Sx}^{max}}\mu + \dfrac{1}{Y_{Sp}^{max}} q_p + m_S}$$

Substitution of the provided data and the kinetic relation for q_p gives:

$$Y_{Sp} = \frac{0.26\,\mu}{2\,\mu + 1.5 * 0.26\,\mu + 0.02}$$

This gives the following results:

μ (h^{-1})	Y_{Sp} (C-mol / C-mol)
0	0
0.005	0.041
0.010	0.060
0.015	0.069
0.020	0.077

It is clearly observed that there exists a relation between μ and Y_{Sp} and that $Y_{Sp} < <Y_{Sp}^{max}$.

11.5   This was quite a difficult question to tackle. We will try to explain the strategy for solving it. We first try to find the effects of temperature on the rate of metabolism in general, but on the rate of product formation (q_p) and maintenance (m_S) in particular.

According to the results from SAQ 11.2, an increase of $T = 300 \rightarrow 310$ K leads to a factor 1.67 increase in maximal production rates (since we assume that enzymes behave similarly and kinetics of metabolism show enzyme-like relationships.

This means that the relationship in SAQ 11.4 $q_p = 0.26\,\mu$ becomes $q_p = 0.43\,\mu$ by raising the temperature 10°C.

We also predict that maintenance metabolism also increases by a factor of 1.67 over the same temperature rise.

Hence m_S becomes 1.67 x 0.02 = 0.033. The values of Y_{Sp}^{max} and Y_{Sx}^{max}, being stoichiometric constants, remain unchanged.

For Y_{Sp} we then obtain:

$$Y_{Sp} = \frac{0.43\,\mu}{2\,\mu + 1.5 * 0.43\,\mu + 0.033}$$

μ	Y_{sp}
0	0
0.005	0.046
0.010	0.072
0.015	0.089
0.020	0.100

Compared to the values of Y_{Sp} calculated for SAQ 11.4 it appears that Y_{Sp} increases by about 15-25%. This is much less than 67% that one might predict from the increases in enzyme activity by a factor of 1.67. This is because the product-rate improvement is counteracted by the maintenance increase.

Suggestions for further reading

Chapters 1-7

Beek, W.J. and Muttzall, K.M.K. (1966) **Transport Phenomena**, Wiley & Sons Ltd. (ISBN 0471061755)

Bird, R.B. Stewart W.E. and Lightfoot, E.N. **Transport Phenomena**, Wiley & Sons Ltd. Chichester, UK. (ISBN 471073954)

Felder R.M. & Rousseau, R.W. (1986) **Elementary Principles of Chemical Processes** (2nd Edition), Wiley & Sons Ltd. Chichester, UK. (ISBN 471837970)

Foust, A.S. Wenzel, L.A. Clump, C.W. & Maus, L. and Anderson, L.B. (1980) **Principles of Unit Operations** (2nd Edition), Wiley & Sons Ltd. Chichester, UK. (ISBN 471047872)

Incropera, F.P. & Dewitt, D.P. (1990) **Fundamentals of Heat and Mass Transfer** (3rd Edition), Wiley & Sons Ltd. Chichester, UK. (ISBN 471517291)

Potter M.C. and Wiggert, D.C. (1991) **Mechanics of Fluids**, Prentice Hall International (ISBN 0-13-571142-8)

Smith, J.M. Stammers, E and Janssen LPBM, (1986) **Fysische Transport Verschÿnselen** 1, Delftse Üitgevers Maatschappÿ bv, Delft, The Netherlands

Welty, J.R. Wicks C.E. and Wilson, R.E. (1984) **Fundamentals of Momentum, Heat and Mass Transfer** (3rd Edition), Wiley & Sons Ltd. Chichester, UK. (ISBN 471886653)

White, F.M. (1988) **Heat and Mass Transfer**, Addison Wesley. (ISBN 0-201-17099-X)

Chapters 8-11

Roels, J.A. (1983) **Energetics and Kinetics in Biotechnology**, Elsevier Biomedical Press, Amsterdam. (ISBN 0-444-80442-0)

Antonie van Leewenhoek (1991) Volume 60
Special issue: **Quantitative Aspects of Growth and Metabolism of Micro-organisms**

Doucet, P.G. and Sloep, P.B. (1992), **Mathematical Modelling in the Life Sciences**, Ellis Horwood, Hemel Hempstead, UK.

Appendix 1

Main symbols used in Chapters 1-7 (with examples of typical units)

A = surface area (m^2)

C = concentration (kg m^{-3})

C_A = concentration of A (kg m^{-3})

C_w = drag (frictional) coefficient

c_p = specific heat at constant pressure (J kg^{-1} K^{-1})

c_v = specific heat at constant volume (J kg^{-1} K^{-1})

D = dilution rate (s^{-1})

D = stirrer diameter (m)

d_p = particle diameter (m)

D = diffusion coefficient (m^2 s^{-1})

e = energy per unit mass (J kg^{-1}) eg e_{fr} = frictional energy

F = force (kg m s^{-2})

F_o = Fourier number $\left(\dfrac{\alpha\, t}{D^2}\right)$

Gz = Graetz number $\left(\dfrac{\alpha\, x}{<v>\, D^2}\right)$

h = partial heat transfer coefficient (kJm^{-2} K^{-1} s^{-1})

h = enthalpy (J mol^{-1}; J kg^{-1}); h_g = enthalpy of a gas; h_l = enthalpy of a liquid

Δh_{ev} = heat of evaporation (kJ kg^{-1})

j_D = Colburn number $\left(\dfrac{Sh}{Re\, Sc^{0.33}}\right)$

j_H = Chilton number $\left(\dfrac{Nu}{Re\, Pr^{0.33}}\right)$

k = mass transfer coefficient (m s^{-1})

k = consistency

k_r = reaction rate constant

l = length (m)

L = overall length (m)

M = mass (g; kg)

M_A = mass of A; M_B = mass of B M_l = mass of liquid phase; M_v = mass of vapour phase

m = partition coefficient

N = stirrer speed (s^{-1})

n = power law index for non-Newtonian liquids

Nu = Nusselt number $\left(\dfrac{hD}{\lambda}\right)$

P = production rate (kg s^{-1}; mol s^{-1}) eg P_{but} = production rate of butanol

p = pressure (Pa)

p_v = vapour pressure (Pa); P^* = saturation vapour pressure

Pe = Péclet number $\left(\dfrac{vL}{D}\right) = \dfrac{\text{convective transport}}{\text{diffusive transport}}$

Pr = Prandtl number $\dfrac{v}{\alpha}$

R = universal gas constant (8.31 J mol^{-1} K^{-1})

R = with subscripts, internal and/or external radius (m)

r = radius

r_A = net production of A per unit volume $\left(\dfrac{Pa}{V}\right)$ (mol m^{-3} s^{-1})

Re = Reynolds number $\left(\dfrac{\rho vd}{\eta} \text{ or } \dfrac{vd}{v} \text{ or } \dfrac{\rho ND^2}{\eta}\right)$

Sc = Schmidt number $\left(\dfrac{v}{D}\right)$

Sh = Sherwood number $\dfrac{kD}{D}$

T = temperature (K)

t = time (s)

U = total heat transfer coefficient (kJm^{-2} K^{-1} s^{-1})

u = internal energy (J kg^{-1})

V = volume (m^3)

v = velocity (ms^{-1})

w = width (m)

z = height (m)

α = thermal diffusion coefficient $\left(\dfrac{\lambda}{\rho c_p}\right)$ (m^2 s^{-1})

ρ = mass density (kgm^{-3})

λ = thermal conductivity coefficient (J s^{-1} m^{-1} K^{-1})

ε = gas hold up also volume fraction

η = coefficient of dynamic viscosity (Ns m^{-2}; J m^{-3} s; kgm^{-1} s^{-1})

η_s = viscosity of a suspension (Ns m^{-2})

δ_L = hydrolytic boundary layer (m)

τ = mean residence time (s)

τ_{xy} = shear stress

λ = thermal conductivity coefficient (J s^{-1} m^{-1} K^{-1})

θ = dimensionless time given by t/τ where t = time, τ = mean residence time

ϕ = flow (mol s^{-1}; kg s^{-1})
   superscript $''$ denotes per unit volume (flux)

ϕ_A = molar flow of A (mol s^{-1})

ϕ_{MA} = mass flow of A (kg s^{-1})

ϕ_p = momentum transport (kgm s^{-2}; Ns s^{-1})

ϕ_q = heat flow (transport) (J s^{-1})

ϕ_V = volumetric flow rate (m^3 s^{-1})

ϕ_{VA} = volumetric flow rate of A (m^3 s^{-1})

Appendix 2

Main symbols used in Chapters 8-11 (with examples of typical units)

A = surface area (m^2)

a = specific surface area = area per unit volume $(m^2\ m^{-3})$

C = concentration $(kg\ m^{-3};\ mol\ m^{-3})$

  subscripts E = enzyme

          $E0$ = initial enzyme

          ES = enzyme substrate complex

          I = inhibitor

          i = intermediate

          p = product

          s = substrate

          x = biomass

  superscripts i = input

          o = output

$\overset{o}{C}$ = rate of change of concentration $\left(\dfrac{dC}{dt}\right)$ $(kg\ m^{-3}\ s^{-1})$

D = dilution rate $\left(\dfrac{\phi_v}{V}\right)$ (s^{-1})

D_s^o = rate of Gibbs energy dissipation $(J\ m^{-3}\ s^{-1})$

D = diffusion coefficient $(m^2\ s^{-1})$

$\overset{o}{E}$ = rate of change of E $\left(\dfrac{dE}{dt}\right)$

ΔG_D^o = standard Gibbs free energy of denaturation $(kJ\ kg^{-1})$

h^o = standard enthalpy $(J\ kg^{-1})$

  subscript c = CO_2

          o = O_2

          p = product

s = substrate

w = water

x = biomass

k = reaction rate constant (units depend on order of reaction)

k_L = mass transfer coefficient (ms^{-1})

k_D = equilibrium constant for reversible enzyme denaturation

k_I = equilibrium constant for inhibitor binding by an enzyme

k_M = Michaelis constant ($kg\ m^{-3}$)

k_S = substrate saturation (Monod) constant ($kg\ m^{-3}$)

m_S = maintenance coefficient based on substrate use ($Cmol\ Cmol^{-1}\ s^{-1}$)

q = metabolic quotient ($kg\ m^{-3}\ s^{-1};\ kg\ kg^{-1}\ s^{-1}$)

  subscripts   A = accumulation

          A_i = rate of accumulation of i

          ATP = rate of accumulation of ATP

          s = substrate

  superscripts   cat = catabolic

          ana = anabolic

r = reaction rates ($kg\ m^{-3};\ mol\ m^{-3}\ s^{-1}$)
  subscripts 1, 2, 3 etc refer to specified reaction

r_A = rate of accumulation ($mol\ m^{-3}\ s^{-1};\ kg\ m^{-3}\ s^{-1}$)

  subscripts   c = CO_2

          i = general case

          n = nitrogen source

          o = O_2

          p = product

          s = substrate

          w = water

          x = biomass

Note, other subscripts are used to denote specific cases for example e = ethanol g = glycerol.

R = gas constant ($8.314\ J\ mol^{-1}\ K^{-1}$)

S^o = standard entropy ($J\ K^{-1}$)

T = temperature (K)

v = velocity of an enzyme catalysed reaction (reaction rate/unit of enzyme) (mol mol^{-1} s^{-1})

V^{max} = maximum velocity of a reaction (mol mol^{-1} s^{-1})

V = volume (m^3)

$\overset{o}{V}$ = rate of change of V $\left(\dfrac{dV}{dt}\right)$ (m^3 s^{-1})

$\overset{o}{X_i} = \left(\dfrac{dX_i}{dt}\right)$ rate of change of intermediate i in biomass (mol Cmol^{-1} s^{-1})

Y = growth yield coefficient

      subscripts   cx = rate biomass produced/rate of CO$_2$ evolution (Cmol cmol^{-1})

                                        Hx = rate biomass produced/rate of heat production (Cmol kJ^{-1})

                                          nx = rate biomass produced/rate of nitrogen source consumption (Cmol N-mol^{-1})

                                          ox = rate biomass produced/rate of oxygen consumption (Cmol mol^{-1})

                                          sp = rate of product made/rate of substrate consumed (Cmol Cmol^{-1})

                                          sx = rate biomass produced/rate of substrate consumed (Cmol Cmol^{-1})

                                          wx = rate biomass produced/rate of water production (Cmol mol^{-1})

Ψ_{Ai} = specific net rate of transfer of i $\left(\dfrac{\phi_{Ai}}{c_x}\right)$ (Cmol Cmol^{-1} s^{-1})

ϕ_{Ai} = flow rate of i into a system (mol m^{-3} s^{-1})

ϕ_H = heat transfer rate (kJ m^{-2} s^{-1})

ϕ_v = volumetric flow rate (m^3 s^{-1})
    superscript i = input   o = output

μ = specific growth rate (s^{-1})

α_i = coefficient in N-balance

β_i = coefficient in water balance

γ_i = degree of reduction

δ_i = carbon coefficient

Appendix 3

Tables of units, conversion factors and physical constants

Table of basic SI-units and derivatives

area			m^2
length			m
mass			kg
temperature			K
time			s
volume			m3
force	N		$kg\ m\ s^{-2}$
energy	J	N m	$kg\ m^2\ s^{-2}$
power	W	$J\ s^{-1}$	$kg\ m^{-2}\ s^{-1}$
pressure	Pa	$N\ m^{-2}$	$kg\ m^{-1}\ s^{-2}$
viscosity dynamic	Pa s	$N\ s\ m^{-2}$	$kg\ m^{-1}\ s^{-1}$
viscosity kinematic			$m^{-2}\ s^{-1}$
conductivity thermal	$W\ m^{-1}\ K^{-1}$	$J\ m^{-1}\ K^{-1}\ s^{-1}$	$kg\ m\ K^{-1}\ s^{-3}$
velocity			$m\ s^{-1}$
rotational frequency			s^{-1}
density			$kg\ m^{-3}$
heat		J	$kg\ m^2\ s^{-2}$
heat capacity		$J\ K^{-1}$	$kg\ m^2\ s^{-1}\ K^{-1}$

Some conversion factors for different units

quantity	unit	SI-unit	
length	inch (in)	0.0254	m
	foot (ft)	0.3048	m
	yard	0.9144	m
volume	UK gallon	$4.55 \cdot 10^{-3}$	m^3
	US gallon	$3.785 \cdot 10^{-3}$	m^3
mass	ounce (oz)	$2.84 \cdot 10^{-2}$	kg
	pound (lb)	0.454	kg
force	pound force (lbf)	4.45	N
	dyn	10^{-5}	N
pressure	atm	$1.103 \cdot 10^{5}$	Pa
	mm Hg (torr)	$1.33 \cdot 10^{2}$	Pa
viscosity	Poise	0.1	Pa s
	Stokes	10^{-4}	$m^2 \, s^{-1}$
energy	BTU	$1.06 \cdot 10^{3}$	J
	kcal	$4.19 \cdot 10^{3}$	J
	kWh	$3.6 \cdot 10^{3}$	J
power	hp (British)	746	$J \, s^{-1}$
	hp (metric)	736	$J \, s^{-1}$

Recommended values of physical constants

Physical constant	Symbol	Value
acceleration due to gravity	g	$9.81 \, m \, s^{-2}$
Avogadro constant	N_A	$6.022 \cdot 10^{23} \, mol^{-1}$
Boltzman constant	k	$1.380 \cdot 10^{23} \, J K^{-1}$
Faraday constant	F	$9.649 \cdot 10^{4} \, C \, mol^{-1}$
gas constant	R	$8.314 \, J K^{-1} \, mol^{-1}$
molar volume of ideal gas of stp	V_m	$2.241 \cdot 10^{-2} \, m^3 \, mol^{-1}$
Planck constant	h	$6.628 \cdot 10^{-34} \, Js$
velocity of light in a vacuum	c	$2.998 \cdot 10^{8} \, m \, s^{-1}$

Appendix 4

Dimensional numbers

This appendix contains some frequently used dimensionless numbers, their definition and their meaning.

symbol	name	formula	meaning	used in
Ah	Arrhenius	$\dfrac{E}{RT}$	$\dfrac{\text{activation energy}}{\text{thermal energy}}$	chemical reactions
Bo	Bodenstein	$\dfrac{vL}{D}$	$\dfrac{\text{convection}}{\text{axial diffusion}}$	Bo = Pe diffusion in reactors
Bd	Bond	$\dfrac{\Delta\rho D^2 g}{\sigma}$	$\dfrac{\text{gravitational force}}{\text{surface tension}}$	Bd = Eo bubbles and drops
Da I	Damkohler I	$\dfrac{k_r D}{v}$	$\dfrac{\text{chemical reaction rate}}{\text{convection mass transfer rate}}$	chemical reactions
Da II	Damkohler II	$\dfrac{k_r D^2}{D}$	$\dfrac{\text{chemical reaction rate}}{\text{diffusive mass transfer rate}}$	chemical reactions
Eo	Eotvos	$\dfrac{\Delta\rho D^2 g}{\sigma}$	$\dfrac{\text{gravitational force}}{\text{surface tension}}$	Eo = Bd bubbles and drops
f	Fanning fricition factor	$\dfrac{d\Delta p}{2\rho v^2 L}$	$\dfrac{\text{shear stress energy at the wall}}{\text{kinetic energy}}$	flow through tubes and channels
Fo	Fourier (mass)	$\dfrac{Dt}{d^2}$	$\dfrac{\text{process time}}{\text{diffusion time}}$	diffusion
Fo	Fourier (heat)	$\dfrac{\alpha t}{D^2}$	$\dfrac{\text{process time}}{\text{conduction time}}$	heat conduction
Fr	Froude	$\dfrac{v^2}{gH}$	$\dfrac{\text{inertia forces}}{\text{gravity forces}}$	
Fr	Froude (rotation)	$\dfrac{DN^2}{g}$	$\dfrac{\text{inertia forces}}{\text{gravity forces}}$	mixing with free surfaces
Gz	Graetz	$\dfrac{\alpha L}{D^2 v}$	$\dfrac{\text{conductive heat transfer}}{\text{convective heat transfer}}$	heat transfer to flowing media
Gr	Grashof	$\dfrac{D^3 g}{v^2}\,\dfrac{\Delta\rho}{\rho}$	$\dfrac{\text{buoyancy forces}}{\text{viscous forces}}$	free convection
Ha	Hatta	$\dfrac{\sqrt{D\,k_r}}{k}$	$\dfrac{\text{mass transfer with chemical reaction}}{\text{mass transfer without chemical reaction}}$	mass transfer with chemical reaction

Le	Lewis	$\dfrac{\alpha}{D}$	$\dfrac{\text{thermal boundary layer thickness}}{\text{mass transfer boundary layer thickness}}$	$Pe = \dfrac{Sc}{Pr}$ combined heat and mass transfer
Nu	Nusselt	$\dfrac{hD}{\lambda}$	$\dfrac{\text{total heat transfer}}{\text{conductive heat transfer}}$	heat transfer
Pe	Peclet (heat)	$\dfrac{vD}{\alpha}$	$\dfrac{\text{convective heat transfer}}{\text{conductive heat transfer}}$	heat transfer in flowing media
Pe	Peclet (mass)	$\dfrac{vd}{D}$	$\dfrac{\text{convective mass transfer}}{\text{diffusive mass transfer}}$	mass transfer in flowing media
Po	Power number	$\dfrac{P}{\rho N^3 D^5}$	$\dfrac{\text{power added}}{\text{power transferred to kinetic energy}}$	stirred vessels and pumps
Pr	Prandtl	$\dfrac{v}{\alpha}$	$\dfrac{\text{hydrodynamic boundary layer thickness}}{\text{thermal boundary layer thickness}}$	heat transfer in flowing media
Re	Reynolds	$\dfrac{\rho vD}{\eta}$	$\dfrac{\text{inertia forces}}{\text{viscous forces}}$	flow
Sc	Schmidt	$\dfrac{v}{D}$	$\dfrac{\text{hydrodynamic boundary layer thickness}}{\text{mass transfer boundary layer thickness}}$	mass transfer in flowing media
Sh	Sherwood	$\dfrac{kD}{D}$	$\dfrac{\text{total mass transfer}}{\text{diffusive mass transfer}}$	mass transfer
T	Thiele modulus	$D\sqrt{\dfrac{k_r}{D}}$	$\dfrac{\text{chemical reaction rate}}{\text{diffusive mass transfer}}$	$T = \sqrt{DaII}$ chemical reactions
We	Weber	$\dfrac{\rho v^2 d}{\sigma}$	$\dfrac{\text{inertia forces}}{\text{surface tension forces}}$	$We = Eo \cdot Fr$ bubbles and drops

Appendix 5

Mathematical tools - differential calculus and integration

Differential calculus: making micro balances

At several places throughout the text we constructed balances over a slab of very small width. The result was a so called micro balance. An important technique from the mathematics of differential calculus is used to arrive at the desired differential equation. Here we will discuss briefly this technique.

Consider a function: $f(x)$

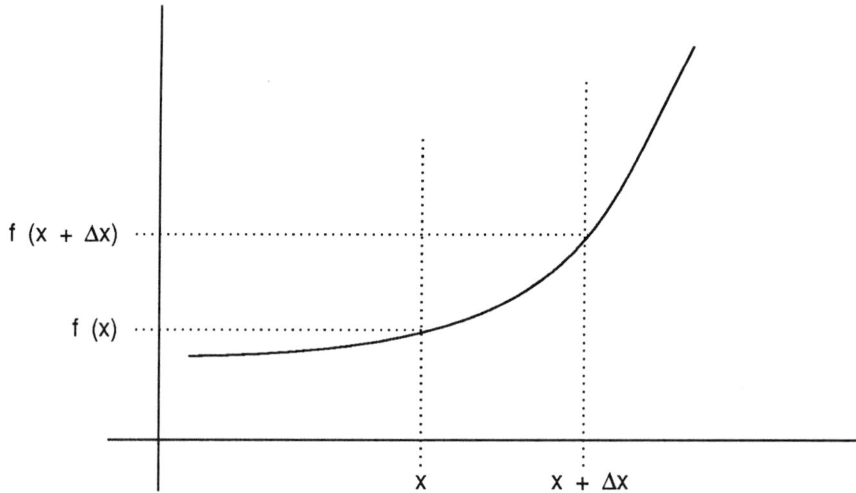

We focus on two points separated by a distance Δx: the value of f belonging to x is $f(x)$ and the value of f belonging to $x + \Delta x$ is $f(x + \Delta x)$. It is clear that the difference between these two values of f vanishes (goes to zero) as Δx goes to zero, since obviously both points 'collapse' to one point. For very small values of Δx of the difference $f(x + \Delta x) - f(x)$ is proportional to Δx:

$$f(x + \Delta x) - f(x) = \alpha \cdot \Delta x \text{ for very small } \Delta x \qquad \text{A.1}$$

The proportionality constant α is just the derivative of f, thus we can write

$$f(x + \Delta x) - f(x) = \frac{df}{dx} \Delta x \text{ for very small } \Delta x \qquad \text{A.2}$$

or alternatively

$$f(x + \Delta x) = f(x) + \frac{df}{dx} \Delta x$$

A.3

A third form that is often used is

$$\frac{df}{dx} = \lim_{\Delta x \to 0} \frac{f(x + \Delta x) - f(x)}{\Delta x}$$

A.4

These rules can be used for instance to simplify

$$\left[\frac{dT}{dx}\right]_{x + \Delta x} - \left[\frac{dT}{dx}\right]_{x} = 0$$

A.5

If we call for the moment $\frac{dT}{dx} = f$, then the first term of the left hand side of A can be written as (for small Δx)

$$\left[\frac{dT}{dx}\right]_{x + \Delta x} = f(x + \Delta x) = f(x) + \frac{df}{dx} \Delta x = \left[\frac{dT}{dx}\right]_{x} + \frac{d\left(\frac{dT}{dx}\right)}{dx} \Delta x$$

A.6

Substitution of A.6 into A.5 gives

$$\left[\frac{dT}{dx}\right]_{x} + \frac{d\left(\frac{dT}{dx}\right)}{dx} \Delta x - \left[\frac{dT}{dx}\right]_{x} = \frac{d\left(\frac{dT}{dx}\right)}{dx} \Delta x = 0$$

A.7

or if we shorten the notation of the second equality of A.7 we obtain

$$\frac{d^2 T}{dx^2} = 0$$

A.8

A similar calculation has to be made to simplify

$$\left[x \frac{dT}{dx}\right]_{x + \Delta x} - \left[x \frac{dT}{dx}\right]_{x} = 0$$

A.9

We can expand the first term on the left hand side of A.9:

$$\left[x \frac{dT}{dx}\right]_{x + \Delta x} - \left[x \frac{dT}{dx}\right]_{x} + \frac{d\left(x \frac{dT}{dx}\right)}{dx} \Delta x$$

A.10

Thus Equation A.9 can be written as

$$\frac{d\left(x \, \dfrac{dT}{dx}\right)}{dx} = 0$$

A.11

Summary

We have seen that in making a micro balance, ie a balance over a very thin slab we can use a simple expansion rule from the differential calculus. This rule will simplify our balance and turn it into a real differential equation.

Differential equations

In the text you came across several types of differential equations. Here you will find some help on solving these equations.

If you want to obtain the solution of the differential equation, you need some extra information on the function you have to solve. Usually this information is given in the form of so called boundary conditions: these are given values of the function or of its derivatives. The boundary conditions are needed to calculate the integration constants that come up in the general solution of the differential equation. Let us look at an example and consider the differential equation.

$$V \frac{dc}{dt} = - \phi_V . c$$

A.12

If we want to solve c as a function of t we have to integrate equation (A.12) one time. Hence, we will end up with one integration constant, which can be determined by one boundary condition. For example the boundary condition could be:

$$t = 0 \Rightarrow c = 0$$

A.13

In the following we will look at several types of differential equations that occur in the text.

Example 1 Consider the differential equation (K is a given constant):

$$\frac{dM}{dt} = K$$

A.14

with the boundary condition

$$t = 0 \Rightarrow M = M_0$$

A.15

Equation A.14 can be separated in its variables M and t:

$$dM = K \, dt$$

A.16

Integration of both side renders (remember that K is a constant):

$$M(t) = K . t + K_1$$

A.17

Here is K_1 the integration constant that has to be determined by substituting the boundary condition A.15 into A.17:

$$M_o = K \cdot 0 + K_1 \Rightarrow K_1 = M_0 \tag{A.18}$$

Thus we find for the solution of differential Equation A.14 with boundary condition A.15:

$$M(t) = K \cdot t + M_0 \tag{A.19}$$

Example 2 Consider the differential equation

$$\frac{dc}{dt} = \frac{1}{\tau}(c_1 - c) \tag{A.20}$$

both τ and c_1 are given constants, with the boundary condition

$$t = 0 \Rightarrow c = c_0 \tag{A.21}$$

The general solution is obtained as follows: write (A.20) in the form (again separation of variables: the result is that the left hand side only contains c and the right hand just t and no c).

$$\frac{1}{c_1 - c}\, dc = \frac{1}{\tau}\, dt \tag{A.22}$$

The left hand side can be written as

$$\frac{1}{c_1 - c}\, dc = \frac{1}{c_1 - c}\, d(c - c_1) = -\frac{1}{(c - c_1)}\, d(c - c_1) \tag{A.23}$$

If we make the simple substitution

$$f = (c - c_1)$$

the Equations A.22 and A.23 can be combined to yield

$$\frac{1}{f}\, df = -\frac{1}{\tau}\, dt \tag{A.24}$$

Note that both sides are also multiplied by -1.

We can easily integrate Equation A.24, since $\int \frac{1}{f}\, df \Rightarrow \ln f$ and $\int -\frac{1}{\tau}\, dt \Rightarrow -\frac{t}{\tau}$. This results in

$$\ln f = -\frac{t}{\tau} + K_1 \tag{A.25}$$

Here again K_1 is an arbitrary integration constant, that has to be determined by the boundary condition A.21. Before calculating K_1 we first substitute back $f = (c - c_1)$. Now Equation A.25 becomes

$$\ln (c - c_1) = -\frac{t}{\tau} + K_1 \qquad\qquad A.26$$

Equation A.26 is the general solution of Equation A.20 (check this by differentiating A.26 to obtain A.20 again). Notice that we need the integration constant also to make Equation A.26 homogenous in its dimensions.

Substitution of the boundary condition A.21 into A.26 yields:

$$\ln (c_0 - c_1) = -\frac{0}{\tau} + K_1 = K_1 \qquad\qquad A.27$$

Thus the solution of Equation A.20 with boundary condition A.21 is

$$\ln (c - c_1) = -\frac{t}{\tau} + \ln (c_0 - c_1)$$

or alternatively

$$\ln \left(\frac{c - c_1}{c_0 - c_1}\right) = -\frac{t}{\tau} \Longleftrightarrow \frac{c - c_1}{c_0 - c_1} = \exp\left(-\frac{t}{\tau}\right) \qquad\qquad A.28$$

Note that now both forms of A.28 are indeed dimensional correct.

Example 3 Consider the differential equation

$$\frac{dc_2}{dt} = \frac{1}{\tau} (c_1 - c_2) \qquad\qquad A.29$$

with boundary condition

$$t = 0 \Rightarrow c_2 = 0 \qquad\qquad A.30$$

but now c_1 is no longer a constant but also a function of time:

$$c_1(t) = c_0 \left[1 - \exp\left(-\frac{t}{\tau}\right)\right] \qquad\qquad A.31$$

This differential equation can not be solved by separation of variables. Now we have to use a different technique. Let us write A.29 like

$$\frac{dc_2}{dt} + \frac{1}{\tau} c_2 - \frac{1}{\tau} c_1 = 0 \qquad\qquad A.32$$

We guess that the solution is of the form

$$c_2 = (\alpha + \beta(t)) \exp\left(-\frac{t}{\tau}\right) \qquad\qquad A.33$$

Here α is a constant, but $\beta(t)$ is yet an unknown function of time. Before we substitute A.33 into A.32 we will first calculate the derivative $\dfrac{dc_2}{dt}$ from A.33:

$$\frac{dc_2}{dt} = -\frac{\alpha}{\tau}e^{-\frac{t}{\tau}} - \frac{\beta}{\tau}e^{-\frac{t}{\tau}} + \frac{d\beta}{dt}e^{-\frac{t}{\tau}} \qquad \text{A.34}$$

Substitution of A.33, A.34 and A.31 into A.32 yields:

$$-\frac{\alpha}{\tau}e^{-\frac{t}{\tau}} - \frac{\beta}{\tau}e^{-\frac{t}{\tau}} + \frac{d\beta}{dt}e^{-\frac{t}{\tau}} + \frac{\alpha}{\tau}e^{-\frac{t}{\tau}} + \frac{\beta}{\tau}e^{-\frac{t}{\tau}} - \frac{c_0}{\tau} + \frac{c_0}{\tau}e^{-\frac{t}{\tau}} = 0 \qquad \text{A.35}$$

Equation A.35 can be simplified to

$$\frac{d\beta}{dt}e^{-\frac{t}{\tau}} - \frac{c_0}{\tau} + \frac{c_0}{\tau}e^{-\frac{t}{\tau}} = 0 \qquad \text{A.36}$$

multiplying by $\exp(-t/\tau)$ gives

$$\frac{d\beta}{dt} = \frac{c_0}{\tau}e^{\frac{t}{\tau}} - \frac{c_0}{\tau} \qquad \text{A.37}$$

We now can separate the variables β and t in Equation A.37:

$$d\beta = c_0\left(e^{\frac{t}{\tau}} - 1\right)\frac{dt}{\tau} \qquad \text{A.38}$$

Integration of both sides renders (check this by differentiation)

$$\beta(t) = c_0\left(e^{\frac{t}{\tau}} - \frac{t}{\tau}\right) + K_1 \qquad \text{A.39}$$

If we substitute this into Equation A.33 we find (also check this yourself)

$$c_2(t) = \left((\alpha + K_1) - c_0\frac{t}{\tau}\right)e^{-\frac{t}{\tau}} + c_0 \qquad \text{A.40}$$

The final solution to problem 3 is now found by using the boundary condition: $t = 0 \Rightarrow c_2 = 0$, hence

$$0 = \left((\alpha + K_1) - c_0\frac{0}{\tau}\right)e^{-\frac{0}{\tau}} + c_0 = (\alpha + K_1) + c_0 \qquad \text{A.41}$$

Thus $\alpha + K_1 = -c_0$ and the solution is

$$c_2(t) = c_0 - c_0 \left(1 + \frac{t}{\tau} \right) e^{-\frac{t}{\tau}}$$

A.42

You can check the validity of our solution by substitution of A.42 into A.29 and by checking that A.42 is in agreement with the boundary condition A.30.

Example 4 Consider the differential equation

$$\frac{dD}{dt} = \alpha D^p$$

A.43

with α, p given constants ($p \neq 1$) and boundary condition

$$t = 0 \Rightarrow D = D_0$$

A.44

Equation A.43 is solved again by separating the variables: all D's to the left hand side, everything that contains t to the right hand side (constants also to the right hand). Then A.43 becomes

$$D^{-p} dD = \alpha dt$$

A.45

Integration of the left hand side of A.45 renders (check this by differentiating):

$$\int D^{-p} dD \Rightarrow \frac{1}{-p+1} D^{-p+1}$$

A.46

From A.46 we also see why we restrict this type of differential equations to $p \neq 1$: if $p = 1$ we would have divided by zero!

Thus integration of A.45 yields the general solution of A.43

$$\frac{1}{-p+1} D^{-p+1} = \alpha t + K_1$$

A.47

As usual, K_1 is the integration constant that has to be determined using the boundary condition A.44. Substitution of A.44 leads to

$$\frac{1}{-p+1} D_0^{-p+1} = K_1$$

A.48

Thus our final solution to differential Equation A.43 with boundary condition A.44 is

$$D^{-p+1} - D_0^{-p+1} = (1 - p) \alpha t$$

A.49

Notice that for the case $p = 0$ we recover the result from Example 1, the case $p = 1$ is treated in Example 2.

Summary

We have seen that a few types of differential equations can be solved by separation of variables. Furthermore, all differential equations have to be completed by boundary conditions. These conditions are necessary to calculate the integration constants that are always present in the general solution of the differential equation. The number of boundary conditions equals the number of times that the entire equation has to be integrated to obtain the general solution. In the cases we considered this was always one.

Index

A

aerobic growth
>and product, 212
>system definition, 212
application of metabolic information, 195

B

basic rate equation, 244
Bingham liquids, 95
biomass yield on a substrate (Y_{sx}), 220
biomass yield on oxygen (Y_{ox}), 220
bioreactor, 5
black box approach to modelling, 184
Buckingham - π theorem, 79

C

C-function, 27
Chilton numbers, 157
closed systems, 168
Colburn numbers, 157
competitive inhibition
>enzyme, 250
concentrations, 13
conduction of heat, 51, 100
conservation laws, 10
conservation principles, 179 , 191
Continuous Stirred Tank Reactors (CSTR), 14
convective mass transfer, 155
convective transport, 48, 155, 205
conversion rate, 246
creep flow, 88

D

degree of reduction, 218
diffusive transport, 48, 148, 206
>unilateral, 143
dilution rate, 22
dimensionless time measurement, 23
dimensional analysis, 66
>generalised technique, 70
dissipation, 40
downstream processing, 5
drag coefficient, 88
drag force, 85
drift flux, 143

E

E-function, 24
eddies, 86
element conservation
>application of, 10, 214
elementary volumes, 170
energy balance, 36
energy conservation, 36, 222
enthalpy, 43
entrance effect, 123
enzyme kinetics, 166, 244, 248
>inhibition, 250, 254, 256
>K_M, 248
>single, 244
>saturation constant, 166
>temperature, 258
>V^{max}, 248
error function, 113
exit age distribution (E), 24
expressions for transfer kinetics, 205
extensive properties, 171

F

F-function, 29
Fick's (or Stefan's) law, 147
Fick's law, 51
film theory, 138
first law of thermodynamics, 36
flow, 10
flow patterns in stirred vessels, 92
flows around obstacles, 87
forced convection, 122
forced diffusion, 131
forced flow around a sphere, 124
form drag, 87
Fourier, 123
Fourier number, 110 , 114
Fourier number and heat penetration, 116
Fourier's law, 51, 100
friction, 85
friction drag, 87

G

generalised model of growth, 212
Gibbs energy, 228
Gibbs energy balance, 228
Graetz number, 123
growth
>system description, 179
growth kinetics, 264
>saturation constant, 265